1007499310

STYLES OF KNOWING

Styles of Knowing

A NEW HISTORY OF SCIENCE FROM ANCIENT TIMES TO THE PRESENT

Chunglin Kwa

Translated by David McKay

UNIVERSITY OF PITTSBURGH PRESS

The translation of this book was made possible by a grant from the Netherlands Organization for Scientific Research (NWO).

Published by the University of Pittsburgh Press, Pittsburgh, Pa., 15260
Copyright © 2011, Chunglin Kwa
All rights reserved
Manufactured in the United States of America
Printed on acid-free paper
10 9 8 7 6 5 4 3 2 1
ISBN 13: 978-0-8229-6151-2
ISBN 10: 0-8229-6151-2
Originally published as *De ontdekking van het weten: Een andere geschiedenis van de wetenschap*, Amsterdam. Copyright 2005 Chunglin Kwa.

Library of Congress Cataloging-in-Publication Data

    Kwa, Chunglin.
    Styles of knowing : a new history of science from ancient times to the present / Chunglin Kwa.
       p. cm.
    Summary: "Inspired by A. C. Crombie's *Styles of Scientific Thinking in the European Tradition*, Kwa offers a full overview of scientific development in cultural and historical context. He introduces readers to the different forms of reasoning used by different sciences. Each chapter examines a different scientific style, illuminating how each style emerges gradually and continues to evolve. Older styles sometimes combine with newer while each also still continues along a solo trajectory. Styles investigated include the deductive, the experimental, the analogical-hypothetical, the taxonomic, the statistical, and the evolutionary. Although primarily designed for use in the classroom, this sophisticated book is also accessible to nonspecialists" — Provided by publisher.
    Includes bibliographical references and index.
    ISBN 978-0-8229-6151-2 (pbk.)
    1. Science—History. I. Title.
    Q125.K83 2011
    509—dc22   2011004071

CONTENTS

Preface vii

1. Introduction: The Six Styles of Knowing 1
2. The Deductive Style of Science 12
3. The Deductive Style in a Christian Context 27
4. From Scholar to Virtuoso: The Renaissance Origins of the Experimental Style 46
5. The Experimental Style II: The Skeptics and Their Opponents 72
6. The Experimental Style III: Alchemy and the New Sciences 92
7. The Hypothetical Style: Analogies between Nature and Technology 134
8. The Taxonomic Style 165
9. Statistical Analysis as a Style of Science 196
10. The Evolutionary Style 221
11. Science in the Twentieth Century 252

   Notes 277

   Bibliography 329

   Index 355

☯ PREFACE

This book has its origin in the former Department of Science Dynamics at the University of Amsterdam. In 1993, a fairly small group of enthusiastic instructors developed a new degree program. When we decided who would teach what subjects, I was very pleased to get the history of science. Soon afterward, Alistair Crombie published his three-volume work called *Styles of Scientific Thinking*. It was a long-anticipated event, and his opus gave me the framework that I wanted. The chapters of the present book originated in my lectures on the history of science, which I revised from year to year.

In the late 1990s, the department was dissolved, and the degree program some years later, due to a reform of the Dutch universities. Science Dynamics was too small to continue as an independent department and too large to be merged with another one. But my course on the history of science was spared, and over the past few years the subject has attracted substantially more students than in the past.

When I wrote the original, Dutch-language version of this book, I intended it for use in the classroom and for a general audience of educated people with an interest in the history of science. The book's first printing sold out before I could even assign it in my own courses. In 2005, some of my colleagues in the history of science convinced me that the book had a significance that went beyond my original objectives. This led me to apply for a translation grant from the Netherlands Organization for Scientific Research (NWO), which I ultimately received.

For a British or American reader, it is probably quite hard to conceive of the difficulties involved in working in a non-English-speaking country where English is the academic *lingua franca*. With ever-rising standards for publication, writing in Dutch has become a risky enterprise. One reason is that European universities are under a steadily increasing pressure to perform on narrowly defined "productivity" scales.

I do have, of course, a scholarly agenda, and here is the obvious merit of publishing this book in English. It is my firm belief that Crombie's study has not received the attention it deserves, especially considering his status as a grand old man of the history of science. Admittedly, he does not make life easy for his readers, with his uncompromising manner of writing and the sheer length of his three-volume tome: some 2,500 pages. Yet however valuable an accessible summary of Crombie's work might be, this book does not provide one. More emphatically than Crombie himself, I have tried to place the history of science in the context of the practice of science as well as of general culture, beginning with intellectual culture. "Forms and styles of painting respond to social circumstances," the art historian Michael Baxandall once remarked, and the same is true of styles of knowing. I am grateful to NWO for recognizing the intellectual importance of the stylistic approach.

For their critical comments on portions of the (Dutch) manuscript, I would like to repeat my thanks to Kees Jan Brons, Christien Brouwer, Gert de Bruin, Christine Delhaye, Leen Dresen, Mieke van Hemert, Frans van Lunteren, Harro Maas, Kees de Pater, Florence Pieters, Ida Stamhuis, Rienk Vermij, Gerard de Vries, and Jan Wattel. I also received valuable advice and assistance from Adrienne Ciuffo, Floris Cohen, Harold Cook, Marijke Duyvendak, Frances Gouda, Margaret Jacob, Rob Kohler, Michiel Korthals, Orpheus Roovers, Sherrill Rose, Wim Smit, and Wouter van Gils. Favorable but sometimes provocative reviews in historical and philosophical journals, such as those by Mieke Boon and Fokko Jan Dijksterhuis, may not have altered my position, but they did lead me to revise and clarify certain points for this English edition. My thanks also go to my translator, David McKay, for a harmonious working relationship.

STYLES OF KNOWING

1

Introduction

The Six Styles of Knowing

A STYLE OF SCIENTIFIC KNOWING is more than a method of scientific practice. This book differentiates between six styles: the deductive (in which science is built on first principles), the experimental, the hypothetical-analogical, the taxonomic, the statistical, and the evolutionary. Each of these styles has its own criterion for good science, the proper way of arriving at "the truth." There is no way to deduce or derive the styles of science from anything else; they form their own justification. The proposal that there are six different styles of science was first made by the historian Alistair Crombie in his magnum opus, *Styles of Scientific Thinking in the European Tradition*.[1] The assertion of six styles is the result of Crombie's taxonomic investigation, of his surveying the many forms in which the sciences have been practiced through history. The foundations of culture and the intellectual capacities of human beings have not been axiomatized to the point where we can prove that they yield these six styles and no others. Perhaps in the future new styles of science will arise, or perhaps they are already here, unnoticed. Technology may constitute an additional style; we will return to this topic later.

This view of science has various implications. In the philosophy of science, it implies that no one style can be regarded as foundational, forming a basis for all the others. In cultural history, it implies that there is no monolith called "science" (or "natural philosophy") that has stood apart from the rest of Western culture since the ancient Greeks. The six styles identified here have their roots in different eras of cultural history and bear the marks of their origins even today. This clearly illustrates that styles are not "paradigms," one succeeding another in strict sequence in the history of science. Once a new style has emerged, it endures, retaining its distinct identity. Yet these styles are not inalterable. Each one has followed its own developmental path. They have entered into various alliances: the deductive and experimental styles, the experimental and statistical, the statistical and evolutionary, and so forth. And there is another way in which this style-based view of science bears on the history of both science and culture: namely, by opening up a wider range of historical contexts for scientific practice.

Crombie sees the stirrings of all six styles in ancient Greece, though most of them did not fully emerge until much later. The Greeks developed the deductive style, in which the only kind of knowledge that qualifies as true science is knowledge derived from first principles that are necessarily true. The view that only the deductive style can lead to genuine scientific knowledge (*scientia*) persisted until the seventeenth century, though by then other styles of science had also taken root, presenting a growing challenge to the ideal of certain and necessary knowledge.

In this respect, the High Middle Ages (1150–1400) constituted a pivotal period for science. The deductive style underwent fundamental changes and, after a long process, was eventually forced to relinquish its claims to dominance. For some time, however, the new, medieval form of deductive thinking remained the only generally accepted style of thought in the sciences. Analogy existed, but was a theological form of reasoning, rather than a scientific one. And while medieval thinkers posed questions about historical development, they couched their answers in the form of deductive systems, as the historian Johan Huizinga observed.[2]

The Middle Ages did not give rise to an experimental style, though a conceptual logic emerged that helped to pave the way for experimentation. The experimental working method did not truly take shape until the Renaissance and the seventeenth century, and even then it took two distinct forms: one associated with the Renaissance virtuoso, and the other with magicians and alchemists. The man of *virtù*, represented by Galileo Galilei, was primarily in search of insights of a general nature, preferably ones that could be expressed in a simple geometric form or as a numerical ratio. The magician, in contrast, represented by Francis Bacon and Robert Boyle, was interested in anything unusual or anomalous.

The Renaissance also produced the hypothetical-analogical style, which was championed around 1600. The beginnings of the taxonomic style also lie in this same period. This book therefore addresses the Renaissance in some depth, though not for the same reason as histories of science that focus on the so-called Scientific Revolution. This revolution is usually said to have begun in 1543, with the publication of Copernicus's theory that the center of the universe is not Earth but the sun. It conventionally ends in 1687, when Newton proved the existence of gravity. Historical narratives based on the Scientific Revolution have well-established merits. But they also have their flaws, such as a rather one-sided emphasis on the "world picture," which is said to have undergone a profound shift, becoming mechanized or disenchanted.

Around 1800, the sciences again underwent such major conceptual changes that one might speak of a second Scientific Revolution, a revolution bound up with Romanticism. Biology established its independence as a scientific field, physics unified a range of natural forces, and history and economics took their place among the sciences. This second revolution is not one of the organizing themes of this book any more than the first one is, but it will be discussed.

Considering the styles all together, we can see that each one opened up new domains of experience to scientific investigation. Accordingly, what this book surveys is not so much the formation of the most influential scientific theories (though they will be included) as the history of empirical scientific practice.

The organization of this book is guided mainly by Crombie's styles of science—or of scientific thinking, as he puts it. This approach has a built-in limitation—namely, that the development of engineering and technology remains largely unexamined (though the influence of these fields on the development of the sciences is discussed). Given the organization of the book, it is important to ask whether engineering and technology represent a distinct style. My provisional answer is yes; one could make a strong case for their separate status as a style characterized by visualization—the use of drawing and computer animation to obtain a concrete (rather than analogical or metaphorical) picture of whatever feat of engineering one has in mind. In our twenty-first-century world, where technological thinking and practice are ever more likely to enter into alliances with various styles of science, it is reasonable to argue that the technological style has the same autonomous conceptual status as other styles.[3] The notion that technology is merely applied science has by now been thoroughly refuted and should be regarded as obsolete.[4]

The view of scientific styles presented here is also inspired by the work of Ian Hacking. Hacking took Crombie's concept of styles of science as his point of departure before it had even been presented to a broad public, and his own explorations of the experimental and statistical styles have pointed the way for other scholars.[5]

This study also draws on a great deal of other British and American scholarship. Historians of science have made great strides over the past thirty years—partly under the influence of the interdisciplinary field of science and technology studies—and, at the same time, strengthened their ties with general history. In the preceding period, the history of science had been plunged into turmoil by Marxist approaches, which had introduced valuable but overly reductive socioeconomic perspectives. Science and technology studies brought quite a few other "continental" perspectives into the history of science, and I hope to do justice to some of them in this book.

Recently, John Pickstone has proposed an interesting threefold distinction between natural history, analysis, and synthesis/experimentalism: three "ways of knowing," along with several "world readings" and present-day technoscience.[6] While Pickstone's natural history can be easily absorbed within the taxonomic style presented in this book, his "analysis" and "synthesis" both fall squarely within the experimental style. Pickstone's distinction between the latter two has the merit of drawing attention to the great wealth of variety within the experimental method. He moreover points to an alliance between analysis and the taxonomic style as practiced in the eighteenth century. All in all, the six styles of knowing presented here are more encompassing than Pickstone's "ways of knowing."[7]

Culture and Science

In 1969, the historian of science Robert M. Young wrote that some of his colleagues were beginning to realize that the history of science had more in common with general history than had previously been assumed.[8] This observation may seem commonplace now, but at the time it was anything but. E. J. Dijksterhuis, for example, in his masterwork *The Mechanization of the World Picture*, described the history of physics from Pythagoras to Newton as an autonomous line of development in which the scientific mind unfolded and awakened over a period of two millennia, virtually without reference to any social and cultural developments.[9] General history, for its part, did not seem to have any need for the history of science. The two subdisciplines were also separated by another factor: the history of science was usually practiced by people with a scientific background, addressing an audience of exact scientists.

When historians of science with a Marxist orientation, such as Young, began to speak out (the Dutch historians Jan and Annie Romein had, in fact, expressed similar views in the 1930s), they generally dismissed their predecessors and opponents as "internalists." These new historians were self-styled "externalists," connecting history to society and demonstrating influences "external" to what had always been regarded as a self-enclosed system. They also imported large parts of the sociology of science into its history.

It was not just the history of science, but the field of history in general, that moved toward a sociological perspective in the 1960s. While political history had once been the norm, social and economic historical writing soon took precedence. This tendency partly reflected the progressive social engagement of a new generation of scholars, but as the historian Lynn Hunt has argued, it was quickly defended on intellectual grounds as well. The French *Annales* school had begun propagating "total history" several decades earlier, and had a growing international influence.[10]

Around 1980, as Hunt demonstrates, the discipline of history began to take what is now known as the linguistic turn. This turn took place, in various guises, within different historical schools and approaches (as well as in philosophy and some social sciences, even playing a role in the formation of new areas of research such as cultural studies). The *Annales* school also shifted its focus to language, deposing social and economic history from their throne. Roger Chartier, one of this movement's later representatives, claimed that "mental structures" are independent of their "material determinations," and further stated: "The representations of the social world themselves are the constituents of social reality."[11] Chartier and other historians of his generation show the unmistakable influence of Michel Foucault, who did not accept any historical category whatsoever as a pretheoretical given. Foucault taught us that madness, the state, and sexuality are not universal concepts that merely take different concrete forms from one historical period to another.

The linguistic turn also led to an anthropological turn, inspired by Clifford Geertz and others, in which the search for historical essences made way for the reconstruction of contemporary systems of meaning.[12] Hunt has pointed out the difficulties resulting from the anthropological turn, particularly its overriding emphasis on shared systems of symbolic forms, which are supposedly organized, within any given culture, into a unified, homogeneous field. This touches on the tension, well-known in the social sciences, between structure and agency, or between structure and event. A historian, like an anthropologist, must have an eye for the differences and inequalities that present cultures and societies with opportunities for change.

Finally, there was also a literary turn, though it encountered more resistance than the other two. Dominick LaCapra used the work of the literary theorist Mikhail Bakhtin to identify multiple voices in a historical text by means of their stylistic features. Hayden White carried on the work of another literary theorist, Northrop Frye, in a book about historical writing itself. White presented compelling evidence that a small set of archetypal literary genres is also identifiable in historical texts, and that each of those genres has a certain rhetorical thrust derived from its narrative mode (its plot structure).[13]

These developments formed a substrate that also absorbed several older cultural traditions of historical writing, such as the art-historical tradition of the Warburg School (whose members included Ernst Gombrich and Frances Yates) and R. G. Collingwood's hermeneutic approach, which had previously been viewed as something of an oddity.

The cultural turn (the product of the linguistic, anthropological, and literary turns) opened up new opportunities for historians, from which I hope to benefit in this book. Essentially, what we now recognize is that science is a form of culture, produced by culture, and linked to other forms of culture. It should be added that science is by no means exclusively a form of "high" culture. There are many episodes in the history of science in which "low" culture was influential. In any case, Chartier notes that in "microsituations" this dichotomy is often irrelevant.[14]

The Term "Style"

Crombie uses the term "style" mainly to refer to the "cultural ecology" of a society: its views, convictions, and sacred cows, as well as its methods of problem solving. In science, he identifies three clusters of convictions. One consists of views about nature and its knowability. Another relates to science itself, specifically the organization of inquiry, argument, and explanation. The third has to do with social conceptions of what is desirable and possible: the moral, scientific, and technical dimensions of human intervention in nature, and the tension between conservation and innovation.[15]

Crombie says nothing about the origins of the term "style" as he uses it. It seems likely, however, that there is a connection with the term "style of thought," coined by the sociologist of knowledge Karl Mannheim.[16] In the 1920s, Mannheim was just about the only scholar to use the term "style" in connection with anything but periods of art history. He was in search of something more fundamental than schools of thought. Schools of thought differ from each other because that they make use of different theories, but the differences between styles of thought have to do with what Mannheim called the *Weltanschauungstotalität* (internal unity of a worldview). [17] Furthermore, each style of thought is associated with a different *seinsmäßige Beziehung* (existential relationship) to the objects of knowledge. Mannheim also used the compound word *Weltwollen* (will to the world) to make this point.[18]

In art history, such styles—Gothic and Baroque, for instance—are a collective matter. Individual artists share a background that provides them with similar "habit-forming forces," as the art historian Erwin Panofsky has argued.[19] A certain style (distinguished from other styles by its conventions) is the result. Individual representatives of that style have internalized the conventions in question. The French sociologist Pierre Bourdieu called a style of this kind a "habitus."[20]

Yet it is highly unconventional to apply the same perspective to the great scientists of the past. Galileo and Newton, for instance, are often described as timeless geniuses for whom the limitations of their age were so much ballast, mere social obstacles rather than intellectual ones. I aim to present scientific achievements such as those of Galileo and Newton as innovations within a style or as combinations of styles. My hope is that this approach accords great scientists the recognition they richly deserve for their trailblazing achievements, while placing them firmly in their historical context.

However, the conception of a style as the embodiment of a collectivity is not free from the usual pitfalls. The art historian Ernst Gombrich was one critic of excessively holistic views that turn a style into an imaginary superartist as a way of attempting to gain insight into the mentality of a period.[21] Looked at in this way, style and *Zeitgeist* are one and the same.

But this puts us in danger of throwing out the baby with the bathwater. Carlo Ginzburg has rightly noted that the belief in a "spirit of the age" is a naive solution to a real problem, namely, how to interpret the connection between different aspects of historical reality.[22] We will return to this point in a moment, but first, note that Gombrich's own conception of style was more modest than Panofsky's. Gombrich analyzed style as a notational system, thus emphasizing the activity of the artist as a maker of visual images.[23] He approvingly quoted the art historian Heinrich Wölflinn, who called art history the "history of seeing."[24]

This view draws a link between style and an important aspect of Thomas Kuhn's notion of a paradigm, which has proved to be the most successful way of conceptualizing changes in science.[25] Kuhn convincingly showed how, for instance, the transition from a Ptolemaic astronomical system to a Copernican one can be seen as a change in worldview. A new theory leads to a new image of reality. In this example, that not only meant that the sun replaced Earth as the center of the cosmos, but also that the Copernicans were open to seeing changes in the heavens, whereas their predecessors and opponents in the field of astronomy believed they saw only an ancient and unchanging universe. Kuhn related such changes in worldview to an emphasis on observations that were problematic for the old theory, even though the supporters of the old theory saw every reason to believe that these problems could be solved. The similarity to style, in Gombrich's representational sense, is striking. Theories are, as it were, different notational systems for reality, just as Impressionists and Cubists represent trees in a landscape in different ways. In both cases, the "truth" of the notation is bracketed.

There is nothing wrong with Kuhn's analysis. But it is possible to expand on it. At the same time that the new experimental style emerged, for instance, the status of scientific knowledge changed substantially. Copernicans had not only a new scientific worldview, but also new aspirations for the cultural significance of that worldview. Even if we assume that science, culture, and society are not a monolithic whole, they are clearly interrelated in countless ways. In fact, they are dynamically interrelated: a change in one component can place pressure on the internal structure of another, and bring about changes there as well. At the very least, we can conclude from this that a style involves more than just a mode of representation.

Crombie brings us closer to resolving this issue through the metaphor of a style as an ecology.[26] He probably did not have in mind the holistic approach of systems ecology, which would not have taken him very far. Crombie was an ecologist before he began his career as a historian of science, and he belonged to what is sometimes called the dynamic school of ecology.[27] An "ecological" view of style in science can help historians studying scientific change to avoid a one-sided emphasis on the representational nature of scientific activity. According to the ecological view, a style is a whole composed of many heterogeneous parts. Art historian Michael Baxandall's concept of "cognitive style" embraces more than just modes of representation. To interpret an image, he says, you need categories, model patterns, and "habits of inference and analogy."[28] A corresponding example from science is that different styles involve different ways of inductive reasoning. But if we return to Mannheim's conceptual model, we can see that the heterogeneity of scientific styles goes even further. Styles draw connections between the criteria for calling things true, rational, possible, desirable, acceptable, and plausible.[29]

The essence of Mannheim's contribution can be captured by the word "ethics."[30] Mannheim contested the "positivist prejudice" that we can free ourselves of metaphysical presuppositions and distinguish between facts and values. Every style of thought has its own way of connecting methodological and metaphysical assumptions. Mannheim drew inspiration for this more comprehensive notion of style from the work of the art historian

Aloïs Riegl, and this makes it noteworthy (though understandable) that Gombrich kept some intellectual distance between Riegl and himself.

Mannheim did not apply his postulate of the cultural relativity of a period's relationship to reality (*Seinsrelativität*) to the natural sciences. He was later criticized for this omission, but according to the sociologist Dick Pels, we must understand that Mannheim's first priority was to free the humanities from the objectivist claims of the logical positivists. He simply never got around to the natural sciences. The job was done later, however, by Mannheim's contemporary Ludwik Fleck.[31]

My analysis of the six styles of science shows that each one involves certain metaphysical assumptions. Purely deductive thinking, for instance, involves optimism about the knowability of the cosmos, which stems from an assumed affinity between the cosmos and the human mind. Such optimism is entirely absent from the hypothetical-analogical style. Though deduction plays a central role in this style as well, the expectations about what it can accomplish are very different. Another example is the experimental style, which could not have developed without the "ethics" of *virtù* and the *vita active* (active life).

A Western History of Science?

A few words of justification are in order. The narratives in this book relate to Western science. The Arab sciences in the Middle Ages are dealt with at some length because of the great influence they had on Western science in that period, but we tend to adopt the perspective of those who receive influence rather than those who offer it. Western science has also absorbed knowledge from many other cultures, in large part as a result of the Portuguese, Dutch, French, and English voyages of discovery in the fifteenth through the eighteenth centuries; this subject is addressed briefly.

The histories of Chinese and Arabic science are independent narratives that unfortunately do not fit within the scope of this book. Science in Orthodox Christian Europe also goes unexamined. I emphasize the relative autonomy of these scientific cultures because, in my view, there is little to be gained by attempting to tie them closely to the history of Western science.

Incidentally, the same exceptions can be made in relation to Greek and Roman antiquity. Between the fall of the Roman Empire and the intellectual reawakening of the West lay a yawning chasm of seven centuries, a long enough period to raise doubts about whether there was any real cultural continuity. Furthermore, our term "science" is much too modern to be attributed to the Greeks.[32] Almost by necessity, the chapter on science in ancient Greece is therefore split between two worlds: on one hand, it emphasizes the aspects of the deductive style that remained highly influential throughout the later history of science; on the other hand, it attempts to reconstruct the deductive style of the Greeks in its original, practical context, and to demonstrate the difference between what the Greeks meant by deduction and the understanding of the term from the Middle Ages onward.

Quite a few attempts have been made to compare Western and non-Western science. For example, Joseph Needham proposed that the absence of a concept of natural law undermined Chinese science (though the Greeks had no such concept either), and that the division of social authority between a variety of ecclesiastical and temporal powers worked in favor of the West. As the argument goes, the fact that the West had no one dominant power center created a scope for science that was lacking in China. Finally, Needham claimed, Europe developed a mercantile class at an early stage, and its merchants became major "consumers" of scientific ideas, whereas their Chinese counterparts only managed to play a minor social role.[33]

In the case of Eastern Europe, it has been suggested that the mystical orientation of the Greek and Russian Orthodox faiths hindered them from developing the practical engagement with the world found in Western science.[34] If this is true, then mysticism was a much greater obstacle in Eastern Europe than in the Western tradition, which includes some mystical movements—Franciscan religious orders in the Catholic Church, for instance, or the Neoplatonic school in the Italian Renaissance—that had profound and constructive impacts on the history of science.

The unsatisfying thing about many histories of non-Western science is that they mainly attempt to explain why such-and-such a place did not give rise to what Western Europe did: science and technology as we have known them over the past few centuries. These "negative" accounts generally tell us more about Western science, and the intellectual and social factors that promoted its development, than about the non-Western systems of science or knowledge that are their nominal subjects. The current tendency in the historiography of Chinese and Arabic science is to emphasize the view "from the inside out," rather than assuming *a priori* that Western science is a universal standard.[35]

Another problem that surfaces when we compare Western and non-Western science is that it is not always clear precisely what achievement is being explained. Is it the birth of Western science as an overarching rational system for understanding and controlling nature? The aim of this book is to plausibly demonstrate that there is not just one form of Western scientific rationality; there are at least six. Although this book gives various examples of the intellectual links between science and technology, the systematic exploitation of science for the purpose of technological advancement did not begin in earnest until the second half of the nineteenth century.[36] And it was not until after the Second World War that science and technology grew so tightly intertwined that "technoscience," a term many recent authors have used, emerged.

Yet the connection between science and power goes back much further. The link between power and knowledge is so close that it might almost be called intrinsic, yet this does not make the two identical. Might it not therefore be preferable to view the relationship between science and power as having a history of its own? Given the distinctions drawn here between styles of science, it even seems probable that each style bears a substantially different relationship to political and social authority.

The idea that scientific progress has led to the constant expansion of our power over nature is a romantic myth. Of course, the development of science and technology has tremendously enhanced our capacity to intervene in the natural and social worlds. But in this context, it is unhelpful to regard "nature" as singular. There is a particular historical period in which this tendency was strongest: the 1950s and 1960s, when the singular was used not only for "nature" but also for "science," and it was said that science should be approached through its (again singular) "method." As Jean-François Lyotard might say, that was a grand story—too grand, in fact. This book aims to tell stories of different natures.

2

The Deductive Style of Science

> *Nature loves to hide.*
> —Heraclitus
>
> *Et ignem regunt numeri.*
> —Jean Baptiste Fourier

THE ANCIENT GREEKS ESTABLISHED the concept that scientific explanation, in the ideal case, means inferring or deducing a natural phenomenon from a higher-order principle. Even today, deduction is the most important form of scientific explanation and is central to our image of good, fundamental science. In the twentieth century, some philosophers of science declared that the one true scientific method was a variant of this Greek form of deductive reasoning: the so-called hypothetico-deductive (or hypothetical-deductive) model. This claim was incorrect—there are other methods and styles in science—but it indicates the high status of the deductive style.

Relative to the original deductive style of the Greeks, the hypothetico-deductive style makes major concessions. Deductive reasoning itself is preserved, but the point of departure is the hypothesis, which the Greeks saw as a lesser form of knowledge. Today's scientists take it for granted that hypotheses have to be tested—experimentally, if possible. In a sense, modern science has reconciled itself to the idea that truth is unattainable, that even the best-supported hypothesis may one day have to be rejected in favor of a better one. Even so, the solidity of deductive reasoning enhances the prestige of the scientific method.

Experimental confirmation would have struck the Greek philosophers themselves as unnecessary and undesirable, because they would have associated experimentation with artifice, which they believed had nothing to tell us about the true nature of things.[1] They saw an obvious connection between deduction and eternal truths that arose by necessity from the Divine. Aristotle presented a precise description of hypothetical knowledge, but added that it does not qualify as true knowledge, which he considered to be the goal of science. Hypotheses are essential tools for explaining some phenomena, yet whenever possible, Aristotle avoided working with them. He wanted to show that whatever is eternal must necessarily be as it is.

Aristotle's most comprehensive account of the deductive style of science was the *Posterior Analytics*. But be warned: when you read Aristotle's work, it matters a great deal in which order you approach it. If you first read the *Physics* and Aristotle's works on biology, ethics, and rhetoric, his philosophy of science may seem to be no more than a convenient didactic overview, a manageable framework for organizing all his detailed studies of natural phenomena. If, however, you begin with his writings on the philosophy of science—the *Metaphysics* and *Posterior Analytics*—then you will come to see his deductive philosophy of science as a method for acquiring knowledge.

Traditionally, the emphasis has been on the deductive Aristotle. Along with Plato's *Timaeus* and Euclid's *Elements*, the *Posterior Analytics* has been one of the most influential texts in the history of Western science; it garnered commentary throughout the Middle Ages and into the early modern period.[2] Descartes's new scientific method was in many ways indebted to Aristotle's text. Some widely used philosophical surveys still claim that if Aristotle had taken the trouble to prepare a definitive edition of his works, he would have begun it with his metaphysical writings.[3] Crombie, too, emphasizes the deductive Aristotle, though he also devotes considerable space to the philosopher's love of observation.[4]

Aristotle starts with a definition: the only knowledge qualifying as scientific is that which can be demonstrated from first principles (*archai*) of which we can be certain. These principles are also the causes of the phenomena that we use them to explain. "We hold," Aristotle writes, "that besides the possibility of scientific knowledge, there is also a definite first principle of knowledge that enables us to recognize ultimate truths."[5] Through the first principles, he continues, all sciences can be connected with one another, and it can be shown that natural sciences, especially those dealing with eternal things, have primacy. In other words, physics is more fundamental than biology.[6]

A number of firm cosmological assumptions underlie Aristotle's metaphysics, the first being that the world is eternal. In a temporal sense, it has no beginning and no end, and never changes. Aristotle believed in a God, but (unlike Plato) not in a creator God. The second assumption is that the world is necessarily as it is, in all its particulars. Goats not only have horns, but could not possibly be without them. Thirdly, the world is a moral order, good just as it is.[7]

If we keep these assumptions in mind, it is clear why Aristotle saw such a close connection between what we now view as three separate domains: logic, causality, and language. Logic specifies what forms of deduction are valid, that is, how you can draw valid conclusions from a set of postulates. Causality describes which causes have which consequences. Language can express a relationship as an elementary proposition in the familiar subject-predicate form: a goat has horns; lead is heavy. Aristotle laid great emphasis on wording definitions properly so that they best expressed the essences of things. If this is done properly, then there is a relationship of mutual implication between cause and effect, subject and predicate, and substance and accidents (perceptible qualities).[8] The world, language, and the structure of reasoning are tightly interconnected. A goat's horns are so much part of it as to be, in a way, a "consequence" of the goat. In *On the Heavens*, Aristotle presented a proof that the heavenly bodies must move in circular paths: the heavens belong to God, and God's movements are eternal and thus circular, because circular movements are eternal. Using similar reasoning, he argued that Earth must be spherical, must be at rest at the center of the universe, and so on.[9] Aristotle also saw circularity in meteorology, in the eternal succession of rain and cloud formation, and in the cyclical succession of sperm, fetus, child, and adult.[10]

In this view, causality is timeless, as is the world itself. All causal connections are linked in causal chains, which are all hierarchically interrelated. The first cause cannot have a cause; from this, Aristotle concludes that there must be a single starting point for the entire world. That starting point is God—Aristotle's God, the Unmoved Mover.[11] According to Aristotle, the entire world arose from this original cause, but it was not a starting point in a temporal sense; instead, it should be seen as a logical process.

Aristotle's system has a number of implications. One is that the most universal concepts are the furthest removed from observation, and the objects of observation are the least universal of all things, and therefore in a sense the least important. This implies a deeper level of reality, a level more fundamental than what we see around us. Heraclitus—a Greek philosopher active around 500 BC, about three generations before Plato—is thought to have said, "Nature loves to hide." In the history of philosophy, this famous maxim has generally been interpreted in a Platonic-Aristotelian sense: the essences of things lie concealed.[12] A second implication of Aristotle's system is that the use of analogy is prohibited because it does not respect the world's hierarchical causal structure. Aristotle called it a category mistake.[13]

Approximately a generation after Aristotle, the mathematician Euclid (who regarded himself as a follower of Plato) accomplished in the realm of geometry what Aristotle had hoped to accomplish for the entire world. Euclid constructed a complete geometric system from a limited number of postulates, which could not be proven and were accepted as a kind of starting point. On these postulates, or axioms, Euclid built the entire edifice of geometry. For more than two thousand years, Euclid's *Elements* was considered the gold standard for knowledge because the deductive system contained therein works so perfectly and because it kept faith alive in the ideal espoused in the *Posterior Analytics*.

Archimedes demonstrated with exceptional force the power of deductive explanation. He performed experiments (or at least thought-experiments), but saw them solely as an intermediate step that could be discarded at the end of the process. Once he had set down his proofs on paper according to the geometrical method, he no longer needed the mechanical method to argue for his conclusions, as he explained in *The Method of Mechanical Theorems*.[14] His work on statics begins with postulates in the Euclidean style: "A balance with two arms of equal length from which two equal weights are suspended is in equilibrium."[15] His work on hydrostatics (which deals in part with measuring the specific weight of objects by immersing them in liquid) is organized in exactly the same way. He does not discuss practical applications of the law that he discovered.[16] Archimedes did not leave any written record of the famed war machines that he designed for the defense of Syracuse against the Romans.[17]

Both Euclid and Archimedes began their analyses (of geometry and statics, respectively), with a set of postulates that was as small as possible. Both based their postulates on knowledge they had already acquired, Euclid through visual experience, and Archimedes by using scales. But each man subjected his experience to such a thoroughgoing process of abstraction that the postulates took on a self-evident quality. A world with different first principles would simply not correspond with the world we know.

As Aristotle said, the first principles of any science cannot be proved. Nevertheless, he was optimistic about the possibility of discovering first principles. His formula was frequent observation, guided by intuition. This may sound like induction, but the truth is much more complex. Aristotle knew very well that inductive arguments, which derive general conclusions from specific facts, or causes from consequences, have no logical validity. "Scientific knowledge cannot be obtained through observation," Aristotle said.[18] But he believed that frequently repeated observation was different because it introduced a level of abstraction; over time, we can learn to ignore the chance properties of individual observations. But in order to acquire certain knowledge, we need intuition (*nous*), which Aristotle regarded as infallible.[19]

Intuition is the explicit subject of only one passage in Aristotle's entire oeuvre, at the end of the *Posterior Analytics*.[20] Here Aristotle says: "No other kind of knowledge except intuition is more accurate than scientific knowledge [. . .] . There can be no scientific knowledge of the first principles [meaning: first principles cannot themselves be proven by yet further principles, . . .] therefore it must be intuition that apprehends the first principles."[21]

Through the centuries, many thinkers have pondered this passage, but it leaves numerous questions unanswered. Yet it is revealing that in another of his works, the *Metaphysics*, Aristotle uses the word *nous* in the sense of spirit, as the active principle of the universe. This suggests a connection between the cosmos and our intellectual capacities.

Plato, Aristotle's teacher, had worked out the nature of this connection in detail. According to the Platonic dialogue *Timaeus*, when the Demiurge was finished creating the world, he took the cup that had contained the World Soul and mingled the dregs with the four elements. This mixture, which contained some diluted World Soul, was then divided into as many souls as there were stars in the sky. The Demiurge took the souls for a ride in a chariot, showing them the universe and explaining its nature.[22] Plato believed that souls were immortal (in *Phaedrus*, he even offers a proof this proposition) and continually reincarnated. It follows that our souls are the same ones originally created by the Demiurge; each one still contains a little drop of World Soul and the imprint of the Demiurge's teachings. This seems to invite the suggestion that all we have to do to understand the cosmos is recall what the Demiurge told us.

In other dialogues, Plato did in fact expound this recollection thesis.[23] But Aristotle rejected it, insisting that the acquisition of knowledge must begin with observation. Nevertheless, Aristotle described the way in which we use our intuition to acquire knowledge of first principles as a process of recognition. This seems to imply, in some sense, that knowledge was already present within us, that the abstract concepts at which we arrive after repeated observation are not of our own making. When we stumble upon them, it seems as if we were remembering them.[24] This is the only clue Aristotle provides as to why intuition should be so strangely unerring. It shows that, like Plato, he believed in a deep connection between the soul and the cosmos. Aristotle thus believed that his first principles must apply to the world of our experience, and so he felt justified in invoking them as the true causes of things and phenomena. At least, this is what his work seems to imply, and many have interpreted it this way.

But the philosopher Wolfgang Wieland has argued that Aristotle discovered the first principles in language, through painstaking analysis of how things and phenomena are normally described.[25] This does not mean that Aristotle thought reality might be different for Greek speakers than for speakers of other languages. Intuition, as he conceived of it, was a universal human faculty. He also had great confidence in human observation and in the integrity of the human senses, and he in no way advocated any form of scientific or cultural relativism. Aristotle told the Sophist Protagoras, who claimed that "man is the measure of all things," that this notion was absurd. It allowed for no meaningful distinction between true and false opinions, yet it was obvious that in a dispute between parties with contradictory views, one of them had to be mistaken.[26]

In developing his well-defined vision of how nature works and how knowledge of it can be obtained, Aristotle built on two earlier Greek traditions. The first was the Ionian school of natural philosophy, which influenced what might be called his substantive ideas about nature. Thales had said that everything was water, and Heraclitus had said that everything was fire. As Aristotle observed, these views are untenable because they impose too many restrictions on our ability to account for natural phenomena. Their value lies in the fact that they find a common thread running through nature, instead of portraying it as a stage on which random, unrelated events are played out. The Ionians proposed material first principles, engaging in a form of cosmological speculation. Aristotle traded in their proposed principles for something more abstract, namely, substance. This more abstract concept enabled him to formulate a philosophical response to the apparently ever-changing nature of the world; throughout all varieties of change, substance remains the same.

A second pre-Aristotelian tradition was that of the Pythagoreans, the disciples of Pythagoras. Their worldview was purely formal—more specifically, mathematical. It is sometimes said that Pythagoras was responsible for liberating mathematics from its practical applications, such as surveying. In ancient Sumer and Egypt, mathematicians had been guides to everyday life, playing a central role in rituals and reestablishing the boundaries between parcels of land after the annual Nile floods. As early as 2000 BC, the Babylonians were masters of astronomy, with series of observations of the positions of the planets organized into clear tables. By about 500 BC, they had identified the regularities in these observations and could accurately predict events such as solar and lunar eclipses.

Theorems and proofs were not part of the style of mathematics practiced by the Egyptians and Babylonians. Such proofs were first used in Greek geometry and went against the grain of Babylonian mathematics, which was largely algebraic. Nevertheless, the Greeks acknowledged the groundbreaking work of their predecessors. Legend has it that Pythagoras traveled in Egypt and Mesopotamia, gathering knowledge about mathematics which he then brought back to the Greek world.[27]

According to a different legend, Pythagoras discovered that the simplest ratios between the lengths of two vibrating strings produce the most fundamental musical intervals, such as the octave (1:2), the fifth (2:3), and the fourth (3:4). The historical Pythagoras, however, was a cult leader who probably knew nothing about math.[28] Many, if not most, of the discoveries usually attributed to him were probably due to the later Pythagorean Philolaos, born around 470 BC, three generations after Pythagoras. It was Philolaos who came to the sweeping conclusion that the entire cosmos is built on simple numerical ratios. This "harmony of the spheres" gave mathematics a significance far beyond its practical applications.

His cosmological thinking also assigned a central role to ideal geometrical figures, such as the circle and the sphere. For clearly speculative reasons, Philolaos described the earth as a sphere, immobile and suspended in space. He also managed to break down the apparently irregular movements of the planets in the night sky into a combination of two circular motions.[29] This remained the standard explanation until about 1600, when Kepler discovered the elliptical form of the planetary orbits, much to his dismay. Like the Pythagoreans, Kepler had strong mystical beliefs, and it was very difficult for him to part with the circle, at least in this case.[30]

Incidentally, the last of the Pythagoreans, Archytas of Tarentum, was the first to use formal mathematical proofs. Archytas and Plato were contemporaries, and Plato incorporated a great deal of the Pythagorean legacy into his philosophical writings. In particular, Plato provided firmer foundations for using math to describe natural phenomena. Unlike Philolaos, he did not claim that the world was identical to numbers and figures, but said that numbers and figures are embodied in nature as well as possible, as part of the rational structure imposed on our unruly world by the Demiurge. Plato believed that science would lead to rational control of the world, and that measuring, counting, and weighing were the foundations of all fields of practical knowledge, such as warcraft, commerce, navigation, construction, and medicine.[31]

Strikingly, the idea of a rational science based on both the systematization of experience and causal reasoning from first principles was also present in the work of the physician Hippocrates of Cos and the Hippocratic writers connected with him. In Plato's *Phaedrus*, Socrates speaks highly of Hippocrates.[32]

Aristotle was opposed to two other schools of Greek philosophy: Sophism and the atomism of Democritus. The Sophists, of whom Protagoras was the best known, were noted for their rhetorical skill. They emphasized the particulars of a case, basing their reasoning on the situation at hand. Plato criticized them fiercely, so fiercely that "sophist" became a term of abuse and for many centuries the Sophists were expelled from the philosophical canon.

The other school, Democritan atomism, had two major advocates after Aristotle's time: the Greek Epicurus and the Roman Lucretius. They developed what we would now call a model of nature and its origins. Democritus's teacher Anaxagoras reportedly said, "Visible things are a glimpse of the unseen." This is surprisingly similar in spirit to Heraclitus's dictum, "Nature loves to hide."[33] But for Democritus, the deeper level of reality consisted of atoms, rather than Aristotle's abstract principles.

Aristotle accused atomists of limiting themselves to the material realm, but what he objected to most was their tendency to think in terms of chance processes. In other words, the atomists' concept of "necessity" (*anankē*) differed from Aristotle's. According to them, necessity arose entirely from the nature of matter.[34] Democritus believed neither in an Unmoved Mover nor in a God responsible for designing the world. Similarly, Epicurus and Lucretius rejected the idea that the traditional, mythical gods played any role in nature. Without denying the existence of the gods, they presented them as indifferent to the world. On the movement of matter, Lucretius said, "The first elements did not arrange themselves in their places with foresight or according to a definite plan, and they certainly did not jointly decide how they would move. But because they have been shifting through the heavens in many ways since time immemorial, harried and buffeted by countless shocks." Lucretius also wrote that the plurality of our nature arose from the first elements, just as poetry can arise from the placement of letters.[35] All in all, he presented a picture of a cosmos that had emerged by chance.[36]

In the ancient world, atomism remained a small philosophical school. (It would be rediscovered in the sixteenth and seventeenth centuries.) Though Greek philosophy was quite diverse in nature, its greatest achievement was the deductive method, summarized most consistently in the *Posterior Analytics*. Nevertheless, the legacy of deductivism was not passed down to the Christian world unaltered.

The worldview of Plato and Aristotle was very formal in nature in the sense that it was based on ideal forms or abstractions. This explains why, in Plato's mathematical creation story, the world is modeled after geometrical figures and the order of the universe is based on numbers. It also explains why Aristotle was preoccupied with logic, with definitions and essences, and with the connection between the general (or abstract) and the specific, which he interpreted as a genus-species relationship.

Nevertheless, Aristotle's philosophy includes enough concrete elements that we can reconstruct his views about the world and cosmos. In the *Physics*, Aristotle defended a teleological principle of explanation that is still generally accepted by biologists today: structures that clearly have a purpose cannot have come into being by pure chance.[37]

Another concrete element of Aristotle's philosophy is his concept of "naturalness"—in motion, for instance. There are two forms of linear motion, ascending and descending. For heavy bodies, only a descending motion is natural; they strive to move downward, and by falling, they come home, as it were. All natural motions, unlike unnatural motions, take place of their own accord and restore the cosmic order, which had been disturbed in one way or another. The true state of the cosmos is one without human intervention or arbitrary human will.[38]

Aristotle imagined the mundane world as the four elements arranged in concentric, nested spheres: earth (of which our Earth is made) in the center, with water around it, air around that, and fire in the outermost sphere. In this system, each element has a natural place; according to Aristotle, this explains facts such as the tendency of fire to move upward. In the Middle Ages, the philosopher William of Ockham pointed out a tension in this system. What Ockham observed was that matter at rest actually tends to keep moving downward, toward the center of Earth, and can be stopped only by other matter. According to Ockham, the result is an inherent instability in Earth's crust and the constant movement of what we now call tectonic plates. He put this forward as an explanation of the fact that Earth is not evenly covered with water, one fortunate consequence of which was the possibility of human life. This clever thought-experiment of Ockham's would never have occurred to Aristotle.[39]

The Normative Context of Science

What role did this rigidly hierarchical causal framework for science play in Aristotle and Plato's thinking, and what made it seem plausible to them? One answer, which still resonates more than twenty centuries later, is that it embodied a kind of mystical beauty. But even if this answer is true for the Pythagoreans who preceded Plato and the Neoplatonists who followed him, it is not necessarily true for Plato himself, let alone for Aristotle. Philosophers who came a few generations after Aristotle, such as the atomist Epicurus and the Stoic Chrysippus, made it clear that, for them, science was in the service of ethics and the central question was how to live. In this respect, they resembled Aristotle and Plato.

Both Plato and Aristotle were teachers. Plato's school, called the Academy, was in a building that had been used for higher education even before his time. Aristotle founded an institute of his own, the Lyceum. After his death, this school remained in operation under the leadership of Theophrastus, the author of books about botany and moral character. After Theophrastus died, it became clear that the Lyceum's future hung by a thread. During a dispute about who would succeed him, the sole copy of Aristotle's lecture notes disappeared from the school. The new owner kept them well out of sight, and they did not resurface until the first century AD. In the meantime, there had been no qualified teachers of Aristotelian philosophy.[40]

The educational programs available in the ancient world were not terribly formal. It appears that no degrees were awarded and schools did not expect their students to work toward any particular qualifications. Completing school was considered a good preparation for a career in politics or public administration. This helps us understand why one of Aristotle's pupils was Alexander the Great. The strong emphasis on science, which sometimes met with protests from the students, was intended to strengthen character and sharpen insight. Ethical and practical questions were considered paramount.

In regard to the deductive style of explanation, the ethical position of greatest relevance is Plato's. He strove to develop an ethical theory modeled after science (*epistēme* or *technē*). This ethics was based on a distinction that had existed since Homer's day, between *tuchē* and *technē*. *Tuchē* is fate, or fortune, whether that fortune is good or bad. *Technē* is the best way to preserve yourself from the vicissitudes of fortune, namely by controlling the world around you. Building a roof over your head and making fire are examples of *technē*.[41]

Plato developed a strict hierarchical system of values, which he also interpreted as a form of *technē*. That was because this system made it possible to decide what to do in the face of various moral problems. According to this approach, the best way to live a good life is to have guidelines for making correct decisions.

In the *Protagoras*, one of his earliest dialogues, Plato identified the maximization of pleasure as the rational principle by which people should be guided. But this does not tell us what he meant by pleasure. In his later dialogues, Plato developed an account of what this pleasure involved. In the

Republic, he argued that the best life a person can live is one of contemplation, the life of a philosopher. This is because the highest possible objective is to achieve certainty and serenity. In ordinary, worldly life, our desires for status, money, and sex often lead to ruin, especially in warlike times such as Plato's. From a philosophical perspective, the best guarantee of a good life is to place one's faith in the existence of an eternal truth in nature, independent of all the uncertainties of human life.[42] In this respect, Platonic ethics are both substantively and procedurally modeled after deductive science.

Aristotle's *Posterior Analytics* is similar to Plato's philosophy and strongly influenced by it, but does not contain one word about ethics. The strange thing is that, around the same time, Aristotle wrote major works about ethics and living nature in which he adopted a very different perspective on the deductive method.[43]

In his ethical work, Aristotle is an outspoken opponent of the deductive method. In his biological work, the situation is more complex; Aristotle concentrates on phenomena as they appear to an observer and seems fairly unconcerned about developing a deductive system to account for them. Although he acknowledges the potential usefulness of such a system, in *On the Generation of Animals* he criticizes those, like Democritus, who construct their theories too hastily.[44] To the extent that Aristotle's biological work strives toward general principles at all, those principles are never far removed from his observations, as illustrated by a passage from *On the Parts of Animals*: "The acetabula [suckers] are set in double line [along the tentacles of] all the Cephalopoda, excepting in one kind of poulp, where there is but a single row. The length and the slimness which is part of the nature of this kind of poulp explain the exception. For a narrow space cannot possibly admit of more than a single row. This exceptional character, then, belongs to them, not because it is the most advantageous arrangement, but because it is the necessary consequence of their essential specific constitution."[45] The "essence" of the cephalopod with a single row of suckers may be a kind of principle, but is far removed from the true archai. Passages like this one pose enduring problems for interpreters of Aristotle. The stylistic contrast between the *Posterior Analytics* and his work in other fields (especially biology) has always given rise to confusion.[46] At first, in the twelfth and thirteenth centuries, Aristotle's more detailed biological work (particularly *On the Parts of Animals*) was ignored entirely, if it was known at all.[47] In the medical field, there were two opposing camps: the followers of Avicenna or Ibn Sina and those of Averroës or Ibn Rushd, two Arab commentators on Aristotle's work. The Averroists championed deductive reasoning and the Avicennists stood up for observation.[48] In the late Middle Ages, the Avicennists gained the upper hand.

Alistair Crombie attached the greatest importance to the conceptual link between Aristotle's work in physics and in biology. But Martha Nussbaum draws a strong connection between Aristotle's biological writings and his ethical work. She also reaffirms the essentially deductive nature of Aristotle's philosophy of science. This does nothing to resolve the tension in his works, which Nussbaum describes as a tremendous problem. In her view, the inconsistency of Aristotle's oeuvre reflects differences in the subject matter. In the fields of logic and the natural sciences, he teaches the conventional wisdom, speaking impersonally. But in his ethical work, he steps into the foreground as an author with opinions of his own.[49]

The Failure of the Deductive System in Ethics

The protagonists of classical Greek stories sometimes face tragic choices without realizing the nature of the dilemma. In Sophocles's *Antigone*, Creon, citing an unambiguous legal rule, forbids his niece Antigone to bury her brother. When she defies him, he is infuriated with her disobedience. Not until too late, after his son has committed suicide for love of Antigone, does he realize that there are other values at stake as well. But by that point he cannot stop his wife from hanging herself.

A second example occurs in Aeschylus's *Agamemnon*. A prophet tells Agamemnon, supreme commander of the Greek forces in the Trojan War, that to gain a favorable wind for the passage of the Greek fleet to Troy, he must sacrifice his daughter. Should he pay heed to the prophesy? Any other father in his situation would hesitate, to say the least, before choosing the public interest over his love for his child. The other Greek leaders are concerned about the girl's fate, yet Agamemnon smothers his child's cries of protest without a moment's doubt and performs the sacrifice—in his own words, as if she were a goat. Agamemnon's inability to perceive the tragedy of the situation makes him an unsympathetic character and leads to his downfall when he returns home after ten years of war in Troy.

Nevertheless, neither Sophocles nor Aeschylus favors surrendering oneself entirely to one's fate (*tuchē*), even in the absence of a method (*technē*) for arriving at a single, ideal moral choice. The ethical stance that they support, and for which Aristotle also argues in the *Nicomachean Ethics,* is not a purely passive one. Instead of placing all the emphasis on controllable actions, they advocate a judicious blend of activity and receptivity, informed by insight into the complexities of a specific situation.[50] It is not only doing the right thing that matters, but also having the right emotion.

For Agamemnon, there was only one hierarchy of values. Lesser goods must sometimes make way for greater ones, and in theory, this does not lead to conflict. What Agamemnon failed to acknowledge is that sometimes there is no clear hierarchy of values and we must therefore make tragic decisions. This makes us vulnerable, and all the *technē* in the world cannot protect us from that.

In a sense, Aristotle's method of ethical deliberation is a return to tradition, building on the insights of the tragic poets. Like them, he realizes that we cannot derive ethical decisions from general laws. On the contrary, we must focus on each specific situation and understand it as a complex whole. We must always be prepared for something new, something unexpected. Some dilemmas will be irresolvable because of their many possible interpretations.[51]

The Platonic virtues of stability and self-sufficiency (modeled after the scientific quest for insight into "eternal" things) are of no interest to Aristotle. For the community as a whole (the *polis*), these are not virtues worth emulating. Aristotle's admiration for the Athenian statesman Pericles illustrates his belief that we are allowed to have ambitions and to stick out our necks for them. In a sense, risk-taking is even admirable, because we know that in our lives different values compete for priority. In interpersonal relationships, we may discover that other people too have desires, and that we ourselves are sometimes the object of those desires. The structure of these desires is part of a world in flux.[52]

A Non-Deductive Approach to Biology

According to Nussbaum, Aristotle's ethical theory of action is, in a sense, founded on his theory of the movement of animals—in other words, on his work in biology. And this biological work, in turn, proves to be surprisingly ethical in nature. What people and animals have in common, according to Aristotle—and what the Unmoved Mover and a stone lack, because they are self-sufficient—is the need, the longing, to reach out for things in the world.[53]

In *On the Parts of Animals*, Aristotle relates that some of his students once complained about the coarseness of this subject matter. He replied that their complaints showed little self-respect, since they too were made of flesh and blood.[54] Aristotle's biological work, like his ethical writings, reflects his wish not to stray too far from the needs of everyday life.[55] Oddly, and in defiance of his own prohibition in the *Posterior Analytics*, he uses an analogy to describe the circulatory system, comparing the body to an irrigated field.[56]

Aristotle characterized his scientific method as the description of phenomena as they appear to us. Nussbaum points out that this term (*phainomena* in Greek) has often been mistranslated as "facts," with all the objectivist connotations of seventeenth-century and later empiricism.[57] Her own preference is for the term "appearances," which clearly invokes the perspective of the observer. Be that as it may, Aristotle's phenomena or appearances stand in contrast to the truth (*alētheia*) as discussed in the Platonic tradition (and in Aristotle's own *Posterior Analytics*), the genuine nature of things which is not accessible to observation. This suggests that Aristotle was being deliberately controversial when he declared phenomena to be the true reality, giving the surface of things priority over the depths.

The observer's role in Aristotle's work must also be interpreted in a cultural context. He frequently prefaces his own observations on a topic with an analysis of how people normally speak about it. As we discussed earlier, Wolfgang Wieland was struck by this tendency, which seems to reflect a kind of conservatism. Yet Nussbaum contends that this Aristotelian method opens the way for new modes of thought in both science and ethics, though we must always take care not to lose sight of the implications of new positions for our lived experience of the world.[58]

Wieland offers an even more radical response to the problem of the two Aristotles, the deductivist and the empiricist. Reading the *Physics*, he notes that Aristotle's first principles are always "principles of something," with no independent existence, and that they are not all that far removed from experience. Accordingly, they are more diverse in character than is generally supposed. We looked at the example of cephalopods and their suckers, but the same point can be made about the first principles in the *Physics*, and about Aristotle's work in the philosophy of science. Wieland believes that the *Metaphysics* and the *Posterior Analytics* should be read only after Aristotle's empirical writings. This leads to an interpretation of Aristotle's work in which the deductive organization of science is not a method of scientific discovery, but a didactic tool for presenting acquired knowledge in an orderly way. If this interpretation is right, then the traditional view of Aristotle is misguided, and he did not, in fact, believe that the world necessarily has a deductive structure, or that science can discover that structure.

It is possible that Nussbaum and Wieland have shown us the true Aristotle, a philosopher who may offer valuable inspiration for us today. But for the purposes of this book, what matters most is Aristotle's reception history—the ways in which his work was interpreted and used in the past. One advantage of Crombie's more deductivist take on Aristotle is that it adheres closely to this reception history. References to Aristotle in later chapters of this volume usually refer to this Aristotle, the deductivist. The tradition in the philosophy of science, from the early Middle Ages to the present, has been to see Aristotle as providing a deductive framework for science. This reading seemed plausible in part because Aristotle's biological works were not available in Latin translation until relatively late in the Middle Ages, when his image was already fully formed. Through a Platonic (or Neoplatonic) lens, the *Posterior Analytics* appears to offer unadulterated deductivism, and Neoplatonism was a widespread school of philosophy from the late Roman period to the sixteenth century. Even after a new Christian theology permanently demolished the ideal of knowledge based on first principles, the pure form of deductivism remained attractive to a variety of natural philosophers and scientists. Descartes, for instance, believed that he could salvage the concept of necessity. The Fourier epigraph to this chapter is a still later expression of an uncompromisingly deductive style, which remained synonymous with real science for many people until well into the twentieth century. Albert Einstein construed relativity theory as a purely deductive system, designating it as a "principle theory," an approach which is reminiscent of Aristotle's thinking in many ways.[59] If it is erroneous to call Aristotle a deductivist, it is not only a persistent error but a very appealing one.

3

The Deductive Style in a Christian Context

> *Thinking is a profession with laws that are laid down in minute detail.*
> —M. D. Chenu

FOR A VERY LONG TIME, deductive explanation of natural phenomena remained the gold standard in scientific inquiry. This standard was tied up with the very definition of science. Thomas Aquinas said that *scientia* was the knowledge of eternal and necessary truths (from which knowledge of more specific natural phenomena could be deduced). Whatever could not be proved was *opinio*, a very broad category (ever broader, in fact, as time went on). Opinions are not devoid of value, Aquinas argued, certainly not if they are so plausible that they are universally believed, especially by the best people.[1] But one aim of natural philosophers should be to bring these fields of knowledge into the domain of true science. Natural philosophers cherished this aim until, and even after, the days of Descartes, whose ideal vision of science was a rational system of a deductive nature. This was largely identical to the conception of science in Aristotle's *Posterior Analytics*, a book that strongly influenced Descartes. The axiomatic structure of the *Posterior Analytics* also showed clearly in the work of Descartes's contemporary William Harvey, the physician and physiologist who discovered the circulation of the blood.[2] Nevertheless, Descartes's and Harvey's deductive ideal for science had a different intellectual context than that of antiquity.

Thomas Aquinas's work went a long way toward cementing the Aristotelian legacy, but could not bridge the many centuries that separated Aristotle's world from his own. Medieval science could no longer claim a direct connection to Aristotle, however much that would have pleased medieval philosophers and investigators of natural phenomena.

Reconquering the Heritage of Antiquity in the Middle Ages

After the collapse of the Western Roman Empire, there was a gap of many centuries in Western Europe's scientific culture. To the east, in the Byzantine world, Greek and Hellenistic culture were carried forward, but the West lost contact with this tradition at least until the time of the Crusades.

The culture of antiquity had to be reconquered, and the place from which that reconquest was launched was the urban society that began blossoming in Europe in the twelfth century. In Paris, Toledo, Naples, and Palermo, the new secular and ecclesiastical elites were aware of the lost civilization in their past. But reclaiming that civilization was no easy task. The language of the church and scholarship was Latin, which had survived in the cloisters in the centuries before the cities emerged as economic and cultural centers. But ecclesiastical scholars, such as Pierre Abélard in Paris, no longer knew Greek. Furthermore, the Romans had read Plato and Aristotle in the original language, and so there were no Latin translations available.[3] Most of the Greek originals had been lost, and all that remained of Plato's work was a fragment of the *Timaeus*. All that was left of Aristotle's work were his writings on logic, which became the core curriculum at schools and universities, constituting the heart of the tradition known as scholasticism.

Under these circumstances, reconquest meant translation, and because in the twelfth and thirteenth centuries the closest civilization was that of the Arab world (most of the Iberian Peninsula remained Muslim, and Sicily had been Muslim until not long before), most translation was from Arabic. Adelard of Bath, the translator of Euclid's *Elements*, travelled to Syria and Sicily to obtain manuscripts.[4] The medical school in Salerno, near Naples, needed instructional materials, and began translating them from Arabic in the eleventh century. Sicily, where the celebrated Holy Roman Emperor Frederick II would establish his court two centuries later, remained an important point of contact with Arabic culture. But in the twelfth century, Toledo was the bustling hub of translation activity. The most prolific translator was the Italian Gerard of Cremona, whose works include Latin versions of Aristotle's *Posterior Analytics* and Avicenna's *Canon of Medicine*.[5] A number of Toledo's twelfth-century archbishops supported translation, and some translators are known to have had church stipends, known as prebends.

These translators were scholars, and many of them held clerical titles.

They produced translations not only of Aristotle and Euclid, but also of Ptolemy's *Almagest* and works by Arabic-speaking scholars such as Avicenna, a physician and commentator on Aristotle's work; Al-Kindi; Alhazen, who had contributed to the field of optics; and Al-Khwarizmi, the author of a book about algebra, a branch of mathematics unknown to the Greeks. Works on alchemy, astrology, and prophecy were also translated.

In a few exceptional cases, translations were made directly from Greek. Archimedes, for instance, was translated by the Flemish cleric William of Moerbeke, who learned Greek before becoming archbishop of Corinth. His accurate translation was less influential than others that were based on Arabic versions but adapted to conform to the prevailing scholastic philosophy, in which geometric proofs were presented as logical arguments. It was not until the Renaissance that the importance of Moerbeke's translation was fully recognized.[6]

The world of the Greeks, as it resurfaced in the West, was no longer simply that of Aristotle—even setting aside the fact that Aristotle had been translated from Arabic. Numerous influences from the Hellenistic period, partly mediated by the Muslim world and partly by early Church fathers such as Augustine, were at work in the new science of the Middle Ages. Arabic scholars, who were themselves heavily influenced by Neoplatonism, made substantial contributions to medieval thought. Finally, it was Christianity that provided the overarching framework. Most universities grew out of ecclesiastical schools and were under the nominal authority of the pope. Many of the instructors were affiliated with monastic groups, especially the Dominican and Franciscan religious orders, which had been founded not long before.[7] But no university was ever dominated by a single order, at least not in the Middle Ages.

Starting around the year 1200, the universities of Bologna, Paris, and Oxford operated as more or less independent educational institutions. Both the students and the instructors were granted all sorts of rights and privileges by both the pope and the Holy Roman Emperor (in the case of Bologna) and by the King of France (in the case of the Sorbonne). The towns where the universities were established also accorded privileges to instructors, such as tax exemptions. The historian Jacques Le Goff painted a rather idyllic picture of thirteenth- and fourteenth-century universities as free associations of teachers and students. The fact is that the universities were never subject to a fully centralized form of authority. The lines of demarcation between the secular and religious authorities, the pope and the bishops, and the towns and the sovereigns, always left room for interpretation and conflict, and hence for a degree of intellectual freedom.[8]

In the thirteenth century, the curriculum took a form that it would retain in most European countries until well into the eighteenth. It began with a liberal education provided by the faculty of arts, consisting of philosophy and the trivium and quadrivium (roughly equivalent to the humanities and mathematics respectively).[9] After that, students could move on to one of the three other faculties: medicine, law, and theology.

Students typically completed the study of the trivium and quadrivium at the age of twenty, receiving the title of *magister* (master). This degree gave them a high social status, comparable to that of a knight, as well as a license to teach, which allowed them to move freely from one school to another. Teaching was a means of subsistence that was always available, even to individuals such as Abélard, who was affiliated with the cathedral school of Notre Dame (which later developed into the Sorbonne) but forced to flee after his love affair with Héloïse came to a tragic end.

The programs of study of the three major faculties, which awarded doctoral degrees, were forms of professional training. Of the three, the medical faculty was most closely linked to the natural sciences. But its curriculum was entirely theoretical, focusing on Aristotle, generally in the company of his Arabic commentators Averroës and Avicenna. Cadavers were very occasionally dissected during the lectures, but for understandable reasons, this was done in the heart of winter and at a brisk pace. The overwhelming emphasis was on textual knowledge.[10] Even practicing physicians did not carry out dissections; surgeon and barbers did so under their supervision. Physicians used only a few techniques of physical examination, and performed no significant operations on the bodies of their patients.[11]

By 1220, Europe's earliest universities had been joined by newer ones in Montpellier, Padua, Naples, and Salamanca. The medical faculties in Montpellier and Padua were especially significant. The number of universities grew steadily, with many new ones founded in the fourteenth century, in such places as Prague (1347), Vienna (1365), and Cologne (1388). In the Low Countries, a university was founded by papal bull in Leuven (Louvain) in 1425.

The Christian God and the Laws of Nature

One revealing illustration of how Aristotle's deductive method was placed in a new, Christian context is provided by Descartes. Though Descartes is classed as a modern rather than medieval figure, his thinking was profoundly shaped by scholasticism.

Descartes asked a question that Aristotle would never have asked: "How is it that I know that one plus two equals three?" To Aristotle, Descartes's answer would have been incomprehensible: "Because God guarantees it, and He would not deceive me." But Descartes also argued—and this is the Christian part of the answer—that if God had wished to deceive him, He could have. In fact, Descartes claimed that, if God wished, He could make one and two unequal to three, and the lines in a circle from center to circumference unequal.[12] This would be impossible for Aristotle's God, who is identical to the notion of necessity. Descartes's God and Aristotle's are emblematic of two very different intellectual contexts for the deductive method. Descartes's God, unsurprisingly, is the God of the Bible. In Western culture, the God of the Old and New Testaments was conceptually dominant, although the role of Aristotle's God never entirely disappeared. And there were also lingering traces of a third god: Plato's Demiurge from the *Timaeus*, which had exerted a formative influence on Augustine.

The God of the Old and New Testaments created the world of his own free will, out of nothingness. This God is not a part of the world himself, but transcendent, in marked contrast to the Demiurge, who is part of the order of the World Soul. The Demiurge created the world and humanity, not out of nothing, but out of matter, and according to a preconceived idea. What is more, the biblical God's act of creation is seen as the historical beginning of the world, while there is no evidence of the Demiurge having created the world at any specific time. It seems natural to regard the Demiurge as a lesser god, a link between the Unmoved Mover and the world, purely logical in nature and not part of any temporal series of events.

Despite the fundamental differences between these conceptions of divinity, from the Hellenistic to the early modern period many theologians, scientists, and philosophers (many of them fell into more than one of those categories) tried to reconcile the Greek and Hebrew traditions, proceeding from a deep conviction that the two *must* be compatible. Each one of them came up with a different solution from his predecessors, and this led to an intriguing intellectual tension running through the history of Western science. For instance, merging the biblical God with the Demiurge (since both were creator deities) made Him seem a little more Greek, more as if He were guided by a rational plan.[13]

Different images of God sometimes collided. In the eleventh century, for instance, the powerful cardinal Peter Damian attacked what he saw as an excessive reliance on logical argument (part of Aristotle's legacy) and argued that God had the power to undo events that had taken place in the past.[14] During the first decades of the Sorbonne's existence, Aristotle was condemned on many occasions. From a long-term viewpoint, these were steps in a process of accommodation, in which the ecclesiastical authorities gradually came to see Aristotle as serviceable to Christian thought. But in the short run, such condemnations could have dramatic consequences for individuals.

Around 1245, Roger Bacon, who had been trained at Oxford, was teaching Aristotle's philosophy in an unadulterated form. But the faculty of theology at the University of Paris (the Sorbonne) grew uneasy about the Aristotelians in the faculty of arts.[15] On the initiative of Bonaventure, the leader of the Franciscan monastic order, the theologians complained to the authorities. Etienne Tempier, the bishop of Paris, responded by condemning a number of Aristotelian doctrines, such as the eternity of the world. Yet the greatest blow, which had far-reaching consequences, came in 1277, when Tempier prohibited teachers at the Sorbonne from teaching 219 separate Aristotelian doctrines.[16] The implicit target was Siger de Brabant, the most prominent philosopher on the Sorbonne's arts faculty. Siger went to Italy, hoping to be rehabilitated by the pope, but he was murdered there.[17] A number of Thomas Aquinas's doctrines were also included in Tempier's condemnation, and it was not until 1322, when Aquinas was sanctified, that the prohibition on those doctrines was lifted.[18]

Siger had a reputation for being an acute and rigorous thinker, in the tradition of the rationalistic Averroës. Dante would later grant Siger a place in Paradise in his *Divine Comedy*, because of the persecution he had suffered.[19] But Siger also believed that Aristotle's death marked the completion of science, and that no one else had contributed anything of worth.[20]

In contrast to the Aristotelian conception of divinity, Damian and Tempier believed in God's absolute power and freedom of will. There had previously been attempts to reconcile this view, known as voluntarism, with Aristotelianism. In 1245, Alexander of Hales united the two in a phrase: *potentia dei absoluta et ordinata*, the absolute and ordained power of God. Thomas Aquinas later clarified this distinction: God's *potentia absoluta* concerns everything that is not self-contradictory, while His *potentia ordinata* relates to the rational order of things. This may be the order not only of our own universe but also of any other conceivable world. Though the concept of the *potentia ordinata* may seem to preserve the essence of Aristotelian thought, it actually involves several departures from it. The *potentia ordinata* is arbitrary in nature, because many orders are possible and the order familiar to us is only one of them. By comparison, Aristotle believed that only one world was possible, namely our own. Alexander's writings pointed to the contingency of the world, a conclusion drawn explicitly by John Duns Scotus and William of Ockham.[21]

Duns Scotus developed an analogy that was, in a sense, entailed by the concept of *potentia ordinata*: ordained power, he wrote, was like a system of laws decreed by God. The source of this analogy is the covenant between God and humankind, God's promise to Noah that there would be no second deluge.[22] By making this promise, Duns Scotus argued, God set a limit on his absolute power and created natural law. Natural law was another concept not found in Aristotle. As we have seen in the previous chapter, Aristotle employed the concept of *ananke* ("necessity") in his philosophy and his logic. The term *nomos* ("law") appears in his work solely in a legal context, and is never metaphorically extended to nature.[23]

The application of the term "law" to nature probably has Hebrew origins, judging by the writings of Philo Judaeus of Alexandria.[24] Philo was the first of numerous philosophers who attempted to reconcile the many gods of the Hellenistic world. Philo's God, however, was primarily the God of the Old Testament, who created the world out of nothing. This God is said to have created the world in accordance with preconceived principles of his own, like an architect carrying out a rational plan. Philo borrowed the term *logos* from the Stoics and applied it to these principles, which he claimed were manifest in the world as laws (*nomoi*) and ordinances (*thesmoi*). "For this world is the great city, and it has a single polity and a single law (*nomos*), and this is the reason (*logos*) of nature, commanding what should be done and forbidding what should not be done."[25] It was through Philo that the concept of natural laws (*naturales leges*) reached Augustine.[26]

In principle, Aristotle's philosophy made deductive reasoning possible through first principles, and natural law made deductive reasoning through laws proclaimed by God. But William of Ockham concluded that there was no alternative to accepting the laws of nature made by God, and no point in trying to understand God's reasons. Hence, "necessity" ceased to be a relevant concept, since laws could have been proclaimed for no other reason than that God apparently wanted them. This greatly reduced the domain of Plato and Aristotle's strict deductive rationality. One advantage of this move was that it brought the fruitless search for first principles to a halt. Laws could be discovered at a much lower level of abstraction than the one at which Aristotle had searched for his first principles. For instance, Roger Bacon declared the regularity he discovered in the refraction and reflection of light to be a *lex naturae universalis* (universal law of nature).[27] The deductive ideal remained intact within each separate domain of research into nature, but there was no longer any conceptual need to unite these domains into a single overarching framework.

The new concept of natural law freed science from its Aristotelian shackles. This was the unintended result of Tempier's condemnation of Aristotle's doctrines. For many years afterward, scholars invoked Tempier in defense of audacious hypotheses, premised on God's absolute power. For instance, in the fourteenth century, Jean Buridan hypothesized that the entire cosmos rotated on its axis.[28] The concept of natural law would reach its height in the seventeenth and eighteenth centuries.[29]

To this day, William of Ockham is remembered for a principle known as Ockham's razor: phenomena should be explained with the fewest possible number of assumptions. What is not as well known is that, according to Ockham, God was not bound by this principle; while we explain a phenomenon by reference to a natural law, God could have produced that phenomenon in hundreds of different ways. This clearly implies that there is no way of discovering how God created the universe.

A second consequence of the new, voluntaristic theology was that it could no longer be assumed that God had created the world for humanity's benefit. This assumption would have restricted God's freedom of action, and therefore had to be rejected as incompatible with God's absolute power. All in all, voluntarism painted an almost existential picture of human life: We find ourselves in a world that we do not understand and that perhaps was not made for us. This bitter fact can lead to the renunciation of worldly things, but also to the insight that we have to make the best of our lot.[30] More than a century later, the Italian humanist Pico della Mirandola argued that the lack of a fixed place and a fixed purpose in the world was the very essence of human dignity.[31]

The development of scholastic philosophy, as sketched here, did not have an equally strong impact on all parts of the church and the universities within its sphere of influence. Duns Scotus was most highly regarded among the Franciscans. Aristotle's impact remained very powerful thanks to the followers of Thomas Aquinas, most of whom were Dominicans. These two fraternal orders left a deep mark on education at the Sorbonne and other universities, with Dominican views sometimes gaining the upper hand, and Franciscan views at other times. The controversy between these two camps went on until the seventeenth century. At the same time, the less influential Capuchin order defended a Christian version of Stoicism.[32]

Natural laws proved to be a very useful way of expressing the operation of forces. The concept of force derived in part from the cosmological speculations of the Stoics, a Hellenistic school of philosophers. This concept had been unknown to Aristotle.

Some time after Aristotle's death, his Peripatetic school lost its primacy among philosophical movements to Chrysippus, the third teacher of the Stoic school. Much more explicitly than Aristotle, Chrysippus drew analogies between the macrocosm (the world, the universe) and the microcosm (the individual). He visualized the world as an organism. This was no metaphor; for the Stoics, the world *was* an organism. Their central concept was *pneuma*, which in reference to a living organism means *breath*.

The pneuma, which fills the cosmos, is made of the elements of air and fire. These two elements bind the cosmos together and make it a unified whole. One inherent quality of pneuma is its tension (*tonos*), and it imparts this quality of tension to the matter with which it unites.[33] Within the Stoic worldview, this is the origin of the concept of force, with which matter can be more or less endowed depending on the workings of the pneuma. Relative to Aristotle's system, however, Chrysippus's philosophy (like Epicurus's atomism) is more speculative. It therefore makes more substantive assumptions about the nature of the cosmos.

The Stoic philosophers were not very influential in the Middle Ages, at least not in mainstream natural philosophy.[34] After the fall of the Roman Empire they went underground, as it were, re-emerging in the late Middle Ages and the Renaissance. The new Stoics, such as Nicholas of Cusa (Cusanus) and Bernardino Telesio, contrasted their forces with Aristotle's forms, and rather than working with hierarchically interpreted genus-species relationships, they focused on the relationship between the whole and its parts.[35] Stoic philosophy was also indirectly transmitted through the writings of Cicero[36] and the physician Galen. Philo and Augustine mingled Stoicism with many ideas from Neoplatonism, which also involved a concept of force.

Augustine was unquestionably the best-known supporter of Neoplatonism to convert to Christianity, and he was also the conduit through which many Neoplatonic ideas entered Christian thought. Neoplatonism was a philosophical movement that enjoyed a great deal of support starting in about AD 200, with schools in Athens, Alexandria, Rome, and elsewhere. It was a rival of Gnosticism, another mystical movement, and of Christianity. Emperor Justinian, whose court was in Constantinople, closed the Neoplatonic schools in the Byzantine Empire in 529 as part of his campaign against heresy.

Neoplatonism's chief representative was Plotinus, author of the *Enneads*. He and the other Neoplatonists attempted to build on the work of Aristotle and Plato and to reconcile the two philosophies whenever they seemed to conflict.[37] In the Middle Ages, Plotinus's writings were not directly available. There was, however, an Arabic manuscript entitled *The Theology of Aristotle* that summarized the second half of the *Enneads*. This work greatly influenced Arabic scholars.

What the Neoplatonists took from Aristotle was the idea of a causal chain beginning with the Unmoved Mover, also known as the One, and ending in the world of observable things. What was inspired by Plato was the idea that the first stages in this chain were entirely immaterial. The One gives rise to Being, which gives rise to the Soul (*Nous*). These are the realms of being that are not perceptible to the senses, and about which we can only gain knowledge by mystical means.[38] Then comes the World Soul, and then Nature, which includes individual living creatures.

The Neoplatonists describe the way in which the One gives rise to the following stages as a current streaming from a source. Their term for this is "emanation." Examples of emanation include light (*lumen*) flowing from a source of light (*lux*), heat from a source of heat, and healing power from a medicine. Light is the best example of emanation, Plotinus says, because it is immaterial. We must imagine the Soul as exactly the same kind of source. The emanation is a result of force (*dunameis*), which diminishes as the distance from the source increases. In Canto 2 of the *Paradiso*, Beatrice uses this theory to explain the dark spots on the moon: because the moon is far from the original source of light, the light received has lost much of its force and is unevenly distributed.

In Neoplatonist thought, a source of light gives light continuously without going out. This may not be true of an earthly source of light, such as a candle, but it is true of the sun and stars, which are eternal bodies.[39] In the religious sphere, this led the Neoplatonists to become sun-worshippers, practically equating the sun with God. Augustine called God the light, and identified light with truth.[40] To Robert Grosseteste, a Franciscan and teacher at Oxford who became the bishop of Lincoln in 1235, light was the medium in which God had created first space itself and then all matter.[41]

For the Neoplatonists, light thus took on a general metaphysical significance, as a model of how a cause could transmit the image of itself to its effects. All bodies on which light falls become sources of light themselves, and in the same way, the entire universe becomes a cohesive unity of forces.[42] In-depth exploration of this analogy led to a physics of light, first in the work of Al-Kindi and Alhazen, and then in that of Roger Bacon. Bacon's ideas were based on those of his Arabic predecessors and Grosseteste, but he developed a more detailed theory of light and the nature of visual perception in the eye.[43] This line of investigation, known to medieval scholars as *perspectiva*, is not only of theoretical interest, but is also noteworthy for its reliance on personal observation and experience.

Toward a Logic of Experiment

Medieval philosophers, university professors, and physicians did not conduct experiments in the modern sense of the word, but some of them did lay the theoretical basis for doing so. Two theorists of particular importance were Grosseteste and Bacon. It should not be supposed that they were radical reformers; Grosseteste did not believe that he was doing or discovering anything novel. He was the author of a commentary on Aristotle's *Posterior*

Analytics, and both he and Bacon tended to regard Aristotle as an infallible authority. Nor did Bacon and Grosseteste see Neoplatonism (which both men largely embraced) as incompatible with Aristotelian doctrines.[44]

Grosseteste acknowledged that natural philosophy would never become certain knowledge, *scientia,* to the same degree as mathematics. But he held that reason, in combination with experience or observation (he used the Latin term *experimentum*) could help to refute a false theory.[45] As we have seen, Aristotle had also made use of observation—a great deal, in fact—but not in the same way as Grosseteste. First of all, there was something peculiar about Grosseteste's choice of the word *experimentum.* Although in this context it meant little more than "observation," it stemmed not from the philosophical but from the magical tradition. Roger Bacon used the same word and made no secret of his interest in magic.[46] Second, Aristotle had used observation to arrive at first principles through an inductive procedure, and once he had arrived at the first principles everything else followed naturally. Grosseteste contended that as a rule, some effects have more than one possible cause; we have only an imperfect ability to determine the correct "first principles" and therefore the true causes of things. He proposed that false theories should be purged both through the faculty of reason (by means such as *reductio ad absurdum,* a procedure used by Aristotle and Archimedes before him) and through observation. Grosseteste's work makes it clear that he arrived at his conclusions mainly through reasoning, and that he conducted no experiments. He did invoke experience, but this probably did not refer to his own personal observations. Nevertheless, experiments had become conceivable.[47]

It should be noted at this juncture that Grosseteste's view of the mathematical structure of the world was quite different from the Pythagorean view. According to Pythagoras, the meaning of mathematics was almost exclusively symbolic. Grosseteste did a much more precise job than all his predecessors of connecting mathematical expressions to observable aspects of reality.[48]

But it is Grosseteste's student, Roger Bacon, who has truly gone down in history as the first empirical scientist, and some early historians of science claimed that Bacon himself conducted experiments and worked to perfect scientific instruments.[49] The historian Eduard Jan Dijksterhuis, however, rejoined that Bacon's work had not led to scientific progress, and that on a technical level Bacon had been more of a Jules Verne–type fantasist than a man of action. Although Dijksterhuis approached Bacon from within the medieval tradition, his assessment of Bacon was not entirely fair.

Bacon's greatest achievement was his synthesis of the *perspectiva* tradition, which was based primarily on Alhazen. In Alhazen's day, there were various opinions about the formation of images in the eye. The dominant theory was that the eye created images by emitting a radiance of its own, like a kind of lantern. In contrast, Alhazen believed that light emitted by objects formed images in the eye. In this theory, the eye was a passive receptor. Alhazen's view brought together three bodies of knowledge; it was mathematical, it was physical, and it took account of what was known about the anatomy and physiology of the eye. Alhazen's experience with physics included some familiarity with the *camera obscura*, which led him to some of his ideas about the characteristics of light (but which he did not regard as a model for the eye). He was also familiar with curved reflective surfaces (concave mirrors).[50] Bacon studied Alhazen's work very thoroughly, and like Alhazen, he pondered theoretical and mathematical issues relating to light and made a major contribution (in the thirteenth-century context) to the understanding of cause and effect.[51] Though Bacon did not investigate the technological applications of his work, it is clear that he was familiar with the camera obscura.[52]

Without fully realizing it, Bacon broadened Aristotle's concept of experience in at least two ways, to include both experience aided by technology (whether instruments or alchemy) and singular experiences (while Aristotle had only recognized everyday experience).[53] In other words, for Bacon, technology and magic were associated with science. This was a departure from the usual medieval view, which was the same as Aristotle's: namely, that alchemy did not belong to the realm of science, but to that of technology.

Bacon considered alchemy a kind of natural magic. The category of natural magic included many innocuous phenomena, such as rainbows seen against the light in the spray of a fountain. But Bacon was undoubtedly also thinking of demons and their power to bring about wondrous effects, though the term "wondrous" should not be understood in any supernatural sense. Although magicians enlisted the services of angels (for white magic) and devils (for black), even angels and devils were considered subject to natural law.[54] Bacon claimed that there was nothing supernatural about magic and astrology, that both could be explained by scientific (that is, rational) means.[55] What interested him most about magic was that it was a vast and fascinating reservoir of knowledge, neglected by philosophers.

In a famous passage, to which Dijksterhuis was alluding when he described Bacon as a Jules Verne, Bacon fantasizes about flying machines and submarine vessels. His inspiration came from tales of travels in the Arab world, where technology was far more advanced than in Europe.[56] It is said that Bacon's beliefs about astrology and alchemy, and about the need to fight the coming of the Antichrist with science and technology, led to his confinement by the leader of his religious order, but the truth of this story is uncertain.[57] It would be an exaggeration to hail Bacon as an early Leonardo da Vinci on this basis, but on the other hand, his ideas were far from irrelevant. In the theoretical domain, they helped to lower the barrier between science and technology.

Some later medieval thinkers shared Bacon's interest in technology, and moreover, they can be credited with some experiments of their own. In 1269, Pierre de Maricourt did some original things with a magnet, though he presented the results in a familiar, Aristotelian theoretical framework. What he did might be described as experimental exploration. Two scholars who carried on the tradition of *perspectiva* research after Roger Bacon, namely Witelo and Theodoric of Freiberg, read about the experiments that Alhazen had performed, and Witelo may have conducted other experiments. For his highly original study of the refraction of light in the rainbow, Theodoric used glass globes filled with water, which he shook and placed on the windowsill of his study on a sunny day. It is thought that his apparatus was, in fact, capable of modeling the behavior of sunlight in drops of water. But Witelo and Theodoric's work did not launch an experimental tradition. This is not to say that optics had no practical implications; it lay behind the invention of spectacles, which came into widespread use in the late thirteenth century.[58]

Quantitative Matters and the New Mathematics

The deductive method and mathematics were connected from the very beginning, and they remained so. Ancient Greek mathematics was limited to plane geometry, with occasional forays into the third dimension, as in the work of Archimedes. Geometric arguments were also taken as models for all forms of argumentation, and the phrase *more geometrico* (literally "in a geometric way") took on the meaning "in a strictly deductive way."

In empirically oriented science it should be possible to choose between two possible causes by quantitative methods, such as counting and calculation. Robert Grosseteste had laid the foundations for this approach.[59]

Counting and calculation fell under the heading of arithmetic, one of the subjects in the medieval quadrivium. Music (harmonic theory), another subject in the quadrivium, was seen as applied math.[60] Since the days of the ancient Greek philosophers, the relationship between geometry and arithmetic had been re-evaluated by Thomas Aquinas (and Augustine before him). Euclid had denigrated calculation with numerals as the stuff of the marketplace, unworthy of the name of mathematics.[61] Aristotle saw numbers as properties of line segments. This relegated them to a secondary role and limited the potential for numeric abstraction.[62] Aquinas, however, accorded metaphysical priority to arithmetic. His reasoning was that arithmetic dealt with discrete amounts and geometry with continuous amounts, and that continuity meant the continuation of something discrete, namely a unit.[63] Nevertheless, the numeral system at the disposal of thirteenth-century Western European scholars was not much superior to that used by the Greeks.

The system used by both Archimedes and his contemporaries in the marketplace had consisted of the letters of the Greek alphabet supplemented with additional symbols, for a total of twenty-seven. These symbols could be used to write numbers less than one thousand. It was a decimal system, but the position of a symbol in the number did not affect the value: ιδ and δι both meant 14 in modern-day notation. But this system allowed the Greeks in the marketplace to count quickly, and apparently it was also useful to scientists—for example, when they compiled astronomical tables. In the Byzantine Empire, the Greek numeral system remained in use until the fall of Constantinople in 1453.[64] Roman numerals, which we still recognize today, were actually less practical than the Greek system, especially for representing large numbers.

Around the year 1000, while travelling in Moorish Spain, the monk Gerbert d'Aurillac (later Pope Sylvester II) stumbled upon an abacus. It was divided into vertical columns corresponding to units, tens, hundreds, and so forth. Gerbert successfully promoted the abacus, but without introducing the Arabic numeral system implicit in its design. For several centuries, European arithmetic consisted in large part of manuals for calculating with an abacus. Fairly complex numerical operations could be carried out, though the tool also had disadvantages. In a calculation with many steps, intermediate steps were lost (for obvious reasons). The abacus was also unsuited to calculations involving both very large and very small numbers. Nor was it quick to work with. The Chinese abacus, which is used to this day, is much faster.[65]

In 1202, the merchant Leonardo of Pisa, known as Fibonacci, published his *Liber Abaci* (*Book of Calculation*). In this book he not only introduced Arabic numerals, but also applied them to numerous algebraic problems, some of which he must have invented on his own, while others were taken from a variety of Arabic scholars whose work went as far back as the ninth century. (Because the Arabs had adopted their numeral system from the Indians, the term Hindu-Arabic numerals is often used.) A book by the best known of Fibonacci's Arab predecessors, Al-Khwarizmi (whose name is the source of the word *algorithm*), had in fact already been translated into Latin in 1145.[66] Fibonacci had probably learned Greek and Arabic at the court of Emperor Frederick II in Palermo and during his travels in the Muslim world.[67]

Yet neither the notation used by Al-Khwarizmi nor Fibonacci's modernized version was adopted by many other scholars. University scholars essentially ignored their work, and Arabic numerals spread at a glacially slow pace. The earliest French manuscript to use them dates from 1275.[68] The new arithmetical techniques were adopted most rapidly in the marketplace, exactly where Euclid had said that arithmetic belonged. In the fourteenth century, they came into use in bookkeeping among Italian merchants, and in combination with the recently introduced double-entry technique, they created the potential for new clarity.[69] It was not until 1494, however, that Arabic numerals alone were used in the financial records of the Medicis in Florence; Roman numerals still had many years of life ahead of them.[70]

The fact that thirteenth-century university scholars did not adopt the new arithmetic is illustrative of the lack of experiment in theoretical fields. There was evidently no demand for clearly organized tables of measurements or algebraic functions describing the relationship between two variables. The practice of systematic measurement originated in crafts, or "arts," such as metallurgy, the extraction of metal from ore. Likewise, time measurement began as a response to practical issues such as the need for an adequate calendar, and entered science by way of astronomy.[71]

At the Sorbonne, Jean Buridan and Nicole Oresme did not engage in any form of measurement, nor did Thomas Bradwardine at Merton College in Oxford. Nevertheless, these scholars were interested not only in *why* physical bodies moved, but also in *how* they moved.[72] Buridan and Oresme developed a variety of mechanics, known as impetus theory, that radically departed from Aristotle's ideas about motion. According to this theory, the act of throwing a projectile gave it a quality called an impetus. This kept the projectile in motion until the impetus was used up. The concept was developed by analogy with heat, another "quality" that could be imparted to a thing. In principle, impetus was quantifiable, but it was never actually described in quantitative terms.

When medieval scholars turned their thoughts from Aristotle's very general deductive methods, they did not concentrate on generalizations at an intermediate level, of the kind that would be developed in the Renaissance, but on individual cases. Where experiments began to take place, they were generally qualitative, based on the systematization of experience. The richest source of these early experiments is medical science.

Medical Science

The oldest urban, as opposed to monastic, program of education for the medical professions was founded in the tenth century in Salerno, a city to the south of Naples. This was not a formal school, but a center that emphasized practical disciplines, such as surgery and obstetrics. In the course of the twelfth century, the manuals and other writings produced in Salerno grew more and more theoretical, no doubt under the influence of translations from the Arabic. The establishment of the universities greatly reinforced this tendency. Students in the medical faculties at Bologna, Padua, Paris, and Montpellier started out in the faculty of arts, where they received philosophical training before beginning their medical studies. The content of medical theory was largely Aristotelian; the influential Arab doctors Avicenna and Averroës had declared that whenever Aristotle's opinion differed from that of the more empirically minded Galen, Aristotle should receive priority.[73]

Among medical professionals, there was a long-standing division between the university-trained physicians and the surgeons. Surgeons generally kept their knowledge to themselves, often transmitting it from father to son, though they did have a place at some universities. Although physicians did not carry out surgical procedures, they did perform a number of practical tests, such as examining urine and checking the pulse. These tests gave them a great deal of information, which they used to make diagnoses and give advice—for instance, on diet, which played an important role in the medieval view of the body.

Throughout the Middle Ages, there were debates about medicine's status as a discipline; Aristotle had called it an art, but the Arab scholars Avicenna and Averroës had both ventured to disagree and label it a science.[74] Averroës was more consistently "scientific" in the sense of employing the deductive method, while Avicenna made a distinction between what counted as knowledge for a physician in his role as a scientist and in his role as a practitioner. Working physicians, Avicenna said, had to be guided by practical questions. Avicenna's medieval followers were more inclined than those of Averroës to take practical experience seriously. One of the most outspoken exponents of this approach was Arnold of Villanova, who taught at Montpellier and served as the personal physician to members of the Aragonese royal family in Barcelona. Arnold probably did not carry out true experiments (except perhaps alchemical ones). But he did employ everything known in practice and in the literature about, say, the effects of a diet or medicine to determine the best possible diagnosis or treatment. In his *Experimenta* and other writings, Arnold stressed the importance of practical knowledge and careful observation.[75] The term *experimenta* was used in standard medical literature from the thirteenth to the fifteenth century to refer to prescriptions that had supposedly proven effective.[76]

Also according to Arnold, the fact that a medicine had a certain effect on one type of animal did not necessarily imply that it would have the same effect on another type of animal, unless this was known from experience.[77] Medieval practitioners tested their medicines, but this "technological" aspect of medical science was seen as irrelevant to the method of scientific inference. Nevertheless, it is reasonable to conclude that in the Middle Ages medicine became a field where science and technology were very closely linked. This was especially true in the realm of diagnosis.

Diagnosis tends to reveal the limitations of the deductive style because a practical decision has to be made about the nature of the illness in an individual case, but the physician has no more to go on than a collection of clues known as symptoms. Experiential knowledge has to be the crucial factor when, for instance, distinguishing leprosy from other skin diseases.

The medical diagnosis exemplifies a model for acquiring knowledge about individual cases that is also found in very different areas, such as law and theology. When theologians were confronted with heresy, they had to identify precisely what form of heresy it was. And in complex legal cases, lawyers had to rely on the case law at their disposal. Case-based legal reasoning is closely related to the technique of medical diagnosis.[78]

The art of making a diagnosis is one of mankind's oldest fields of knowledge, older than the deductive method, from which it differs fundamentally. As the cultural historian Carlo Ginzburg has observed, diagnosticians move from observation to observation like a hunter moving from one pawprint to the next. Their goal is not to devise a theory or arrive at an abstract generalization. There would be no point in that, because their object of scrutiny is an individual or an individual case. When making a diagnosis, or tracking an animal, what matters is to reconstruct the story, how things came to be this way, what happened, and in what order. Medieval doctors were trained in the deductive method because a degree in philosophy was required for admission to a university medical program. But they also preserved their tradition of non-deductive thinking.[79]

While we may be tempted to judge the Middle Ages by its failures, such as the absence of experimental natural science and even mathematical physics,[80] that would be to evaluate the period by the anachronistic standards of Galileo Galilei's *nuova scienza* (new science). What we can conclude is that there is no continuous research tradition linking Oresme and Bradwardine to the scientific upheavals of the seventeenth century. The medieval optics of Bacon and Witelo did remain influential in the early modern period; there is a continuous line of author-reader relationships to Kepler.[81] But optics played only a modest role what is normally called the Scientific Revolution of the sixteenth and seventeenth centuries.

On the other hand, a number of tendencies with medieval origins continued in later periods. One was the development of the concept of natural law, which freed deductive thinking from Aristotle's all-embracing framework and permitted generalization at lower levels. Another was the dawning recognition of the role of experience in science—the beginnings of a "logic of experiment," as Alistair Crombie calls it.

Among historians of science, the debate about whether we find continuity or rupture between the Middle Ages and the Renaissance is still being waged. In the 1930s, Alexandre Koyré strongly emphasized the uniqueness of the Scientific Revolution, which he interpreted as a shift in theoretical outlook (primarily due to the application of mathematics) and in worldview. Koyré objected to what he saw as an overemphasis of the role of experiment in science. To this day, most historians of science have accepted Koyré's appraisal of the Scientific Revolution, though their views on experiment have often been more sophisticated.[82]

But Eduard Jan Dijksterhuis made compelling arguments in favor of continuity, arguments later presented even more forcefully by Crombie.[83] This gave a powerful impetus to research into the development of science in the Middle Ages,[84] although Crombie has been faulted for portraying Roger Bacon as too much of a "modern" experimental scientist, a direct predecessor of Francis Bacon.[85] I would argue that this criticism is unjustified, at least with regard to Crombie's later work. Today, there is a consensus among historians of science that Roger Bacon should be approached by way of the medieval tradition, but the controversy about precisely what intellectual and other contexts are most relevant is unlikely to be decided any time soon. One of the principles on which this book is based is that in the history of science, individuals are never "ahead of their time." But there are people whose practices—ideas and methods—have a historical impact, and Bacon was certainly one of those people.

The continuity between the Middle Ages and the Scientific Revolution is most visible in connection with the concept of natural law. In the hands of Descartes and Newton, this became one of the most potent concepts in seventeenth-century science. Crombie's argument is that Roger Bacon, William of Ockham, and others developed this concept on the basis of their voluntarist view of God's absolute power, which holds that God is free to ordain any natural laws he chooses (and that He subsequently abides by them). This chapter largely reflects Crombie's view on this point. Recently, however, it has been argued that while Descartes adopted some elements of Bacon's view, his own conception of natural law also involved many new aspects.[86]

Steven Shapin has commented, "there is no such thing as the Scientific Revolution" (even though he wrote a book about that very topic).[87] What he meant was that the period in question cannot be understood in terms of one central idea. The historiographic situation has grown even more complex as, over the past twenty-five years, our view of the Scientific Revolution has been stripped of its anachronistic trappings. We can no longer view seventeenth-century scientists as our contemporaries. Paradoxically, this has made the Middle Ages more relevant than ever to our understanding of the Scientific Revolution.

4

From Scholar to Virtuoso
The Renaissance Origins of the Experimental Style

When one goes too near [to great nobles], one gets burned; when one goes too far away, one freezes.
—Johann Joachim Becher

ARISTOTLE MADE OBSERVATIONS, BUT did not perform experiments. Archimedes performed experiments, but kept quiet about them, at least in his formal arguments. In Greek and Roman antiquity, experiment was not regarded as a valid path to knowledge. Experimentation, generally in the form of trial and error, was the province of the arts and crafts, and of engineering. Science studied nature, which was understood to exclude human intervention. In the Middle Ages, some scholars posed questions that lent themselves to experimental investigation, but in practice this led merely to a slight expansion of the scope for personal experience and observation.

By about 1600, this situation had changed completely. Francis Bacon, Galileo Galilei, and Marin Mersenne, the leading scientists of this period—or natural philosophers, as they were known in those days—attested to the importance of experiment in science through their words and actions. Bacon believed that scientists had to force nature to divulge its secrets. The way to exercise this force was by applying the thumbscrews, "twisting the lion's tail," forcing nature to do something it would not normally do—in short, experimentation. Though Bacon did not carry out any significant experiments of his own, his experimental philosophy had a profound influence on the century that followed, especially in England and the Netherlands, two countries that were to become major centers of experimental science.

Marin Mersenne represented a different aspect of the new experimental philosophy. A moderate Skeptic, he placed more confidence in the practical results of an experiment than in theoretically proven theses about nature. But it was Galileo who truly introduced experimentation into science. Galileo was active in the same period as Bacon and Mersenne, but was not influenced by them (though he did influence them). His most significant experiments had to do with the law of falling bodies, which states that the distance traveled by a falling object is proportional to the square of the time elapsed and that, in principle, there is no difference in the speed at which light and heavy objects fall. Galileo deduced from this law that the path of a projectile must be parabolic. In other words, Galileo did two things at once; he carried out experiments, but within the framework of a mathematical physics of his own design. "The book of nature," he famously said, "is written in the language of mathematics."[1]

Where did he get this idea? The obvious answer, which turns out to be difficult to prove, is that experiment entered science by way of engineering. There is an undeniable conceptual link: both engineering and scientific experimentation involve interfering with the natural course of things, making phenomena bend to your will.

The historian of science Edgar Zilsel drew a connection between engineering and the origins of modern experimental science. According to Zilsel, in fifteenth- and sixteenth-century Italy a kind of cross-fertilization took place between three different groups. University scholars, humanists, and artisan-engineers crossed paths at the countless academies in the cities of Italy.[2] The early Italian academies were social clubs under the patronage of a ruler; in some cases, "academy" was no more than the name for a one-time intellectual gathering at the sovereign's table.[3]

No one denies the profound importance of the Italian humanists, but the same cannot be said of the artisan-engineers. Their influence on science is more difficult to demonstrate, and for this reason, Zilsel's thesis has often been challenged. In the sixteenth century, no branch of engineering had made even a rudimentary attempt at quantification, according to the historian of science Rupert Hall, who argues that it was not until well into the eighteenth century that engineering was put on a more or less scientific footing. This seems to rule out any possibility that engineering contributed to the development of science. Hall argues that the experimental turn was an achievement of theoretical scientists, an autonomous development produced by science itself.[4] Likewise, the quantitative aspect of experimentation, the key element of measurement, which was very clearly present in Galileo's work, could not have come to science from engineering. Hall claims that, if anything, it was the other way around, and in making this claim he revives an argument originally made by the French historian Alexandre Koyré: theoretical science solved the problems that engineering could not satisfactorily address. [5]

In Hall's writing, Galileo is the personification of theoretical science, and his law of falling bodies proved the engineering tradition wrong. Hall writes that Galileo did not need engineers to come up with the experimental method; his experiments were so straightforward that no technical skill was needed to carry them out. And the results of Galileo's measurements were so perfect that there was some doubt as to whether he actually performed experiments at all. During Galileo's life, Marin Mersenne aired his suspicions that Galileo was presenting thought-experiments as if they were real ones. Galileo himself fueled these suspicions with a well-known remark in his *Dialogo sopra i due massimi sistemi del mondo* (*Dialogue Concerning the Two Chief World Systems*, 1632), in which Salviati, Galileo's mouthpiece, says, "Without experiment I am certain the result will happen as I say because it is necessary that it should happen that way."[6]

Yet contemporary historians have become convinced that Galileo did in fact perform experiments, and did so with considerable skill. This skill was due in part to his roots in the social circle of the artisans, who influenced him more than Hall supposes. Galileo's experimental background was shaped by the liberal arts (mathematics and music), rather than the less respectable crafts, or mechanical arts. But the importance of this distinction should not be exaggerated. As we saw in the previous chapter, time measurement and bookkeeping were regarded as practical techniques. Though the practitioners of bookkeeping, called abacists, had certain mathematical skills, they were equal in status to craftspeople and usually came from humble origins; for example, Niccolò Tartaglia's father was a postman. It may seem counterintuitive, but in the Renaissance, many areas of mathematics were more closely linked to craftsmanship and engineering than to the sciences. Likewise, the mathematically trained artist-engineers known as *virtuosi* were socially and intellectually tied to the artisan class.

In addition to the abacists and engineers, there was another group of mathematicians, who had much stronger ties to the high culture of humanism, including the poet Torquato Tasso. Some of them came from the nobility, such as Guidobaldo, Marquis del Monte; they all admired the elegance of Archimedes's mathematical work and sought to reconstruct his lost work. But upper- and lower-class mathematicians agreed that mathematics had practical value. They were interested in the same issues as the engineers. Guidobaldo also did research in mechanics, which in that period still meant the study of machines, and he held up the prospect of human "control of the realm of nature."[7]

I believe that Zilsel's thesis is, for the most part, defensible. While Zilsel uses the term "artisans," it is important to keep in mind that these were fairly well-educated people, who had stronger ties to city elites than Zilsel suggests. Many of them took an interest in natural philosophy.[8] The group was fairly heterogeneous; Galileo, Mersenne, and Bacon were, each in his own way, products of this group's historical influence. Francis Bacon greatly admired the alchemists and drew inspiration from them in many respects. Alchemy did not exist in ancient Greece; it entered medieval Europe through the Arabic world. Its methods were unmistakably experimental. But the question is how humanists and natural philosophers came to accept experimentation not merely as an alchemical practice, but as a means of obtaining knowledge within their own domain of natural philosophy (or science).

A similar question could be asked about the transition from the applied mathematics of the bookkeepers and *virtuosi* to the use of mathematics in science. As a result of his origins and early career, Galileo himself had close ties with *virtuosi*, teachers at art schools, and high-born humanists with an interest in mathematics.

The universities could not keep pace with all the new developments and so they retained an organization based on the outdated hierarchy of disciplines. The experimental sciences gravitated to a new type of institution: the academy. The defining feature of early academies, especially in Italy, was that they fell directly under the patronage of a sovereign.

The Humanists

In the fourteenth and fifteenth centuries, Europe underwent a gradual shift away from Arabic and toward Byzantine culture. Spain ceased to be a major point of contact between Arabic and Jewish culture in 1492, when Granada was captured by the Catholic monarchs Ferdinand and Isabella. Granada had been the last remaining European city in the hands of the Moors, and the Spanish Inquisition began that same year.

Arabic science also declined in importance under the influence of Italian and Dutch humanists, who advocated a return to the original Greek sources and saw translations from the Arabic as a form of typical medieval corruption. Many of the Arabic texts had not been translated directly from the Greek, but were based on Syriac (that is, Aramaic) versions, and it was therefore concluded that the Latin translations must be full of errors and imprecisions. The Arabic contribution also faded somewhat into the background. The new University of Alcalá in Spain stopped using Avicenna's

Canon as a medical textbook in 1560, though the book remained in use in Salamanca. For some time, the medical curriculum at the University of Padua in Italy remained focused on Averroës, but Aristotle eventually came out on top. At Leiden University in the Netherlands, Jacob Gool held chairs from 1629 to 1667 in both Arabic and mathematics, a combination which he saw as useful and productive.[9]

Let us turn back to Italy, where the intellectual emphasis lay on reclaiming the culture of antiquity. There was direct contact (mainly through Venice) with the Byzantine world, the continuation of the Eastern Roman Empire, where there had always been a high degree of continuity with Hellenistic Greek culture. Though the schism between the Catholic and Orthodox churches had led to a strict religious division, the influence of Byzantine culture was felt in Italy.

The Italian humanists formed a secular group that was not affiliated with the monastic orders and had no strong ties to the universities. This group had emerged from officials in the service of Italy's city governments.[10] When the medieval Italian mini-republics became small absolutist principalities (with the exception of Venice, which remained a republic), a characteristic relationship emerged between humanists and princely patrons. Humanists often served princes as diplomats or in other capacities, and this experience of public service inspired a new intellectual ideal, the *vita activa* (active life) rather than the *vita contemplativa* (contemplative life).[11] Their translations and publications were not intended for use in university education but were addressed to a broader public of doctors, lawyers, and merchants.[12] In France and the Netherlands, more than in Italy, a few successful authors became less dependent on a patron. Erasmus went through a long series of benefactors, but later established a direct relationship with his readers through leading printer-publishers.[13]

Some Eastern Orthodox scholars moved to Italy, swelling the ranks of the humanists there. In 1397, Manuel Chrysoloras began teaching Greek in Florence. The best-known of these Byzantines was Johannes (or Basilius) Bessarion, the former archbishop of Nicea, who brought about reconciliation between the Eastern and Western churches and was invested with the rank of cardinal by the pope. Bessarion took up residence in Italy in 1439, and after the capture of Constantinople by the Turks in 1453, he decided to remain in Italy permanently. He had brought hundreds of Greek manuscripts with him, including many that dealt with mathematics. His house in Rome became a kind of academy, frequented by Cusanus, the astronomer Johannes Müller von Königsberg (better known as Regiomontanus), and many others.[14] In 1463, Regiomontanus discovered the *Arithmetica* of

Diophantus, who had developed a system of algebra around AD 250 that had been adopted in the Arabic world but never in Western Europe.[15]

In sixteenth-century Italy, knowledge of Greek deteriorated. Jacob Burckhardt, the Swiss historian who shaped our understanding of the Renaissance to an important degree, commented that it was fortunate that northerners such as Erasmus had mastered Greek by then.[16] Erasmus presented himself in no uncertain terms as the successor to Italian humanism, and many thought that he had assumed that mantle from Rodolphus Agricola.[17] But as late as 1502, Nicolaus Copernicus learned Greek in Padua, and this enabled him to read old astronomical works in the original.[18]

The fifteenth century was a golden age for translation from the Greek. The Italian humanists were capable and critical philologists. They wrote magnificent Latin, inspired primarily by Cicero, and they held the crude Latin of the scholastics in contempt. The Florentine humanist Marsilio Ficino had a particularly strong impact on fifteenth- and sixteenth-century science with his translations of Plato and Hermes Trismegistus. Hermes was a mysterious Egyptian sage mentioned in the writings of Augustine. When Ficino had just begun translating Plato, his patron Cosimo de' Medici the Elder sent him a manuscript attributed to Hermes and found by an Italian monk in a cloister in Macedonia. Ficino first translated Hermes and then returned to Plato, finishing the latter in 1468. Hermes was thought to have been a contemporary of Moses and to have influenced Plato. His writings were widely read. More than a century later, the Swiss humanist Isaac Casaubon, who resided in England, used philological methods to unmask Hermes as a Neoplatonist of the first or second century AD. [19]

Many humanists were, above all, literary scholars. Lorenzo Valla showed that the so-called Donation of Constantine (a document on which the popes based their worldly authority) was an eighth-century forgery. Valla and his method of indirect proof later served as an important model for Erasmus.[20]

The philosophical and scientific impact of the new translations and editions was tremendous. The humanists' predilection for Plato and Neoplatonism contributed to the study and use of mathematics by such figures as the painter Piero della Francesca, the musician Franchino Gafori (a professor in Padua), and the mathematician Luca Pacioli. Ficino's student Giovanni Pico della Mirandola developed a hermetic and Neoplatonic philosophy that later became a major source of inspiration for John Dee's program of alchemical research in England.[21]

The humanists very consciously experienced the Renaissance as a break with the scholastic philosophy of the Middle Ages, which was still being taught at many universities. Nevertheless, practically all of them read the work of the philosopher who was most central to medieval thought, namely, Aristotle. In fact, many more new translations and editions of Aristotle's work were published between 1400 and 1600 than of Plato's. Most of Aristotle's readers were at universities or in monastic orders such as the Society of Jesus.[22] The humanists (who include the Jesuits in this context) focused on two specific Aristotelian works, the *Topics* and the *Rhetoric*, in which the emphasis is not on unshakeable certainty but on defending an opinion (*opinio*) on reasonable grounds, with reference to experience.[23]

The fifteenth and sixteenth centuries also saw the rediscovery of many other scholarly works and philosophical systems that had not played a direct role in the intellectual life of the Middle Ages. Lucretius's didactic poem *De Rerum Natura* (*On the Nature of Things*) was rediscovered in 1417 and printed in 1473. Ptolemy's *Geographia* (*Geography*) came into circulation in 1406, having been brought from Byzantium to Western Europe by Chrysoloras and Giacomo d'Angelo da Scarperia, who made their own Latin translation.[24] In 1475, Diogenes Laërtius's collection of biographies—of Lucretius's teacher Epicurus, the Skeptic Pyrrho, and countless other Greek philosophers—became available in translation; it is still a valuable source of information today. Galen's philosophical writings, first published in translation in 1490, provided information about Stoicism. And from the late fifteenth century onward, the newly invented printing press brought translations and editions of classical works to a much larger public than ever before.[25]

The astronomer and mathematician Regiomontanus was a typical representative of a movement with a purely humanistic objective: to recover and restore the works of classical Greek authors, such as Diophantus, Archimedes, and Ptolemy.[26] Renaissance mathematicians, humanists who often enjoyed the patronage of a like-minded ruler, were largely independent from the universities. In their own intellectual hierarchy, Greek mathematics was elevated high above the "corrupt" medieval mechanics and philosophy that enjoyed preeminence in the university system, and Archimedes was unquestionably the author they most revered. For much of his life, Regiomontanus was under the protection of Cardinal Bessarion, and he later spent time at the court of King Matthias Corvinus of Hungary, the first humanist monarch outside Italy. Through the intercession of his royal patron, Regiomontanus became a professor of mathematics at the Academia Istropolitana in Pressburg (now Bratislava) in 1467.

Regiomontanus was primarily an astronomer. Since the thirteenth century, astronomy had been taught from a text by Sacrobosco (John of Holywood) using a simplified and incomplete presentation of the Ptolemaic system. To those who wished to calculate planetary orbits, it was no help at all. Regiomontanus's teacher, Georg von Peurbach, had written a far better astronomical manual, which Regiomontanus improved still further. But what most occupied Regiomontanus was his search for more complete and accurate manuscripts of Ptolemy's work to serve as guides. The manuscripts that Regiomontanus had found were full of errors made by poor copyists. He was able to correct some of the inaccuracies through observation, inferring from experience what the manuscript should have said. Regiomontanus's work was greatly valued by astronomers. But, in a telling illustration of the gulf between the world of the humanists and that of the universities, Sacrobosco's text was still in use more than a century after Regiomontanus's death.[27]

Meanwhile, sixteenth-century humanists were still working to recover the work of Ptolemy. Francesco Maurolico was an eagle-eyed editor of ancient Greek texts with a compendious knowledge of geometry and astronomy. In the year of Maurolico's death, his books came into print, including one that he had hoped would ultimately replace Sacrobosco's, as it contained the best possible reconstruction of Ptolemy's system. Maurolico has become infamous for his remark that Copernicus—whose book *De Revolutionibus* (*On the Revolutions*) had been published in 1543—deserved to be lashed for his errors. Apparently, however, it is wrong to see this as a denunciation of Copernicus's heliocentrism. Instead, we should see Maurolico's remark as a case of one humanist grumbling about another, accusing him of sloppiness.[28]

In many respects, Copernicus was firmly within the humanistic tradition. Though he was a conscious innovator, he showed his loyalty to Ptolemy by very closely following the order of presentation used in Ptolemy's treatise known as the *Almagest* (*Great Compilation*). Copernicus was born in a town on the border of Poland and Prussia in 1473 under the name of Nicolaus Koppernigk. He studied mathematics and astronomy in Kraków, at a university with humanist leanings where he encountered the work of Peurbach. After that, he went to Bologna to study canon law, and then to Padua to study medicine. When he returned to Poland, his undemanding church appointment enabled him to devote himself to astronomy. This involved making some observations, but the importance of this aspect should not be exaggerated. In any case, it was not observation that put Copernicus on the path to heliocentrism, but the inspiration that he drew from his studies of ancient Greek and Arabic works.[29]

It was not simply a love of old books that led the humanist mathematicians to recover ancient Greek works on mathematics and astronomy. Ficino linked his view of mathematics to the practical orientation of the arts and crafts, which allowed humans to control their world. "We call arts the sciences using the hands; they owe their acuity and perfection in the first place to the mathematical faculty, that is counting, measuring, weighing, which stands forth as the greatest of all those of Mercury [that is, trade] and reason."[30]

The utilitarian bent of humanistic mathematics is also apparent in the humanists' writings on the history of math. Regiomontanus (in a 1464 work) and Polydore Vergil (in 1499) traced the origins of geometry to practical problems of land measurement in ancient Egypt and to the mathematics of the Phoenicians, a well-known nation of traders.[31] This discourse about the practical value of mathematics was eagerly embraced by the mathematicians' princely patrons, who educated themselves and their children in math and who elevated a number of mathematicians and military engineers to the nobility. The French king Charles VIII had brought a new type of cannon into use during his invasion of Italy in 1494, and since that time, every Italian principality had been within one day's march of an army that could unleash devastating cannon fire. The only way for the Italian city-states to protect themselves was by using techniques from geometry to dramatically improve the design of their bastions. A new type of bastion was developed around 1500.[32] Hall has argued that the geometrical knowledge required was not terribly advanced or complicated, but the new design did win great admiration, and humanist mathematicians successfully laid claim to the domain of fortification. Their knowledge was valued in other countries too. In 1547 Italian engineers designed Fort Rammekens near the town of Vlissingen (Flushing) in Zeeland at the request of Mary of Austria, Governess of the Netherlands. They were also responsible for Antwerp's city walls and citadel, in 1540 and 1568 respectively. These engineers were well compensated for their work.[33]

In contrast, the field of ballistics retained its low social status, despite its practical value. Tartaglia's fictional dialogue between a scholar and a gunner shows, through its literary form, an ambition to elevate ballistics in the eyes of the humanists. Yet it remained the preserve of the lower-status engineers and mathematicians in the artisan class.[34]

Belief in the tremendous usefulness of mathematics had an additional dimension. Mathematics was thought to make you a better person and, partly for that reason, it was assigned a central place at many levels of education. In the early sixteenth century, Philipp Melanchthon introduced educational reforms at Protestant universities in Germany that strengthened the role of mathematics.[35] Pierre de la Ramée (Petrus Ramus) strove for similar reforms at the Sorbonne, as did Joseph Scaliger and later Rudolf and Willebrord Snel in Leiden, and John Dee in England.[36] In Italy the Jesuits, led by Christopher Clavius, took the lead.[37] Ramus advocated the unification of geometrical mathematics with his own deductive system and the mathematics of artisans and workshops. It was his belief that this unification would lead to prosperity.[38] Nevertheless, at most universities, mathematics remained low in status.

The Artist-Engineers

Ficino had adopted a maxim of Plato's from the *Republic*: "numbers have the power of leading us to truth."[39] The humanists passed on their Neoplatonic mysticism about numbers to artists and engineers, who turned it to practical purposes. Some of them developed into polymaths: Filippo Brunelleschi and Leon Battista Alberti were draftsmen, soldiers, engineers, and architects. Brunelleschi designed a new type of cargo ship, for which the city of Florence granted him a three-year patent, the first patent in Western history.[40] The architects Alberti and Brunelleschi designed every detail of their structures before they were built. The master builders of the Gothic period had never done that. Lorenzo Ghiberti was a goldsmith, sculptor, and painter. All three of these men were well trained in mathematics; Alberti was also a linguist and an impressive athlete.[41] We see the same versatility among the great Italians of the following century. Michelangelo was a painter, sculptor, and architect; Leonardo da Vinci was an engineer, painter, and much more. All of them embodied the ideal of the virtuoso. With one or two exceptions, however, these men did not have university educations, and because of their rudimentary Latin, they preferred to write in the vernacular.

Brunelleschi's first characteristically Renaissance works were the Ospedale degli Innocenti, a hospital whose construction began in 1419, and a chapel that he designed for the Pazzi family in 1430, both in Florence. His work was a radical departure from all previous architecture, and especially from the dominant Gothic style. While these works, and especially the chapel, have an undeniably "classical" look, they do not resemble Roman architecture.[42] Brunelleschi and Ghiberti carefully studied the Roman architecture and classical sculpture that they were able to find, just as the humanists studied ancient texts. They also read books by the Roman architect Vitruvius. But their own work departed from these models in significant ways. For instance, Ghiberti believed that the study of the classics led to new inventions.[43] Of all the intellectually active groups in the fifteenth and sixteenth centuries, the artist-engineers had the strongest sense that they were moving beyond the achievements of the ancient world. Many humanists, in contrast—like the universal scholar Gerard Vossius—believed even as late as the seventeenth century that everything significant in the field of physics had already been said by Aristotle.[44]

Leonardo, as shown by remarks at various points in his famous notebooks, brought nature and technology together; as he saw it, both were subject to laws that linked cause and effect "in the shortest possible way." He saw mathematics as the key to science, and mechanics as the way to reap the benefits of mathematics. If we had complete knowledge of causes in nature, he wrote, then we would have no need for experience, but we are usually well advised to let experience guide us. Leonardo investigated both nature and technology by means of thought-experiments, and sung the praises of the speculative mind. He often spoke of the mechanical arts and painting in the same breath.[45]

Leonardo's contemporaries saw him as more of an engineer than a painter; his present fame as the painter of the Mona Lisa goes back no further than the late nineteenth century.[46] He was always appointed in his capacity as an engineer—at the court of Ludovico Sforza, Duke of Milan; for Cesare Borgia; and, toward the end of his life, in the service of King Francis I of France. The great bulk of his activity was in the field of hydraulic engineering. The background of his painting Madonna dei Fusi (1501) shows the regulated flow of the River Arno, one of the projects to which Borgia assigned Leonardo.

One cannot bring up Leonardo without mentioning his technological projects, which still inspire the imagination: the aircraft, the automobile, his many and varied machines, and so forth. In most cases, however, Leonardo was playing with concepts that had already been familiar to earlier engineers, especially Francesco di Giorgio Martini, an engineer from Siena. Leonardo also seems to have known about Roger Bacon's technological speculations, judging by the striking similarities between their views.[47]

But according to the historian of technology Bertrand Gille, Leonardo's most significant achievements were his investigations of smaller-scale construction problems, such as the use of beams and arches above windows and doors to prevent cracks from forming. Leonardo used numerical data for this purpose, which he must have gathered through experimentation. He generally searched for methods that would allow him to use one type of solution in a variety of situations. This is visible in his studies of riverbeds to ascertain the causes of river meanders and thereby stabilize the courses of rivers more effectively.[48]

Alberti and Valla were responsible for fleshing out the Renaissance ideal of *virtù*. The word *virtù* is difficult to translate; its core meanings include virtue, courage, and strength. Valla disagreed with the concept of virtue put forward by Aristotle, who emphasized prudence and the golden mean. In Alberti's hands, *virtù* became a philosophy of life that combined scientific, artistic, and moral aspects. The *virtuoso* was independent of the whims of fortune and used his intellectual powers to gain total mastery of the situation. He gave shape to his actions in both private and public life the way an architect gives shape to a building. The ideal of *virtù* stressed a combination of activity and mathematical rationality that was a perfect fit for the artist-engineers of the Renaissance, who were highly receptive to the humanists' ideas about the utility of mathematics. Their strong emphasis on the *vita activa* probably contributed to a scientific mentality aimed at sweeping aside obstacles, making decisions, and then taking action, rather than focusing on consensus, like the medieval scholastics. For humanists, it was the *will* that mattered.[49]

The starting point for the discovery of linear perspective was Ptolemy's *Geography*, which introduced a method of map projection involving an abstract geometric grid based on a single observer's point of view. Brunelleschi appropriated this technique for a different purpose. In 1425, he demonstrated linear perspective in Florence with a picture he had painted of the Florence Baptistery in the Piazza del Duomo. Viewers were invited to look at the building from the very same point of view used in the painting. The illusion of three-dimensional reality is said to have been very convincing.[50] In 1435, Alberti published a treatise on linear perspective, the earliest written European work on painting.[51] Another decade later, Ghiberti sought to explain the psychological effects of perspective by referencing the writings of Alhazen and Roger Bacon. Ghiberti asked how an observer could estimate the size of an object that was a specified distance away. In this work, he presented Alhazen as an experimental mathematical physicist.[52]

After Alberti's treatise, the perspective technique was further refined by Piero della Francesca, Albrecht Dürer (whose vernacular German work *Underweysung der Messung* dates from 1538), and Leonardo da Vinci. Dürer's book includes the well-known illustrations in which a grid is hung between the draftsman and his subject matter, so that he can reproduce the correct proportions on paper, detail by detail. The technique of perspective, and the new drawing techniques based on it, brought about a revolution in communication. A great deal of information could be conveyed by a drawing of the inner workings of a cannon, for instance. Human anatomy also became much easier to grasp. Botanical drawings, such as Leonardo's, led to a stable classification system for plants throughout Europe.

The very act of seeing was profoundly influenced by perspective. Galileo could make out mountains, craters, and seas on the moon thanks to the new art of three-dimensional interpretation of images. And his familiarity with chiaroscuro, the painterly rendition of contrasts between light and dark, helped him to recognize the shadows on the moon.[53] In contrast, Dante had interpreted the dark spots on the moon as parts that reflected less sunlight, rather than as indications of a three-dimensional structure. In 1611, when Galileo's observation was presented to a committee of four Jesuits from the Collegio Romano, three of them declared themselves convinced. The renowned old mathematician Clavius remained true to the traditional interpretation, but did nothing to stop Galileo.[54]

Most of Piero della Francesca's paintings can still be found in the area where he lived and worked, on the border of Tuscany, Umbria, and the Marche. In Urbino, there is a painting by Piero of the flagellation of Christ, dating from 1465, which uses strict perspective and is full of Neoplatonic numerical mysticism, including a reference to squaring the circle. In the foreground is an astrologer-mathematician.[55] In the small town of Monterchi, the birthplace of Piero's mother, there is a fresco of the pregnant Madonna. Mary is standing in front of a simple round tent. Round tents were often used in problems in arithmetical manuals that involved calculating the surface area of a cylinder. Piero was the author of one such manual. Of course, contemporary viewers of Piero's painting were not asked to carry out such a calculation on the spot, but the merchants among them may have felt a sense of familiarity when they saw the shape of the tent.[56]

One of Piero's mathematics students was Luca Pacioli, whom Piero trained as a Christian Neoplatonist. Pacioli and Piero both stayed at the ducal court of the Montefeltris in Urbino in the same period. The court was then a center of science and technology with a renowned library.[57] There is a painting by Jacopo de' Barbari, now in Naples, in which Pacioli is depicted teaching mathematics to the young duke Guidobaldo. Pacioli probably met Piero thanks to Alberti, who also introduced Pacioli to the papal court.[58]

Pacioli's *Summa de Arithmetica* is an encyclopedic survey of all mathematical knowledge. Much of it is devoted to the double-entry bookkeeping technique and to taking inventory. The relevance of bookkeeping to science is that its techniques allow you to organize a large quantity of data without losing track of anything, providing a quick visual overview of the situation. In the sixteenth century, Pacioli's books were translated into many vernacular languages. Pacioli's *Summa* did not present much new information, but adhered closely to its sources, such as Fibonacci and Witelo's *Perspectiva*. It is fair to say, however, that Pacioli rediscovered Fibonacci; in the sixteenth century, most mathematicians mentioned the two of them in the same breath.[59]

Pacioli wrote the book *De Divina Proportione* at the Sforza court in Milan, where he stayed from 1496 to 1499. Leonardo was there too, and produced the geometrical illustrations for Pacioli's mystical Platonist book. A variety of other philosophers and mathematicians, such as Cusanus, were in the service of Duke Ludovico Sforza, and were collectively referred to as an academy. Leonardo and Pacioli are said to have spent a great deal of time together, and Pacioli praised Leonardo copiously in his book. But their friendship was interrupted when the French took Milan. Leonardo lost his job (and his rank) as ducal engineer, and Pacioli had to leave as well. Both then spent some time in Florence.

Pacioli's connections with various princely courts do not imply that he belonged to the scientific or cultural elite. The true elite mocked him for the confused mixture of Italian and Latin in his writings. For a short time, from 1501 to 1502, he taught at the University of Bologna, but he spent most of his life as an itinerant teacher of business math. Pacioli's "low birth" was not unique. Leonardo had received his education as the youngest servant and dogsbody at a large painting and sculpting workshop, where he also learned arithmetic. In Florence, there was a single guild for the non-university-trained abacists and the land surveyors, while in other towns, abacists were grouped with teachers.[60] The members of the artists' guild also retained the status of artisans. It was not until the death of Michelangelo in 1564 that the guild in Florence decided to display its social ambitions with a funeral fit for a king.

The generation that followed Pacioli's included one of the greatest Italian algebraists of the sixteenth century, Niccolò Tartaglia. Like his predecessor, Tartaglia was mocked for his linguistic shortcomings. This was hardly surprising, since his mother had been too poor to send him to school. Tartaglia had taught himself mathematics and started making his own living as a teacher of business math around the age of eighteen. He was catapulted to fame in 1535, when he prevailed over another mathematician in a major competition. Competitions of this kind were an important way of drawing attention, and Tartaglia took part regularly, relying on his victories to ensure his success as a private teacher. Such events remained popular until well into the seventeenth century.[61]

In the 1535 contest, Tartaglia and his rival had to solve cubic equations (used in practice to calculate volume) by algebraic means. Pacioli had thought this to be impossible and used several circuitous geometric methods as substitutes. Tartaglia's opponent knew how to solve one type of cubic equation, but Tartaglia managed to solve them all.

This resounding victory attracted the attention of the famous Gerolamo Cardano, who was eager to learn Tartaglia's secret method and contacted him personally in 1539. Cardano was a doctor and astrologer, and a very well-known if somewhat controversial figure in Milan. His father, a lawyer, had been a friend of Leonardo's.[62] In those days, Cardano earned his living as a mathematics teacher at the Scuola Piattine, a large private institution of higher education. Later, when his medical career took off, he passed on this teaching position to one of his students.

Cardano's attention must have been very flattering to Tartaglia. He disclosed his method to Cardano, who promised to keep it secret and to introduce him to the governor of Milan. But when Cardano found out that Tartaglia's earlier opponent had independently arrived at a technique for solving some cubic equations, he no longer felt bound by his promise and published his work on the subject. Though Cardano did acknowledge Tartaglia's discoveries, this did not appease Tartaglia in the least, and he vented his anger about Cardano's actions at some length in his subsequent book.

The introduction to the governor had probably not been of much value to Tartaglia, and it has been suggested that this was the main reason he felt cheated.[63] Despite his great talents and ambition, he was unable to improve his social status through either mathematics[64] or his non-mathematical activities. For Cardano, in contrast, algebra was just one of many activities that bolstered his reputation as a doctor. (He was also actively investigating the design of a seat that would remain stable on a ship's deck, the fields of astrology, meteorology, and geology, the operation of the senses, and hallucinogenic substances, and he is credited with the invention of the Cardan joint used in automobile drive shafts today.)[65] Later mathematicians granted him a place of honor in the history of algebra.

Italy was not the only place where mathematics was associated with business calculation. Simon Stevin began his career as a bookkeeper in Antwerp, until he left for the northern Netherlands to serve Prince Maurice as an engineer and personal mathematics tutor.[66] Later, the States General appointed him to the post of quartermaster-general of the army. Stevin invented the modern notational system for decimal fractions, and the notation for exponents also derives from his work.[67] Another interest of his was defensive works; his book *De Stercktenbouwing* (*Fortification*) dates from 1594. This theoretical, mathematically sophisticated work was used in a degree program for military engineers established in 1600 at Leiden University.[68]

For one of the other major algebraists of the sixteenth century, François Viète, mathematics was just a sideline. Viète was a legal scholar who worked for the French kings Henry III and Henry IV. His knowledge of algebra helped him to decipher coded messages from King Philip II of Spain. Viète introduced the modern plus, minus, and equals signs, and improved and systematized the theory of equations.[69]

The University Scholars

In seventeenth-century Italy, the University of Bologna had a good reputation in the field of mathematics, with sometimes as many as eight professors teaching the subject, and a history that went back to the twelfth century. Yet the significance of this university should not be overstated. The most prominent Bolognese algebraist, Rafael Bombelli, was an engineer unaffiliated with the university. Bombelli made algebra more abstract, less focused on practical problems, partly under the influence of his study of Diophantus.[70]

Since about 1300, mathematics had been losing ground within the university quadrivium to physics, and hence to philosophy as well. At this juncture, a few remarks about terminology may be useful. Until well into the eighteenth century, the word *scientia* had a restricted meaning: namely, knowledge that could be proved entirely from first principles. The dominant terms for what we now call the sciences were *philosophia naturalis* (natural philosophy) and, in later periods, *philosophia experimentalis* (experimental philosophy). But the significance of this way of speaking went beyond terminology. The natural sciences were part of the domain of philosophy, rather than that of mathematics. In the Middle Ages, Jean Buridan and Nicole Oresme had developed and used mathematical techniques for their theories of impetus, but they had remained philosophers, with all the rights that philosophers enjoyed. In the hierarchy of the medieval universities, philosophy had a much higher status than mathematics, and that did not change until well into the seventeenth century.[71] Thomas Aquinas, as well as the scholastic tradition that stemmed from his work, had held that mathematics could not make any statements about true causes.[72] One exception was optics, which had the status of a "mixed science," in which mathematicians could draw conclusions about physical phenomena (in this case, the path of light rays); this could be traced back to Aristotle's own confusion about whether optics fell under mathematics or physics. The status of optics was insecure, however, and permanently open to debate.[73]

In the Renaissance, the distinction between natural science (philosophy) and mathematics was still sharp. For instance, statics, the measurement of weight, was regarded as a branch of mathematics, and dynamics, the study of motion, as a branch of philosophy.[74] This distinction seemed natural to most philosophers, and to most mathematicians, much like the twentieth-century distinction between physics and chemistry, which was not only theoretically defined but also institutionalized in the form of two separate disciplines. All the same, the borders between mathematics and philosophy were occasionally crossed by scholars such as Tartaglia. In the seventeenth century, Galileo's law of falling bodies would close the gap between the two fields. But just a few decades earlier, in 1577, the mathematical theorist Guidobaldo del Monte declared this gap unbridgeable, and excoriated Tartaglia for trying to cross it.[75] In practice, the division between mathematics and philosophy was partly a way of allocating intellectual rights and responsibilities; mathematicians were expected to steer clear of philosophical issues.

Even in the Middle Ages, instructors in faculties of arts did not tend to stay there long. Ambitious individuals moved on to one of the larger faculties.[76] In sixteenth-century Italy, mathematics professors were usually appointed for just a couple of years. Furthermore, philosophy professors earned six to eight times as much as mathematics teachers.[77] A renowned medical professor might earn even more.

Astronomy fell under mathematics, to the extent that it aimed to calculate or predict the positions of heavenly bodies. Initially, astronomers and arithmeticians differed in status, and a firm intellectual distinction was made between the two disciplines. Astronomers in sixteenth-century Italy—in Padua, for example—were generally appointed to the faculty of medicine, where they taught subsidiary courses in astrology.[78] Business mathematics was, as one might imagine, quite a bit less prestigious, and not merely because it was taught at less illustrious commercial institutes and private schools like Tartaglia's. Astronomy dealt with heavenly things, rather than earthly ones. During the sixteenth century, these two branches of mathematics grew closer together, and at some universities the same professor was appointed to teach both—again, Padua is a case in point.[79] But the interpretation of the structure of the cosmos was the territory of philosophy, and astronomers were expected to keep quiet about it.

Not all mathematicians accepted these constraints. Galileo managed to escape being pigeonholed as a mathematician by acquiring the status of philosopher in an unusual way. The Jesuits, Galileo's opponents in a different context, also broke free of university traditions to create their own educational institutions. Their mother institution was the Collegio Romano, founded in 1551. Inspired by humanism, the Jesuits laid a strong emphasis on mathematics education.[80]

The science of motion fell within the domain of philosophy, and in the sixteenth century it was still being taught at universities in the medieval style and tradition. In general terms, the approach was scholastic and Aristotelian, though without adhering to every letter of Aristotle's doctrines. There were, however, several competing theories about the phenomenon of motion. Renaissance thinkers showed a growing tendency to conceptualize this phenomenon in terms of two interrelated topics: the paths of projectiles and the behavior of falling bodies, whether falling freely or along a inclined plane. (This problem had already attracted great interest in the Middle Ages.) The most important of these approaches was the impetus theory mentioned above. Galileo was taught this theory at the University of Pisa and included it more or less unchanged in *De Motu* (*On Motion*), his first, unpublished book, in 1592. Impetus theory deals primarily with the paths of projectiles and seeks to explain why an object that is thrown or fired follows an "unnatural" path rather than dropping directly to the ground. Aristotle himself had suggested that the motive force was imparted to the projectile by the medium through which the object moved (usually the air), but impetus theorists located the force in the projectile itself. It was in many respects a satisfying theory, much better than Aristotle's. But one of its shortcomings was that it was difficult to quantify. This had not been a serious problem in the Middle Ages, but it became one in the Renaissance.

A second approach was based on the treatise *Quaestiones Mechanicae* (*Questions of Mechanics*), which was attributed to Aristotle himself in the sixteenth century, but is now thought to have been written several generations after Aristotle's death by a member of his Peripatetic school. The book resurfaced in the Renaissance—the first Latin translation was published in 1497—but even in the Middle Ages, some people must have been aware of it.[81]

The author of the *Quaestiones Mechanicae*, now often referred to as Pseudo-Aristotle, based much of his discussion on the workings of a balance, trying to use its moving arms to analyze the relationship between an object's velocity and its distance from the fulcrum. From there, he developed an analysis of circular motion, which (being a true Aristotelian) he divided into two components: natural downward motion and unnatural lateral motion. In mathematical terms, the *Quaestiones Mechanicae* contains many errors and fails to produce a satisfactory explanation of motion, but its ideas long remained influential.[82]

In the early thirteenth century, a man named Jordanus, about whom not much more is known than that he came from the German lands, tried to achieve a synthesis between Pseudo-Aristotle's dynamic analysis of the balance and Archimedes's static analysis. Jordanus's problems included a ball rolling down an inclined plane; the steeper the incline, the faster the ball rolls. His approach, too, was riddled with errors and internal inconsistencies, but it also contained the seeds of useful ideas. Most significantly, Jordanus suggested that movement along an inclined plane has two components: a straight-line downward motion, which is natural, and a sideways motion, which is forced.[83] In the fourteenth century, the impetus theory of

Buridan and Oresme had offered a qualitative interpretation of the movement of the ball in such cases.

Up to that point, the problem of motion had been studied by university scholars, credentialed natural philosophers. But it was the autodidact Tartaglia who, by returning to Jordanus's theory, arrived at a mathematical account of projectile motion. In his first book, *Nova scientia* (*New Science*, 1537), he based his theory on Aristotle's, describing the path of a projectile as having three components: a first, horizontal part (the forced motion), a final, downward vertical part (the Aristotelian natural motion in its pure form), and in between, as a transition, a section of a circle.[84] It is important to realize that the idea that a cannonball's path is perfectly horizontal at first was somewhat plausible. In Tartaglia's day, gunners were usually just a short distance away from their target, at point-blank range. Tartaglia observed that a cannonball travels farthest when shot at an angle of 45 degrees and claimed to have reached this conclusion experimentally.[85] Given Tartaglia's practical involvement with artillery forces—he developed an instrument for estimating the height of distant objects, and another for making a cannon barrel adjustable in height[86]—it seems highly probable that he was telling the truth.

In his book *Quaesiti et inventioni diverse* (*Diverse Problems and Inventions*, 1546), Tartaglia corrected his theory of projectile motion, writing that the trajectory did not begin with a horizontal section, but was curved throughout. How did Tartaglia arrive at this insight? In theoretical terms, he drew on the work of Pseudo-Aristotle and Jordanus, adapting and improving the latter's approach to inclined planes. In a typically Renaissance style, he adopted Archimedes's Euclidean mode of presentation; that is, he organized his arguments (to some extent) by proceeding from initial postulates to his conclusions. Jordanus's presentation had been much sloppier. But Jordanus's basic idea must have been what made Tartaglia realize that two forces, one natural and one artificial, can act on a cannonball simultaneously, resulting in a curved trajectory.[87]

Tartaglia did not solve the problem of projectile motion; his mathematical proof was not convincing enough. Hall speaks dismissively of the mathematical window dressing with which he embellished his theories.[88] Tartaglia's contemporaries were equally scornful and struck out in other directions; Giovanni-Battista Benedetti, his former student, devised explanations involving the impetus theory, while Guidobaldo wanted to base everything on Archimedes. They both believed that any adaptation of Jordanus's theory was barbaric.[89] (Galileo would later prove them wrong about this.)

Tartaglia's work was also unusual in other respects. One practical problem in ballistics is that projectiles encounter quite a bit of air resistance, which influences their trajectories. Tartaglia appealed to Plato, however, arguing that the truth of mathematics was not limited by its imperfect realization in the material world.[90] This strikingly modern position could not be reconciled with the conventional view, which was based on the writings of Thomas Aquinas. Galileo later became the leading advocate of this new vision of mathematics, which in his day was also supported by several prominent Jesuits.[91]

A few decades after Tartaglia concluded that the path of a projectile is curved over its entire length, Guidobaldo del Monte determined that its shape was parabolic. He did so in two ways: first, through an incorrect analogy with a catenary, the curve formed by a hanging chain, and then through a remarkable experiment, throwing a ball covered with ink over a vertical tabletop and observing the trail it leaves behind. His explanation was drawn from Pseudo-Aristotle, and cited the mix of "violent" and natural motion in changing proportions. But Guidobaldo did not provide any mathematical evidence for the parabolic trajectory.[92]

Guidobaldo's relationship to dynamics was complex. As an orthodox mathematician, he believed that dynamics did not lend itself to quantification of any kind. Dynamics was the domain of Aristotle (or Pseudo-Aristotle), and mathematics was the realm of Archimedes. Each had its own area of application, and the disciplinary boundary between them had to be respected.[93] It wasn't the low status of mathematics that led Guidobaldo to take this point of view. He was completely independent of any university hierarchy; his father had been ennobled by Guidobaldo II, Duke of Urbino, for his accomplishments in military architecture. Guidobaldo had studied mathematics under a renowned teacher with the young Duke Francesco Maria II, and was married to the duke's half-sister, an illegitimate daughter of Guidobaldo II.[94] Guidobaldo emphasized his prestige as a humanist at court by writing only in Latin. He was powerful enough to act as a patron in his own right, and his leading protégé was Galileo, who was eighteen years younger. Guidobaldo arranged a professorship for Galileo in Pisa, followed by one in Padua.

Guidobaldo's main project was to provide a solid mathematical foundation for mechanics and the science of machines; to him, "mathematical" meant based on Archimedes.[95] He was interested solely in simple machines, such as the pulley, that were based on Archimedes's law of the lever. Since this was a static, rather than dynamic, principle, Guidobaldo had to divide mechanics into two separate sciences. In his disgust with the "barbaric" contributions to mechanics made by Jordanus and Tartaglia, he tended to use statics to solve problems unsuited to that approach.[96] But dynamics was a very different domain than Archimedean mathematics, and that may have been why Guidobaldo was prepared to take an exploratory, experimental approach. He also performed other experiments in dynamics, some involving balls on inclined planes, and corresponded with Galileo about them.

Galileo Galilei

In his *Dialogue* (1632) and especially in *Discorsi e dimostrazioni matematiche intorno a due nuove scienze attenenti alla meccanica* (known in English as *Two New Sciences*, 1638), Galileo published the first almost-modern theory of falling bodies and projectiles.[97] At that point, Galileo was an old man, but he had collected most of the material dealt with in *Two New Sciences* during his time as a professor in Padua, before 1609. In other words, even though Galileo had the original insights around 1604, it was not until 1638 that he published the law of falling bodies in its modern form—the distance traveled by a body in free fall is proportional to the square of the time elapsed—and stated the inference that the path of a projectile is parabolic in form. The question is, how did he discover the quadratic relationship that this law expresses?

Dijksterhuis describes Galileo's development in purely intellectual terms, as the gradual dwindling of the "medieval" aspects of his thinking (especially impetus theory, which was dominant during his time in Pisa) and a growing emphasis on inquiring into how things worked and trying to describe them in mathematical terms. Dijksterhuis does not question the idea that Galileo performed experiments to measure the motion of a falling object, but he describes these experiments as the mere verification of insights obtained by other means. In support of this view, he quotes Galileo himself: "I have made the experiment, and even before that, natural reason had firmly persuaded me that the effect had to happen the way it indeed does."[98] For Dijksterhuis, it is of little relevance that Galileo's initial deductive argument for the correct law of falling bodies was based on an incorrect premise.[99] The important thing is that theoretical intuition came first for Galileo, and experiment came second. Dijksterhuis challenges the "Galileo myth," the idea that he discovered the quadratic law of falling bodies as a result of taking measurements, through experimentation.

The above quote from Galileo is significant, in that it expresses an attitude still found frequently among scientists. The deductive style, in the form it has taken since the seventeenth century, incorporates the procedure of experimental testing. And there are times in Galileo's life in which he undoubtedly adopted that working method. But in the specific case of the law of falling bodies, we can assign a larger role to experiment than Dijksterhuis does. Dijksterhuis cannot be faulted for his conclusions, given that he was writing in 1950.[100] But from 1970 onward, the historian Stillman Drake spent a period of decades studying Galileo's notes, and his findings and reconstructions compel us to rethink Dijksterhuis's view.[101]

During his time in Pisa, before 1592, while writing *De Motu*, Galileo performed experiments with balls of various weights on an inclined surface. These experiments, largely based on impetus theory, failed to yield any useful results. In 1602, he referred to this problem in his correspondence with Guidobaldo, who was performing experiments in the same area. According to Drake, it was in 1602 that Galileo started experimenting with balls in grooves, without any idea of where this line of investigation might lead him.[102]

Galileo's stay in Padua had a highly practical purpose, as we shall see. Around 1601–1602, Galileo was writing another work that would remain unpublished, *Mechanica* (*Mechanics*), in a style markedly less theoretical than that of *De Motu*. For instance, in *De Motu*, Galileo does not regard a horizontal plane as perfectly flat, but makes a minute correction for the curvature of the earth, while in the *Mechanica* a horizontal plane is simply horizontal.[103]

In 1604, Galileo performed the decisive experiment, investigating the law of falling bodies by observing a bronze ball as it rolled down a groove in a gently sloping ramp. There were marks along the groove, and he measured how many units of time had passed as the ball reached each of these marks. Drake has reconstructed Galileo's method of measurement. Galileo ran gut strings over the groove in such a way that the ball made a sound whenever it passed over one of them. He then adjusted the position of the strings until the sequence of sounds corresponded precisely to the tempo of a melody. This implies that Galileo sang during his experiment. He had no watch, of course, and there were not yet any reliable pendulum clocks at that time. But Drake discovered that the ear is capable of detecting deviations from a rhythm small enough to correspond to the tiny degree of error in Galileo's measurements. This explains why Galileo's measurements seemed too accurate to be believed. It is important to note that Galileo's father and brother were musicians and he himself was an accomplished lute player.

These days, the typical way of phrasing the law of falling bodies is that the distance traveled increases with the square of the time elapsed. In other words, after 1, 2, 3, 4, 5, 6, and 7 units of time, the ball has traveled a total of 1, 4, 9, 16, 25, 36, and 49 units of distance, respectively. But what Galileo measured was not the total, but the additional distance traveled in each unit of time. For the units of time 1 through 7, this yielded the series 1, 3, 5, 7, 9, 11, and 13.[104]

Galileo did not immediately recognize the quadratic relationship underlying this sequence of odd numbers. Furthermore, he did not put the strings in the "right" places from the start. These facts suggest that he had not arrived at the correct law for theoretical reasons before carrying out the experiment.[105] It is true that Galileo was searching for a simple mathematical relationship, but he did not know the nature of that relationship in advance.

After his experiment with strings, Galileo devised still more ways of accurately measuring intervals of time, such as collecting water in containers and measuring its weight.[106] Hall's claim that Galileo's experiments were easy is untenable. It should be added that, from his youth, Galileo was accustomed to conducting experiments.

Vincenzo Galilei, Galileo's father, was a mathematician and a musician. In 1588, he had performed experiments with strings of equal length under differing degrees of tension. It was a long-established fact that when two strings were under equal tension and the ratio of their lengths was 3:2, they produced two tones at a musical interval of a fifth. Many people assumed that the same ratio applied to the degrees of tension when the strings were of equal length. But Vincenzo showed experimentally that the ratio of tensions needed to produce a perfect fifth is 4:9. It is more than probable that Galileo helped his father with these experiments.[107] It is also noteworthy that Vincenzo, like Galileo in later years, arrived at a simple, elegant mathematical ratio, but not one predicted by an existing theory.

Galileo spent his studies and the first years of his career surrounded by *virtuosi*, including his father. In 1581, he went to the University of Pisa to study medicine, but never earned a degree in the subject. His interest in mathematics dates from 1583–1584, when he took private lessons from Ostilio Ricci, a friend of his father's. Ricci was a military engineer and former instructor at the Accademia del Disegno in Florence, which had been founded in 1562 by the painter and architect Giorgio Vasari.[108] In addition to applied mathematics, Ricci taught pictorial perspective. Under his tutelage, Galileo read Euclid and Archimedes.[109] Around 1587, Galileo was an independent private teacher of mathematics and also had a one-year teaching position in Siena. In this period he attracted Guidobaldo's attention with an ingeniously designed problem, inspired by Archimedes, about a balance with unequal weights. It was the start of a friendship that profoundly influenced Galileo's career.[110]

Galileo never received a degree in mathematics, either, before he began teaching it, first in Siena, then in Pisa, and then in Padua. This is indicative of the low status of the field, comparable to that of engineering. For an appointment in philosophy or medicine, a doctoral degree was the minimum requirement.

During his time as a professor in Pisa and Padua, Galileo taught not only mathematics but also mechanics and the construction of fortifications. He visited the shipyards at the Venetian Arsenal, absorbing the working methods and mentality of the shipbuilders.[111] He investigated methods for controlling the flow of rivers (the same problem Leonardo had tackled in the service of Cesare Borgia) and was paid to give advice about defensive works. In Padua, in 1594, he received a patent on an improved version of the Archimedean screw used to move water to a higher level. He improved Tartaglia's instruments for gunners in 1596 and supplemented his salary by manufacturing them, selling them at a profit, and offering private instruction in their use to students. He had little choice but to take measures like this, as he was constantly short of money.[112]

Galileo was well acquainted with Tartaglia's work on ballistics, and unlike his mentor and benefactor Guidobaldo, he appreciated its value. This work sparked his interest in the law of falling bodies more than ten years before he solved the problem.[113]

Galileo did not publish his formulation of the law of falling bodies, or at least not in 1604. As mentioned previously, he published his work in 1632, and the final version, with his measurements and calculations, did not appear until 1638. One possible explanation is that Galileo was a mathematician in 1604 and a philosopher in 1632 (and a disgraced philosopher in 1638, which explains why *Two New Sciences* was published in the Netherlands but not in Italy). We have seen why it was important to be a philosopher rather than merely a mathematician. Galileo's law of falling bodies took him into the domain of philosophy. The same was true of his second major program of research, in the field of astronomy. This time, Galileo's ambitions went further than making observations and calculations. He also hoped to draw conclusions about the truth of Copernican cosmology. If Galileo had resigned himself to his role as a mathematician, he could have made use of Copernicus's work without anyone trying to stop him.

When we take in the entire sweep of Galileo's career, three moments stand out. The first is when he obtained the protection of Guidobaldo del Monte, who offered him professorships in mathematics, first in Pisa in 1589 and then in Padua in 1592. The second and most crucial step in his career came in 1610, when he was appointed as a mathematician and philosopher at the court of the young Grand Duke of Tuscany, Cosimo II de' Medici. Galileo dedicated the moons of Jupiter, which he had discovered, to the Medici family, and for a while the Medicis actually used them as the heraldic emblem of their dynasty. For more than ten years, Galileo lived the good life at their court, enjoying power and influence. His connections gave him influence over the appointment of professors of mathematics and philosophy at the University of Pisa, which was on Tuscan soil. The position of court mathematician and philosopher may seem a peculiar one, but it was primarily a way for the Medici to show off Galileo and delight in his intellectual skills—for instance, during debates at the duke's dinner table, in which he was contractually obligated to participate.[114]

The third major step in Galileo's career was born of necessity, and he paid a heavy price for it. Cosimo II died in 1621 at the age of thirty, and his successor Ferdinand II, a mere child, could not be expected to actively promote Galileo's interests. Two years later, a friend of Galileo's, Maffeo Barberini, was elected pope, and took the name of Urbanus VIII. In those days, the popes were mighty rulers, wielding absolute power over the Papal States, which covered most of central Italy. Galileo sought and received the protection of the new pope.

In a sense, Rome was dangerous territory for Galileo, not because it was so Catholic (so was Florence, and Galileo himself was quite devout), but because he had so many rivals there. He had been admitted to Rome's social circles long before, in 1611, when invited to become a member of the Accademia dei Lincei. But the Accademia's patron, Prince Federico Cesi, died in 1630, and the Jesuits of the Collegio Romano were a powerful force in the city. Galileo and Clavius had always been on friendly terms, but Clavius had died many years earlier, and Galileo, regarding the Collegio's younger generation as his juniors, was too proud to court their favor. Galileo was undeniably the leading individual scientist of his day, but the Jesuits insisted that their Collegio Romano was the leading scientific institution, a fact that led them to confidently assert their scientific priority over Galileo.[115] Christopher Scheiner took credit for discovering sunspots, and Galileo was embroiled in debate with Orazio Grassi about the nature and origin of the comets that were visible to all the world in the skies in 1618.[116] In short, Galileo's sole power base in Rome was the pope himself.

But after a while, Pope Urbanus began to suspect that Galileo was deceiving him. He thought that Galileo had used illicit means to procure the license to publish his *Dialogue*, which argues for the superiority of the Copernican system. It may have been the Jesuits who gave Urbanus this idea,[117] but the fact that Galileo gently mocked Urbanus in his book certainly did not help.

In any case, Urbanus did nothing to stop Galileo's spectacular fall from grace. By the time Galileo produced exonerating evidence (a document given to him by the late Cardinal Bellarmino), the damage had already been done, and his actions only made things worse. For the pope to save face, Galileo had to be convicted. At the infamous trial of 1633, he was forced to abjure the Copernican system and sentenced to a term of imprisonment. At this point, the Medici family took Galileo under its patronage again, seeing to it that he was placed in house arrest in relatively comfortable conditions and could spend the final years of his life in Tuscany.

It was the dynamics of the patronage system that cost Galileo his freedom. He was the fallen favorite, a familiar character in those days.[118] Urbanus himself never anticipated that Galileo's fall would be interpreted as an attack on science by the church. On the contrary, the pope was a great supporter of the sciences, and Galileo's successor in Rome was a rising star in the world of scholarship: the prolific and many-talented Athanasius Kircher, a Jesuit father.[119]

The patronage system evolved during the seventeenth century (starting in Italy), and rulers began maintaining a greater distance from their favorites. Prince Leopoldo de' Medici, the younger brother of Grand Duke Ferdinand II, established the Accademia del Cimento in 1657, fifteen years after Galileo's death, and presided over it until 1667, when it was discontinued. Leopoldo refrained from comment on the views of the members of his academy, as long their claims were not too extreme. The Accademia's dedication to the experimental method may have been prompted by the desire to keep its members from making bold claims with philosophical implications.[120]

In his career as a mathematician, Galileo united the two traditions of Renaissance mathematics: that of the *virtuosi* such as Tartaglia, and that of Guidobaldo's humanism. Like Tartaglia, Galileo needed benefactors, but unlike Tartaglia, he managed to obtain them. He would not have come far with Ricci, who had to sit at the servants' table when he went to court. Guidobaldo sat at his ruler's table, just as Galileo did from 1610 onward. Before Galileo could introduce mathematics and experimentation into science, he had to win the status of philosopher. No university could grant him that status, but a princely patron could. When Galileo broke through into the Medici court's inner circles, it was also a breakthrough for mathematical and experimental physics.

⌘ 5

The Experimental Style II

The Skeptics and Their Opponents

Life is a material and corporal motion, an action imperfect and irregular of its own proper essence . . .
—Michel de Montaigne

THE FINAL GREEK PHILOSOPHER that Renaissance humanists reclaimed for the Western tradition was Sextus Empiricus. We do not know much about him; he lived around AD 180 and was a disciple of Pyrrhon, the great Skeptic of the ancient world. Sextus probably did not attract a great deal of notice during his life, since there were other philosophers with similar views. Coincidentally, however, he is the only radical Skeptic whose writings have survived. Before the rediscovery of Sextus, Diogenes Laërtius had provided some information on Pyrrhon, and Cicero was known to have advocated a much milder form of Skepticism. When Sextus became available in Latin translation for the first time in 1562, he immediately drew considerable attention. A century later, Pierre Bayle, the publisher of the Rotterdam-based philosophical review *Nouvelles de la république des lettres*, proclaimed Sextus the father of modern philosophy.[1] How did Sextus achieve such renown more than fourteen centuries after his death?

Skepticism was a frontal attack on the validity of the deductive style of thinking. If we survey the whole turbulent reception history of Skepticism in the sixteenth and seventeenth centuries, we can see that while it did not annihilate the truth claims of the deductive style, it did limit their scope dramatically. The domain of certain knowledge (*scientia*) again lost ground to knowledge that was plausible but fundamentally uncertain (*opinio*).[2] In addition, Skepticism cleared the way for an exploratory, experimental style that was not tied to the deductive style of thinking; in other words, a less constrained experimental style than Galileo's. Yet Skepticism won its first battles in the field of religion. That was also what made it useful; the rediscovery of Skepticism coincided with the rise of the Reformation. At first, Skepticism was mainly used to argue for tolerance, particularly by Desiderius Erasmus. Later, Protestants and Catholics began to use Skepticism to undermine each other's positions, and it became a dangerous double-edged sword, a potential threat to public order. Descartes was the most formidable opponent of Skepticism, and his reasoning was both theological and scientific. Likewise, Galileo and Kepler wanted nothing to do with Skepticism. As adherents of the Copernican system, they could not support a philosophical view with a noncommittal approach to important questions.

Sextus, like his teacher Pyrrhon, espoused an extreme and uncompromising strain of Skepticism. Descartes mounted a defense that was just as uncompromising, and ultimately untenable, but one of his contemporaries, Marin Mersenne, developed a moderate form of Skepticism that put aside the question of a theory's ultimate truth, giving free rein to the method of experimental exploration. For Mersenne, good technique (along with the criterion of effectiveness derived from it) was the hallmark of good science.

Radical Skepticism

The Greek Skeptics opposed all other schools of philosophy, especially those of Plato, Aristotle, and the Stoics. They challenged the idea of certainty, rejected dogmas of all kinds, and distrusted sensory experience. Above all, they contested the syllogism, the gold standard of logical reasoning, as well as the claim that there is a criterion for determining which of two or more views is correct. A Skeptic would advise suspending judgment.[3]

Is Skepticism a negative dogmatism, then, based on the dogma that there are no dogmas? No, because it suspends judgment on this question too. This makes it a difficult position to maintain in a psychological sense; the Skeptic perceives no underlying regularity in the world, no firm footing. Sextus himself claimed that accepting this situation would bring peace and serenity, but the famous Skeptic Montaigne expressed feelings of agitation and anguish. Montaigne could not find peace, not even deep within himself, and that is striking because Calvin and Descartes described the inner self as the basis of certainty. Sextus would have said that Skepticism is also the best remedy for this type of agitation. Montaigne sought a balance, temporary if necessary, among the patterns of his life, without aspiring to universal solutions. In this respect, he and Descartes were polar opposites.[4]

The Skeptics' arguments against the criterion of truth and the syllogism played a central role in sixteenth- and seventeenth-century debates on religion and science. Skeptics punctured the criterion of truth by showing that it led to an infinite regress. Imagine a discussion about whether or not there is a criterion of truth. To bring that discussion to a successful conclusion, you need a criterion on which to base a decision. Is there any such criterion? And so on.[5]

The syllogism had been developed by Aristotle and formed the cornerstone of university education in philosophy throughout the Middle Ages. Skeptics discredited it with the following syllogism, which in formal terms is beyond reproach:

> 1) All men are stones; 2) All stones are animals; Conclusion: Men are animals.[6]

A true conclusion can thus be deduced from false premises. It was this argument, above all, that shook the foundations of science. The Jesuits used it against Galileo, who had asserted that some of the phenomena he had observed, such as the phases of Venus, were more compatible with the heliocentric Copernican system than the Ptolemaic model. According to the Jesuits, this was a typical example of deducing a true conclusion (the phases of Venus) from a false premise (the Copernican system). The argument underscored the sheer impossibility of having true premises at one's disposal.

Erasmus used Skeptical arguments, such as those against the criterion of truth, to combat a religious crisis: Luther's attack on the authority of the church. While Erasmus sympathized with much of what Luther said, he did not approve of Luther's dogmatic faith that he was in the right. It did not take long, however, before Skepticism had set off an intellectual crisis of its own, which spread from religion into the domain of science.

Protestant Anti-Skepticism and Catholic Skepticism

In 1517, when Luther nailed his ninety-five theses to the church door in Wittenberg, he was claiming that the church was in error by its own standards. His more fundamental challenge followed two years later, when he argued that the authority of the church, its pope, and its traditions has no basis. Religious truth can only be found in what the conscience must accept on reading the Gospels. That was Luther's criterion of truth, but on what grounds could it be accepted? The Church of Rome's counterclaim was unsurprising: if everyone were to follow only the dictates of his own conscience, the inevitable result would be anarchy. In works such as *The Praise of Folly*, Erasmus confronted Luther with a mild form of Skeptical doubt, claiming that efforts to arrive at absolute truth were futile. Luther, however, did not understand how a Christian could build his faith on Skepticism.[7]

A generation later, in France, John Calvin sharpened Luther's arguments. In 1559, Calvin wrote that the authority of faith must reside in scripture rather than the church, because the church based its authority on a scriptural passage (Matthew 16:18: "Thou art Peter, and upon this rock I will build my church"). But how can we know for certain that scripture is the Word of God? What is the criterion for this certainty? Calvin's answer was that the Holy Spirit instills in us (or, at least, the chosen among us) absolute inner conviction, which is so overwhelming that it serves as a full guarantee of our faith. The elect are recognizable by this inner conviction.

This is circular reasoning: God's truth gives us confidence in our inner conviction, and inner conviction gives us confidence in God's truth.[8] Even so, for those in search of firm foundations, there is no other path. The only alternative is Skepticism, a view that Protestant theologians often ascribed to Catholics. And indeed, moderate forms of Skepticism formed the core of Catholic theology. Yet many Protestant philosophers in the humanist tradition subscribed to the same sort of "Ciceronian" Skepticism; a case in point is Philipp Melanchthon, Luther's right-hand man in the field of education at *gymnasia* (pre-university grammar schools) and universities in German-speaking Protestant countries. Philosophers too could propound radical Skeptical views without fear, as Montaigne was startled to observe during his travels through Italy in 1580. In the seventeenth century, Cardinals Richelieu and Mazarin acted as patrons of Skeptical philosophers.

The Catholic defense of tradition is based in part on a Skeptical argument. If everything is subject to debate, why not just respect tradition? This inherently conservative view can be extended to the realm of science; if the choice between Ptolemy and Copernicus cannot be made on the basis of observation (as was true for a very long time), why not accept the version that seems to have biblical support? If conclusive evidence emerged in favor of Copernicus, the traditional interpretation of the Bible would have to be reconsidered, but as yet there was no reason for that. This was what Cardinal Bellarmino, a Jesuit, reasoned in 1615 when he advised Galileo to be wary of the Copernican system.[9]

Unlike Protestant denominations, the Church of Rome recognized sources of authority other than scripture. When the Vatican declared the Immaculate Conception of Mary to be official church dogma in 1854, it invoked oral tradition, even though prominent theologians such as Thomas Aquinas had argued against this belief on rational grounds. Since the Council of Trent, the official position of the church had been that it would investigate doctrines critically, in a humanistic way, before accepting them. Yet we have seen that it based its rejection of heliocentrism on more than just the literal text of the Bible.

In their fight against Protestantism, Jesuits made use of a moderate form of Skepticism: the tradition of dialectics established by the fifteenth-century humanist Lorenzo Valla.[10] The Dutch thinker Rodolphus Agricola further developed this approach in his widely distributed book *De inventione dialectica* (*On Dialectical Invention*), published in 1515 mainly through the efforts of Erasmus. In his own work, Valla took Cicero as his model, advocating a rhetorical style in which one presents one's viewpoint with the greatest possible force. Presenting deductive evidence is part of dialectics as Valla describes it, but it is only one of many techniques. Valla advises us to abandon the quest for utter certainty and base arguments on probability (in the sense of plausibility).[11] Humanist Skeptics in this tradition included Melanchthon and Ramus in the sixteenth century and Mersenne in the seventeenth. Radical Skeptics like Montaigne, however, believed that this probabilistic view was too weak and led to dogmatism.[12]

Specifically, probabilism could lead to conservative conclusions because it justified the invocation of authority. In *Les Provinciales* (1657), Blaise Pascal is harshly critical of this style of reasoning among Jesuits, especially when applied to moral issues. In debate with a Jesuit priest, Pascal says, "I do not want what is probable, I want what is certain." The Jesuit then points out to Pascal that if a reliable person told him about an event he had witnessed in Rome, Pascal would take him at his word. Analogously, the priest continues, we can believe wise, pious men in issues of conscience, because their authority lends great plausibility to their opinions. When Pascal raises the possibility that wise men might disagree about some issues, the Jesuit replies that we may then choose the answer that we prefer, even if it is the less plausible of the two. This dialogue of Pascal's greatly contributed to the Jesuits' reputation for defending whatever conclusion suited them best.[13] Pascal's scientific work was strongly rooted in the deductive tradition, in keeping with his desire for certainty in moral issues, and he tended to use experimental results more as illustrations in the course of deductive arguments.

In defense of the Jesuits, one might argue that they are in fact selecting the answer best suited to the specific case at hand. They called this method casuistry, and it more than superficially resembles the Anglo-Saxon common-law system, in which case law plays a pivotal role. In this regard, the Jesuits were carrying forward the sixteenth-century humanist tradition.

There was a striking parallel between the Jesuits' views on moral and scientific issues. According to the historian John Heilbron, the Jesuits as a group were the most important scientists of the seventeenth century.[14] In particular, they excelled in the experimental sciences. Discoveries such as diffraction (of light, when reflected by a scratched surface) and electrical repulsion were made by Jesuits. Galileo's two greatest rivals and opponents, Cristoph Scheiner and Orazio Grassi, were outstanding experimenters.[15] Giovanni Battista Riccioli conducted experiments on the law of falling objects before Galileo's work on the subject was published, and in 1651 he provided detailed experimental confirmation of Galileo's measurements. Riccioli had his fellow Jesuits sing in chorus in time with a pendulum. This made it easy to keep track of the number of swings, which was essential for measuring time.[16]

However, Jesuits avoided explaining the phenomena they discovered in a unified theoretical manner. If asked why a new star had appeared in the sky, one Jesuit would give fourteen possibilities, including a novel explanation of his own, without stating a preference. He was content to leave open the question of what the true theory might be.[17] The priests' distaste for unified theory was in large part due to their role as defenders of theological orthodoxy; in 1632 they officially condemned atomism, declaring it incompatible with transubstantiation, the transformation of bread and wine into the body and blood of Jesus during the Eucharist.[18]

The only case in which the Jesuits did support one theory at the expense of another was the one that nearly cost Galileo his life, namely the debate about whether the sun or Earth was the center of the universe. Yet this case too confirms the general pattern. The professors of the Collegio Romano endorsed Tycho Brahe's hybrid system, which drew on both Copernicus and Ptolemy: Earth is in the middle, the sun revolves around it, and the other planets revolve around the sun.

Another aspect of the Jesuits' working method was a team approach, cooperation based on a shared institutional identity. The Collegio Romano did not act independently, but was the leading institute in a network of European colleges founded by Jesuits. Finally, the Jesuits can be considered the inventors of peer review. In their theological work, they developed the practice of having articles checked by colleagues prior to publication. They then extended this principle to their scientific research.

Even the scholars at the Accademia dei Lincei, who were the arch-rivals of the Jesuits, adopted some of their methods. The members of the Accademia were more or less celibate, and without exception, they were not allowed to publish anything until it had been read and approved by three other members.[19]

In 1541, a final round of negotiations between Rome and the Protestants ended in failure. Pope Paul III convoked a council in the northern Italian city of Trent (Trento) in 1545. The Vatican exercised firm control over the Council of Trent, which continued to meet, with interruptions, until 1563. This is where the Counter-Reformation was forged. In Trent, the church put its internal affairs in order, rationalized religious doctrines, and centralized its administrative structures. This council gave birth to the Inquisition and the index of prohibited books. In 1557, Erasmus's entire oeuvre was added to this index; this included his Latin grammar books, to the irritation of the Jesuits, who used them at their schools.[20]

The philosopher Stephen Toulmin sees the Council of Trent as the first step in a process of political, religious, and philosophical polarization that put an end to tolerant Skepticism in religion and science by about 1600. Philosophers such as Descartes began to seek certainty, and science followed in philosophy's footsteps.[21]

Toulmin has a point, but it should not be overstated. To be certain, religious tensions mounted. In the Netherlands, for instance, doctrinaire Calvinists rejected Arminian views at the Synod of Dort in 1618. As a result, the jurist Hugo Grotius had to flee the country and spent many years in exile in France. Toulmin also cites the assassination of Henry IV of France in 1610 and the Thirty Years' War in Germany from 1618 to 1648. Starting in 1624, however, when Cardinal Richelieu took control, France became a little more tolerant again. Furthermore, the church in France was effectively beyond the control of the Vatican, making the influence of the Inquisition very slight. Marin Mersenne developed a new, highly influential form of Skepticism in this period.

Despite Mersenne and the relatively benign cultural climate in France, it is fair to say that Europe became divided. In the seventeenth century, the modern, Cartesian version of the deductive style became dominant in France, and in some senses it has remained so until today. In the Protestant countries of England and the Netherlands, an experimental empiricism emerged that was congenial to moderate Skepticism.

Ptolemy and Copernicus

Skepticism was not an entirely novel philosophical position. In the Middle Ages, Averroës had used it, albeit in a more diluted and selective form, to argue against the Ptolemaic system. Ptolemy had been forced to make three conspicuously inelegant assumptions to reconcile his observations of the motion of the planets with the idea that they were orbiting around Earth. A planet's apparent motion, which can be tracked with the naked eye, follows a strange, meandering course relative to the background stars (which together trace a circle of their own). According to Ptolemy, the center of the planetary orbits was not Earth, but an imaginary point near it. He referred to such orbits as eccentric. The planets were also said to revolve around a second type of imaginary point, which in turn revolved around the Earth: these small circles moving along larger ones were known as epicycles. Furthermore, the Ptolemaic system could only explain why the planets appeared to move at variable speeds by referring to a third imaginary point, known as the equant. For the special, difficult case of Mercury, Ptolemy developed yet another mathematical device.

Averroës observed that this formal system did not prove the existence of epicycles and eccentric orbits in reality, and Thomas Aquinas agreed.[22] The renowned thirteenth-century physician Arnold of Villanova and many other scientists from the Middle Ages onward held an even stronger version of Averroës's position, contending that planets do not move in epicycles or anything of the sort.[23] As they saw it, epicycles and equants were just mathematical devices that bore no relation to reality. To use a well-known phrase, these were seen as techniques for "saving the phenomena," which could in principle be explained by another hypothesis. This shows the low status of mathematics, which was deemed irrelevant to true causes.[24]

In technical terms, there was nothing wrong with the Ptolemaic system. It served the purposes for which it had been designed: making calendrical calculations, predicting eclipses of the sun and moon, and so forth. Yet it had other flaws that tarnished its claim to represent the physical universe. Ptolemy himself had become inclined toward a Skeptical view, because the motion of some planets, such as Mars, could be accounted for using either an eccentric or an epicycle. He chose to use both devices for all the planets to maintain maximum homogeneity among them, even though simpler solutions would have been available if he had distinguished between the outer and the inner planets. But of course, he had no theoretical grounds for any such distinction.[25] Another problem was that the path of a planet with an eccentric orbit could cross that of a planet moving in epicycles. This was a problem because, in the Middle Ages, planets were not seen as moving freely through space, but as embedded in crystalline spheres; if two orbits crossed, the rotating spheres would have to intersect, which is physically impossible. Nevertheless, Georg von Peuerbach accompanied his fifteenth-century reconstruction of the Ptolemaic system with an illustration of precisely this situation. Apparently, it did not bother him.[26]

Although Nicolaus Copernicus was a mathematician himself, he had higher ambitions for his heliocentric system. In *De Revolutionibus* (*On the Revolutions*), he praised Hermes Trismegistus in his role as a sun priest. It seems very much as though Copernicus, who was in the employ of the Catholic Church, saw the sun as a god who deserved a place in the center of the cosmos. This aspect of his system does not seem to have struck his contemporaries; in any event, they did not comment on it. Whatever the case may be, Copernicus believed in the truth of the heliocentric system and assumed (probably correctly) that Pope Clement VII would be pleased with his work.[27] Although Copernicus too had to work with epicycles and eccentric circles (dispensing solely with the equant), he made sure there were no intersecting spheres in his model.[28]

By the time *De Revolutionibus* was published in 1543, there was a new pope, Paul III, and Copernicus had missed his opportunity to obtain the church's approval. (Even so, Copernicus had dedicated the book to Paul III.) For Catholic theologians in the sixteenth century, Copernicus's greatest sin was not heliocentrism in itself, but the fact that his view seemed to be based on mathematical arguments. He was not accused of heresy.[29]

Nor did Copernicus have better prospects in the Protestant country of Germany. Without his knowledge, a preface had been added to the book, stating that the system it outlined was merely a hypothesis intended to simplify calculation, without any claim to truth. The book had been printed through Copernicus's connections in Wittenberg's Lutheran circles. Copernicus's disciple Joachim Rheticus was a professor of mathematics in Wittenberg and a strong supporter of the new system.[30] However, Rheticus had delegated coordination with the printer to Andreas Osiander, a confidant of Philipp Melanchthon, the Lutheran humanist.

Osiander was the author of the anonymous preface. His aim had probably been to forestall objections from Melanchthon, who was opposed to expressly backing a heliocentric system. It should be added, however, that Melanchthon did not have a strong opinion on the issue; the spirit of Erasmian tolerance was still at large and polarization had not set in.

Whatever Osiander's motives may have been, readers took his preface to heart. Philosophers and theologians concluded that Copernicus was not so different from Ptolemy; astronomers were still mathematicians, whichever system they used. Regnier Gemma Frisius, a cartographer and professor of mathematics in Louvain, claimed the reason he had adopted the Copernican system was that it provided the most convenient method of calculation, but he was probably genuinely sympathetic to it as well.[31] In 1554, the German astronomer Erasmus Reinhold used Copernicus's book to draw up his "Prutenic," or Prussian, tables, a kind of almanac showing the expected positions of planets in the sky. These tables were more accurate than their predecessors, but their author did not feel compelled to view this as evidence for heliocentrism, and neither did their many grateful users.

In fact, it proved possible to reconcile Reinhold's tables with the Ptolemaic system. In 1582, they played an important role in the development of the new Gregorian calendar; Christopher Clavius, who led the project, and Pope Gregory XIII, who authorized it, seem not to have seen any harm in this. The Jesuits at the college in La Flèche taught Copernicus's system to Descartes, perhaps because it made calculations somewhat easier. And Pope Urban was astonished when Galileo insisted on the absolute truth of the Copernican system, for he could quite happily live with a selective Skepticism that did not require a choice.

Many others backed Ptolemy or Copernicus for probabilistic reasons. In Mersenne's view, the balance swung toward Ptolemy at first, and later Copernicus, but neither system was ever definitively proven. A variety of considerations thus played a role in this period. Around 1600, for instance, a Jesuit might prefer the Copernican model to Ptolemy's for instrumental reasons, because it was easier to use and more accurate, but this mathematical argument did not otherwise lend status to the Copernican system. In a physical sense, probabilistic reasons were thought to militate against a heliocentric system and in favor of a geocentric one (in other words, Ptolemy's). The views of the church played an important role in this probabilistic decision, but that was due to its role as an authority; preferring geocentrism for this reason was different from choosing it on the basis of personal faith. Probabilistic argument did not unequivocally support the Ptolemaic view, however, since it was implausible that epicycles and other such devices corresponded with physical reality.

This was a manifestation of the age-old distinction between philosophy (including physics) and mathematics (including astronomy). The philosopher-physicists at the Collegio Romano, as well as philosophers and theologians among the Dominicans and other monastic orders, believed in enforcing this distinction and were active defenders of orthodoxy. However, Jesuit mathematicians, particularly Clavius, took a different view. They believed that astronomy, and hence mathematics, dealt with real causes.[32] Galileo held the same opinion, though he advocated a different theory.[33] It may seem surprising today, but when Christopher Scheiner, the leading Jesuit astronomer, discovered that Venus and Mercury revolved around the sun rather than Earth, his faith in geocentrism was unshaken, even though he considered his discovery a true "physical" fact and not a mere mathematical fiction.[34]

Around 1612, Galileo tried to show that the Copernican system was more in harmony with the Bible than Ptolemy's. These attempts were censored, however.[35] A 1616 edict prohibited any use of the Bible in support of the Copernican system. The works of Copernicus were added to the index of censored books, with the qualification "until they are corrected."[36] The church objected only to a few minor, largely technical points, however, and did not ban books that made use of Copernicus's ideas.[37] That same year,

Galileo came under the scrutiny of a suspicious Inquisitor, but escaped punishment thanks to Cardinal Bellarmino's protection. Galileo heeded Bellarmino's admonition that Copernicus could not be endorsed until there was clear and compelling physical evidence for his views. Later, Galileo thought on several occasions that he had found such evidence, or was on the verge of doing so.[38]

The edict of 1616 was a sign that the terms of the debate were changing. Since about 1600, the hunt for heretics and the legacy of the Council of Trent had led the church into an increasingly orthodox Aristotelian (or "Peripatetic") stance. Galileo was not the only one who suffered the consequences. Francesco Patrizi, who held the chair of Platonic philosophy at La Sapienza, a university in Rome, was prohibited in his old age from teaching that there is only one heaven, instead of the many described by Dante and other authors, and that the stars are in an infinite space.[39] By this point, the church could not maintain its moderate Skepticism, even though this position held the most appeal even for Urban himself. The mood had shifted, and the church was expected to take sides. One important reason that Galileo advocated Copernicus so zealously was his fear that the Jesuits, and the church as a whole, would throw their weight behind Tycho Brahe's system instead.[40] In any event, the Ptolemaic system inspired indifference among Galileo's opponents.

While the church became ever more apprehensive about truth claims for the Copernican system, Galileo grew more and more insistent about the truth of theories expressed mathematically, if they corresponded to observed reality. In 1615, he stressed the importance of acknowledging that epicycles and eccentric orbits really existed.[41] Galileo never acknowledged the work of Johannes Kepler, who ascribed elliptical rather than circular orbits to the planets and dispensed with epicycles and all the rest.[42] What is even more striking is that Galileo maintained that epicycles truly corresponded to reality. Galileo, Kepler, and Clavius were the first scientists to interpret a theory as a pictorial representation of reality, each in his own way.[43]

Anti-Skepticism

Descartes most clearly expressed the vision that Clavius, Kepler, and Galileo had developed: understanding the world means having an accurate picture of it, and that picture is located in the mind.[44]

Historians and philosophers have debated whether Galileo was a Platonist or Neoplatonist, an Aristotelian, or neither of those but a modern scientist. The evidence for the last view includes the fact that he spurned reliance on authority, instead choosing to trust observation. Supporters of this view might argue that Platonic elements in Galileo's thinking, such as his use of mathematics, and Aristotelian ones, such as deductive argument, were merely convenient tools that he used when it suited him.[45] However, this glosses over the fact that there was not yet any such thing as a modern scientist; it was a role and attitude that had yet to be created.

To be sure, Galileo's work had Aristotelian elements, which are most evident in his anti-Skepticism, his unshakeable faith that he could apply mathematics to observed reality and gain certain knowledge of the world and cosmos.[46] For Galileo, it was "impossible to believe that nature was not constrained by necessity" and just as impossible *not* to believe in the "true constitution of the universe," which "could not possibly be otherwise."[47] In Galileo's eyes, that true condition was the Copernican system. A Jesuit at the Collegio Romano presented him with the familiar Skeptical argument that various causes can have the same effect. But in Galileo's treatise on the motion of the tides (*Discorso del flusso e reflusso del mare*, 1616), a phenomenon which he saw (incorrectly) as a confirmation of the Copernican view, he tried to show that cause and effect mutually entailed each other, and that causes could be derived from effects in a mathematically necessary fashion.[48] Galileo may not have been a thoroughgoing Aristotelian (and certainly his physics was a clean break with Aristotle's), but in these passages one hears more than a faint echo of Aristotle's *Posterior Analytics*.

Galileo took a pragmatic approach in cases where more than one causal explanation of a phenomenon was possible, and his position could border on Skepticism when it suited him—as in the case of the comets, one of the many controversies in which he was involved.[49] But he usually took the stance that it was sufficient to eliminate the incorrect explanations one by one, leaving only the correct one. What he overlooked is the possibility that future scientists would come up with new explanations. Galileo, like Descartes after him, always believed that it was possible to determine true causes. His anti-Skeptical vision of scientific knowledge also applied to observation. Galileo consistently stressed reliability in observation and optical instruments. His perspective was a practical one; he developed procedures for keeping the number of observational errors in astronomy to a minimum, and arrived at the fundamental principles of what would become error analysis a century later. By indicating the margins of error in his measurements, he tamed the unreliable senses.[50]

Johannes Kepler was a mathematician in the service of Holy Roman Emperor Rudolf II, whose court was in Prague. There Kepler obtained very accurate observations of the planets, made by Tycho Brahe over a period of many years. In 1609, Kepler published the first two laws that bear his name, one on elliptical orbits and the other on conservation of angular momentum in planetary motion (the law of equal areas). In 1619, he added a third, on the relative speed of the planets in relation to their distance to the sun.

Kepler attacked Skeptics for the lack of ambition implicit in their argument that there was no rational choice to be made between the Ptolemaic and Copernican systems. Skeptics contended that because both "hypotheses" were compatible with observations of the planets, they were in that sense "equivalent." For Kepler, this was too severe a constraint on theory selection. Unlike Descartes, he accepted that both systems were hypotheses, but he was optimistic about the possibility of eliminating incorrect hypotheses. One theory might also be preferable to the other on "physical grounds."[51] What Kepler meant by this was that an incorrect hypothesis will betray itself if it is applied to parts of the natural world for which it was not originally intended. In other instances, Kepler argued that Copernicus's system was superior because he could explain what to Ptolemy was purely coincidental.[52]

René Descartes studied with the Jesuits at La Flèche college for more than nine years,[53] but later came out against their scientific method. He wrote in his *Discours de la méthode* (*Discourse on Method*, 1637), which was intended for a general public, that scholars at La Flèche preferred plausibility (*vraisemblance*) to causes (*les raisons*). Students were trained to become lawyers, rather than judges.[54]

Descartes's mathematical work took an approach very unlike the Jesuits'. Descartes consciously viewed himself an innovative thinker. He abandoned the mathematical restoration project of the Renaissance, attributing a much more abstract meaning to mathematics than the humanists had with their discourse of utility. At the Jesuit school, Descartes had received the impression that mathematics was the handmaiden of "the mechanical arts," as he wrote in the *Discourse on Method*, but he rejected this view.[55] Descartes had begun writing this *Discourse* in 1633, when he heard about Galileo's condemnation. Shocked by the news, he set out to ensure the survival of Galileo's "method" and the Copernican system along with it.[56]

Descartes's chief philosophical work, *Meditationes* (*Meditations*, 1641), is intended as a refutation of Skepticism from the first page to the last. Its rhetorical structure shows how influential Skepticism was. The first sentence of the first meditation begins with an apparent concession to Skeptics. The unreliability of the senses, which Descartes illustrates with a few vivid examples, leads him to decide that nothing can be taken as certain. He is even compelled to question the existence of God. But then Descartes discovers something that is certain after all. Even though he may not yet know *what* to think, he becomes aware *that* he thinks, and this introspective observation is the basis of his discovery that he exists. *Cogito, sum*: "I think, I am."

Many people are familiar with the version of this phrase that appears in Descartes's work for a broader public, the *Discourse on Method*: *Je pense donc je suis*; "I think, therefore I am." The "therefore" in this version can be misleading. No deductive inference is being made from "I think" to "I am." On the contrary: "I am" is a more general truth than "I think," which is why in the *Meditations* the two phrases are linked solely by a suggestive comma. Strictly speaking, Descartes makes an inductive argument.[57]

The problem with inductive arguments is that the conclusion is not guaranteed to be true. In the seventeenth century, inductive chains of reasoning came to be interpreted in probabilistic terms, but that was definitely not what Descartes had in mind. This was no ordinary induction. Descartes required a guarantee that the inference *cogito, sum* was true. His concept of clear and distinct ideas provided this guarantee: a necessary connection that appears entirely clear and distinct to the mind must be true. Apparently, what was needed was the firm inner conviction that an idea was correct. Inner conviction? The historian Richard Popkin has observed that Descartes could have borrowed this criterion from Calvin.[58]

Richard Rorty has argued that Descartes gave shape to the modern concept of the mind unintentionally, through his reflections on certainty in science and what is needed to obtain it.[59] With this argument, Rorty turns the conventional history of philosophy on its head. For philosophers who followed Descartes, especially from Kant onward, "I think" was the cardinal discovery of Cartesian philosophy. From that point onward, philosophy was "modern."

For Descartes himself, however, the *cogito* was no more than an intermediate stage, a step on the way to proving God's existence. In the *Meditations*, he goes on to consult his own mind again, and finds the ideas of perfection and infinity. An imperfect and finite being could never have such ideas on its own, he reasons; they must have been placed there by God. Having thus established the existence of God, Descartes can invoke Him as the guarantor of the accuracy of his observations, his sense impressions, and no longer has to worry that there might be a "malignant demon" deceiving him.

This provided the foundation of a science based on observation, and vanquished the Skeptics. God was also a guarantee that Descartes's laws of physics were absolutely certain (especially with regard to motion), while for an atheist they could be no more than hypotheses.[60] In *Principles of Philosophy*, Descartes wrote:

> And certainly, if the principles which I use are very obvious, if I deduce nothing from them except by means of a Mathematical sequence, and if

what I thus deduce is in exact agreement with all natural phenomena; it seems [to me] that it would be an injustice to God to believe that the causes of the effects which are in nature and which we have thus discovered are false. For we would then be accusing Him of having made us so imperfect as to be liable to make mistakes, even when correctly using our reason [which He has given us].[61]

In other words, Descartes believed he could use the criterion of clarity and distinctness to assess specific scientific concepts. By current standards, he was sometimes remarkably on target, but in other cases he was extremely wide of the mark. The source of Descartes's criterion of clarity and distinctness was the notion of a mental image and the extent to which one could be vividly evoked in the mind.[62] This theory went back to Aristotle,[63] as well as Quintilian, who concentrated on the qualities that made such an image convincing to others. For this purpose, the mental image (*phantasma*) had to be transformed into its rhetorical equivalent (*eikon*).[64] Descartes, in contrast, set out to convince himself first, and others only afterward. The original meaning of the Latin word *evidentia* is perspicuity, or vivid representation, and vivid representation was the goal.[65]

For Descartes, the scientific criterion that ideas must be clear and distinct was a way of making scientific discoveries and solving mathematical problems. It was not a foundation for knowledge, but a God-given guarantee that newly acquired knowledge was correct. The same applied to the *cogito*. Descartes placed knowledge that had already been acquired in a deductive framework. This framework was not important to him as a method for acquiring knowledge, but ultimately the certainty provided by clear and distinct ideas was expected to apply to that entire deductive edifice.[66]

Pierre Gassendi remarked that "everyone thinks that he clearly and distinctly perceives the truth that he champions."[67] Letting go of the criterion of clear and distinct ideas leaves nothing but the hypothetical method. Most of Descartes's followers went down this path; in 1646, Descartes severely berated his Dutch disciple Henri Regius, a medical professor in Utrecht, for doing so.[68]

Descartes's reason for concern was the trial of Galileo, which made a deep impression on him and filled him with dread. A guaranteed truth was the only thing that could not be brushed aside as a mere hypothesis. Yet despite all the difficulties that Cartesianism encountered in France and the Netherlands, Descartes had no need to fear that he would be condemned like Galileo. And what Cartesianism lost in guaranteed certainty, it compensated for with the cogency and plausibility of what Descartes had been trying to protect: mechanism, a topic we will return to later in this book.

The only thing we need note here is that mechanism retained its appeal, especially in France, almost throughout the eighteenth century.

Descartes gave a new impetus to Aristotelian deductive thinking. Although contemporary Skeptics were not at all convinced, he did a great deal to ensure that the deductive, "rationalistic" style would remain dominant in France for many centuries. The period after 1633 is marked by the downfall of humanistic values, and the direction in which Descartes steered philosophy gave priority to the timeless and universal above everything that is bounded in space or time.[69]

Marin Mersenne

Kepler, Galileo, and Descartes had lashed out at all forms of Skepticism. Mersenne attacked only Sextus's radical Skepticism, proposing to replace it with a moderate form. While he owed much to his Jesuit teachers, he also introduced some new perspectives. Like the Jesuits, he was a probabilist; one amusing example of this tendency is his proof of the existence of God. He did not stop at a single proof, but gave thirty-six, none of which were entirely convincing. His hope, in doing so, was that the sheer number would have an overwhelming effect. In essence, he tried to "pitch" his ideas, believing it was impossible to provide conclusive proof.[70]

Mersenne concluded that it was not possible to learn the true nature of things, reasoning from the fact that we cannot know what God's intention was in creating the universe. In our ignorance, we see creation as an arbitrary structure that could have been otherwise.[71] What could we possibly learn about that? Mersenne thus believed in modesty: first principles such as Aristotle's were beyond our grasp. At the same time, however, he countered radical Skepticism with the claim that one type of knowledge was possible: knowledge of phenomena. While conceding to the Skeptics that there can be no certain knowledge about the nature of things, he added that this was not a problem. What is possible is subjective knowledge about the way the world appears to the researcher, who can formulate hypotheses about it, draw connections between events, and make predictions about future observations. Popkin regards Mersenne as a pragmatic positivist *avant la lettre*.[72]

In *Questions théologiques* (*Theological Questions*, 1634), Mersenne wrote: "It seems that the capacity of men is limited by the outside and surface of corporeal things, and that they cannot penetrate further than quantity with complete satisfaction. That is why the ancients could not give any demonstration of what appertains to qualities, and restricted themselves to numbers, lines and figures, if one excepts weight, of which Archimedes spoke in his *On the Equilibrium of Planes* [Isorropiques]."[73]

Initially, Mersenne had treated mathematical physics (like optics and music) as an exception to these Skeptical considerations, but in his later work he no longer set these fields apart. Even in music and optics, according to Mersenne, unprovable claims are made about the nature and characteristics of light and sound. His system excluded all of physics from the realm of *scientia*, demonstrable knowledge.[74] In a wide range of fields, Mersenne was unable or unwilling to choose among competing theories. For instance, he considered the question of whether light traveled at a finite or an infinite speed to be undecidable on the basis of the evidence available to him. Descartes and Christiaan Huygens were none too pleased with this standpoint.[75]

Yet there was compensation for this loss of certainty. In this respect, too, Mersenne drew on the tradition of the humanists, including the Jesuits, and specifically their discussion of the practical value of mathematics. Mersenne not only used this as a pragmatic argument against radical Skepticism, but also derived an epistemological insight from it. Making something, said Mersenne, is the proof of true knowledge.[76]

In other words, experimentation and technology were the only road to truth. Mersenne was a meticulous experimenter; he repeated Galileo's experiment with a ball rolling down a groove, but unlike Galileo, he described how he had carried it out and what materials he had used for the groove and the ball. His measurements of free-fall confirmed the ratios of the distances traveled in successive time intervals, but not the distances in absolute terms, and Mersenne did his utmost to determine what units of distance Galileo had used. (Mersenne was not as specific about how he measured time. It is assumed that he used a pendulum, a device on which he had done a good deal of experimental research.[77])

Mersenne also took an experimental approach to the argument from the unreliability of the senses, which was the pride of the radical Skeptics. In doing so, he built on his knowledge of music theory, expanding his research to include physiological aspects of human hearing.

In the generation before Mersenne's, music had changed utterly with the advent of polyphony and other new developments, and this had repercussions for music theory. The subject interested many well-known scholars, in part because music was in the traditional quadrivium of higher liberal arts. The authors of works on music theory included Simon Stevin and Isaac Beeckman, Kepler, Descartes, and Christiaan Huygens.[78] It was in fact the field to which Mersenne devoted most of his scientific work.

The new music theory was confronted with a knotty problem. In traditional music theory, which went back to the Pythagoreans and Plato, the only pure intervals were those expressible as harmonic ratios of open-string length to stopped-string length on a taut musical string. Harmonic ratios are those involving only whole numbers between one and four. The ratio for an octave is 1:2, for a fourth 3:4, and for a fifth 2:3. But in 1558, Gioseffo Zarlino, the *maestro di capella* at St. Mark's Basilica in Venice, increased the number of pure intervals by allowing ratios of whole numbers between 1 and 6. This gave him the major third (4:5), the minor third (5:6), and the major sixth (3:5).[79] At this point, a serious problem arose; the intervals do not add up properly. For instance, the notes in two series of intervals whose fundamentals are separated by a fourth do not correspond precisely. In polyphonic music, this could lead to tuning problems. Zarlino tried to deal with this new situation by developing a severely constrained theoretical system. However, this drew sharp criticism from his former student Vincenzo Galilei (the father of Galileo), who presented the empirical argument that in practice musicians adjust the series to one another in such a way that the ear is not bothered by the discrepancies. This process is known as tempering.

Mersenne endorsed Vincenzo's view without reservation, contending that how we hear in practice takes precedence over the mathematical precision of a musical scale. While this admission of the imperfection of human senses might seem to be a concession to the Skeptics, Mersenne also tried to investigate what the senses *were* capable of. This inspired a series of experiments investigating sound and its perception by humans. His discoveries included the speed of sound and the part of the ear where the physical sensation of sound is produced. Mersenne conducted a similar program of research into light and vision, with results that included a method for measuring differences in the intensity of sunlight.[80]

In the 1640s, the Englishman Thomas Hobbes spent time in Paris. He soon gained admission to gatherings at Mersenne's Minim convent. Although Mersenne rendered significant services to Hobbes (such as publishing his work on optics), Hobbes remained critical and later recalled that Mersenne had dominated the meetings. Whoever wanted to demonstrate an experiment or discuss a problem addressed himself primarily to Mersenne, who then passed judgment.[81]

Hobbes's first objection to the experimental method, which he formulated more precisely in his debates with Robert Boyle, was the role of the master, whose opinion on the relevance and reliability of an experiment carries more weight than that of someone with less experience and status. The validity of a deductive proof, in contrast, can be checked by anyone with enough intelligence. Hobbes also noted that experiments require a space in which they are performed, which despite all the rhetoric is never entirely open to outsiders, because the number of observers who could potentially oversee the process in a useful way is extremely limited. Experiments require instruments that are affordable only to a small minority of people—the pump with which Boyle created a vacuum in a bell jar has been referred to as the seventeenth-century version of Big Science. (The telescope, microscope, vacuum pump, and barometer were the quintessential seventeenth-century scientific instruments.) Hobbes's final complaint was that experiments require a community of people who are involved in carrying forward an experimental method. A community of this kind has a social structure to which individuals are forced to adapt.

It was Robert Boyle, in England, who first explicitly stated the conditions under which scientific experimentation could take place. The standards he set are still considered definitive of good experimental science. Boyle followed the example of Mersenne, whom he admired, by creating an informal network with himself at the hub. But he added new elements, some borrowed from Francis Bacon and others of his own invention. Bacon had considered it absolutely crucial to the success of experimental science that the results of experiments be made public. It was probably due in part to this ideal of public disclosure that, in the course of the seventeenth century, scientists came to see themselves as forming an international community. Within this community, they could develop trust in each other's work—an important goal, since they could not always repeat each other's experiments.[82] Trust was built step by step.

Steven Shapin and Simon Schaffer have described Boyle's contribution to experimental science as the creation of a new "form of life," a term coined by Wittgenstein. More particularly, they refer to Boyle's material, literary, and social technologies. We have already touched on the material technologies involved in his vacuum pump experiments. The term "literary technology" may have an odd ring to it, but it neatly captures the style of writing that is still standard in the natural sciences. Boyle was the first to use this style consistently. He forbade polemics and *ad hominem* arguments, insisting that experiments be reported in the driest possible manner, without frills.

Boyle also stressed the importance of keeping disputes about the results of experiments within the community of experimenters, a principle that can be classified as social technology.[83] His aim was separation from the rest of society, and the protector of the Royal Society, King Charles II, was willing to provide it. The king promised the scientists the power and authority to work together "without any molestation, interruption or disturbance." In return, the Society pledged that it would not meddle (in the words of Robert Hooke) with "Divinity, Metaphysics, Moralls, Politics, Grammar, Rhetorics, or Logic."[84] This pledge should not be seen as a form of submission. It was in perfect harmony with the moderate Skepticism that formed the quasi-official philosophy of the Royal Society's founders, and Boyle heartily endorsed the sentiment.[85]

6

The Experimental Style III

Alchemy and the New Sciences

> *I suppose that the Philosophers stone is in the North West passage.*
> —William Watts
>
> *For man by the fall fell at the same time from his state of innocence and from his dominion over creation. Both of these losses however can even in this life be in some part repaired; the former by religion and faith, the latter by arts and sciences.*
> —Francis Bacon

SURVEYING THE WHOLE RANGE of seventeenth-century sciences, we can identify two broad categories. Thomas Kuhn called the first group—mathematics, astronomy, harmonics, optics, and statics—the classical sciences because of their roots in classical antiquity. A new group of sciences had joined them, including metallurgy, magnetism, chemistry ("chymistry"), and the study of heat. Kuhn called these Baconian, after the English statesman and philosopher Francis Bacon.[1]

The new sciences were not taught at universities; they were "lower" sciences, and their practitioners were generally amateurs (or physicians). In the seventeenth-century Republic of Sciences, these amateurs were, in Kuhn's words, second-class citizens, except in England and the Netherlands. These sciences had their roots in the knowledge of skilled craftsmen such as dyers and metallurgists, who were a step below even painters, architects, and abacists in the intellectual hierarchy (though if they owned large workshops, they could be accepted into the upper middle class).

The practical utility of the new sciences was not the only thing they had in common (and geometry had practical uses, too). These Baconian sciences also required active intervention in the normal course of nature, usually with the aid of a device or instrument. A chemical experimenter, for instance, makes something happen that would not occur in the same way in nature. The same applies to a scientist like Robert Boyle, who conducted experiments under a glass bell jar from which he had pumped out the air. Placing a bell inside the bell jar, Boyle demonstrated an unprecedented phenomenon: when it rang, even people standing nearby could not hear a sound.[2] Francis Bacon described this type of intervention in nature as the task of the scientist, who in his opinion had to take an active role. Bacon mustered a broad range of metaphors in support of his argument that scholars should seize control of the forces of nature.

This style of experimental science came into being as a variety of the mathematical experimental style practiced by Galileo and others.[3] But even more than the mathematical engineering tradition associated with Galileo, the new sciences required the practitioners to get their hands dirty. Moreover, they were practiced within an alchemical, magical, or astrological framework, and in various combinations of and variations on these systems. Remarkably, this form of experimental science remained largely the same in character throughout the seventeenth century, and historians' estimates of the date at which magic and alchemy started to become irrelevant have moved ever further toward the present.[4]

As far as we know, Bacon himself did not perform any alchemical experiments, but he had made a thorough study of the writings in the occult tradition.[5] Two generations later, in the latter half of the seventeenth century, the eminent scientist Boyle was still seriously in search of the philosopher's stone, and Newton's new concept of gravity grew out of mystical notions inspired by alchemy. Johann Friedrich Schweitzer (also known as Helvetius), the personal physician to William III, claimed to have transmuted half an ounce of lead into gold at his home in The Hague in 1666 and to have had the metal assayed by a goldsmith. A foreign stranger from North Holland had given Helvetius a small piece of a philosopher's stone after hearing that he was skeptical about alchemy.[6]

In 1675, the physician and mathematician Johann Joachim Becher transmuted lead into silver. He had the silver made into a medallion with an inscription, which is still on display in Vienna.[7] Becher, who had the ear of the German emperor, received permission three years later from the governing council for Holland and West Friesland to start a company that would produce gold from sand. He found private investors and had sand brought from the seaside near The Hague. By the time the project failed, he had vanished. This episode smelled more than a little fishy, despite Becher's excellent credentials as the brain behind many projects for German princes—and even the emperor himself—to use metallurgy and other technology for the public good. He had often held up the Dutch spirit of trade and enterprise as a model for the Germans.[8]

This episode cannot have shaken the general faith in alchemy all that much; Herman Boerhaave wrote in the early eighteenth century that there was gold concealed in lead.[9] Over the course of that century, the magical and occult aspects of alchemy gradually moved out of the scientific mainstream. But they survived in popular culture, and scientists are still pictured today as brewers of mysterious potions in oddly shaped vessels.[10]

Alchemy came to Europe in the twelfth century, mainly from Arabic sources.[11] But natural philosophers at universities were generally uninterested, with a few major exceptions such as Roger Bacon. For many centuries it was physicians, like Arnold of Villanova, who had the greatest exposure to various occult practices. Throughout the sixteenth century (if not longer), astrology was a required subject in their program of study. Marsilio Ficino, the translator of Plato and Hermes Trismegistus, was a medical practitioner (though he had never entirely completed his medical studies), and Cardano, Paracelsus, and Jan Baptista van Helmont held medical degrees. Compared to natural philosophers, physicians had easier access to court circles, where there was much interest in the subject of magic. And a doctor who became the personal physician to one of the many Italian princes earned much more than a natural philosopher at a university, enough to dabble in occult matters at his own expense. Some physicians helped their princes to build collections of objects with magical properties.[12]

For physicians, astrology was more than a method for predicting the future (for example, by casting a patient's horoscope). It was the core of a theory in which the properties of planets, plants, metals, magical objects, human body parts, and temperaments (such as melancholy) were all part of a large, interconnected system of similarities, sympathies, and antipathies.[13]

By 1600, however, natural philosophers were very active in the field of alchemy. Giordano Bruno, who was burned at the stake that year, was a Dominican and a natural philosopher. The English alchemist John Dee was a mathematician. In seventeenth-century England, it was primarily mathematicians who gave explanations derived from the magical traditions. Some of these mathematicians also practiced alchemy,[14] despite its drawbacks, which were numerous. For one thing, alchemy was a good way to lose your money, given the large investments required in supplies and equipment. Seventeenth-century painters in Flanders and the Northern Netherlands, such as Adriaen van de Venne and Adriaen van Ostade, produced many scenes of alchemists' workshops with this moral message.[15] Secondly, the practice of alchemy was prohibited in many countries for both political and religious reasons. Because alchemical experiments were thought to rely on assistance from devils, the experimenters were distrusted and feared, and the authorities and the public were not usually willing to take alchemists at their word when they claimed to practice white rather than black magic. Popular images of alchemists were strongly influenced by the character of Doctor Faust, portrayed by the English playwright Christopher Marlowe and others. As is well known, Faust made a pact with the devil. But there were also good magicians in literature, such as Prospero in Shakespeare's *The Tempest*.[16] For a short while, the Roman Catholic Church was sympathetic to ideas drawn from alchemy and magic. But that changed after the Council of Trent, where these practices were condemned as heresies.[17] Most secular authorities were also hostile to the independent practice of alchemy, but court alchemists were often allowed to display their skill for the greater glory of their royal patrons. Emperor Rudolf II, who had Kepler at his court, was a great patron of both the arts and alchemy.[18] And Emperor Leopold I prized the ability to make gold and appointed a number of alchemists to lucrative positions in the mining industry.[19] They must have done well there, because German mines were seen as a model for all Europe.[20]

Alchemy, by nature, was a form of esoteric knowledge only accessible to insiders, and that never changed. Newton kept very quiet about his alchemical pursuits, which therefore remained undiscovered during his lifetime and for two hundred years afterward. When this aspect of Newton's work was first brought to light in 1946, thanks to the economist John Maynard Keynes, historians of science ignored it.[21] What did change over the centuries was that alchemists became more pragmatic, gradually losing interest in elaborate theoretical frameworks and ties to other bodies of esoteric knowledge, such as astrology. In England around 1660, alchemy, like other forms of occult knowledge, provided inspiration for the experimental study of more and more domains of the natural world. At least in that country, occult knowledge coexisted peacefully with modern systems of thought such as mechanism and atomism.[22]

In the seventeenth century, alchemy and related occult systems certainly had their opponents, and in hindsight, it is tempting to interpret the resistance to them as foreshadowing the triumph of a non-occult scientific worldview. But that view is only partly correct. One famous controversy pitted Johannes Kepler against Robert Fludd and his "hermeticism" (a philosophy based on the work of Hermes Trismegistus). Kepler criticized Fludd's non-mathematical method, particularly his belief that he could understand the world through pictures and diagrams in the hermetic tradition.[23] But Kepler himself was far too deeply entangled in a network of mystical and magical ideas for us to view him as a spokesperson for a modern worldview.[24] As late as the 1660s, serious alchemist/chemists such as Becher continued to cite Hermes Trismegistus as an authority, even though Causabon had by then shattered the myth surrounding the hermetic writings.[25]

Marin Mersenne also criticized Fludd, making arguments rooted in Skepticism. In contrast, Mersenne's main objection to alchemy was that it promised a form of salvation outside the Church.[26] He had no objection to alchemical experiments, as long as they were performed openly.[27] Around the same time, the chemist Andreas Libavius disparaged the alchemist Oswald Croll for his "vulgar" views, meaning his Paracelsianism. But Libavius himself believed in the transmutation of lead into gold.[28]

Finally, pure mechanists such as Galileo, Descartes, and Huygens were consistently opposed to obscure causes in scientific reasoning, except when seemingly obscure causes could be explained mechanically. One could even say that they firmly established the causal explanation of obscure qualities as a central part of the scientific method.

But the mechanists proved to be too ambitious. Neither Newton nor his successors ever found a mechanical explanation for gravity. The first authors to oppose alchemy in terms that are familiar to us today were the authors of the *Encyclopédie*, that monument of the French Enlightenment.[29] But that was not until 1770 or thereabouts.

The reception history of alchemy, and of occult knowledge in general, can be separated into two strands. The ways in which alchemy influenced scientific theories in biology, chemistry, and physics are discussed in the next chapter. The magical aspect of the occult sciences also had a profound impact on scientific practice, because it gave new meanings to terms such as "observation," "experience," and, above all, "experiment."

Before it could exert such influence, alchemy had to gain intellectual respectability. In the Middle Ages, there were a few alchemists who strove for recognition of their theoretical ambitions, but even so, alchemy never rose above the status of a craft. Not until the Renaissance did alchemists receive the recognition they desired, and not from university scholars, but from humanists motivated by beliefs that no modern scientist would acknowledge as scientific. Nevertheless, it was alchemical ideas that led to the enrichment of natural philosophy with new experimental methods.[30]

Marsilio Ficino finished his translation of the writings of Hermes Trismegistus in 1463, just in time to present it to the dying Cosimo de' Medici the Elder.[31] Magic and alchemy had always aroused suspicion, as Roger Bacon had learned to his detriment. But the more innocuous varieties of magic had gained a firm foothold in the courts of the Italian princes, for whom it was a matter of prestige to acquire rare objects with special powers. For instance, Philip the Good, the Duke of Burgundy, owned a series of artfully designed mechanical apes with parts that moved automatically.

His son Charles the Bold had no fewer than six "unicorn horns"; only the highest nobles could afford these coveted objects, which were thought to possess miraculous healing powers.[32] And before they undertook any major enterprise, princes always had their horoscope cast. In short, the occult was already part of court culture. At the universities and in the Church, however, the situation was very different. Avicenna had argued against alchemy, and his works were widely read. Alchemy had also been denounced by Pope John XXII. Theologians disapproved of occult knowledge, and natural philosophers in the Aristotelian tradition could not fit it into their theories about how knowledge should be obtained. As noted above, physicians formed an exception.[33]

Yet Ficino significantly enhanced the intellectual status of occult doctrine. Thanks to him, the magical tradition became woven into the fabric of scholarly life in an unprecedented manner. Not that scholars had been utterly ignorant of it; Hermes Trismegistus was a familiar figure. Cicero was thought to have mentioned him; Augustine had criticized him and Lactantius, another father of the Church, had praised him. Aquinas quotes him approvingly, in his *Summa theologica*. In a handbook of magic translated from the Arabic, with the Latin title *Picatrix*, Hermes was spoken of with reverence. Ficino had made Hermes's writings accessible at last, and he cloaked their author in the aura and authority of an Egyptian sage who long predated Plato and Aristotle. In 1488, a marble mosaic of Hermes was laid in the floor of the cathedral in Siena. It shows him handing Moses a book, and the accompanying text describes him as a "contemporary of Moses."[34] Another picture of Hermes with Moses is found in the Borgia Apartments at the Vatican.[35]

The supposed antiquity of Hermes's writings was doubly significant for humanists such as Ficino. For one thing, it played into their general tendency to move forward by looking back to the classics. But Hermes's antiquity was especially significant for quite a different reason: a magician did not acquire his knowledge through personal experience or investigation, but from his teacher. The original source of the knowledge was revelation, more or less as in the case of the Bible. In this scheme, Hermes Trismegistus was a *priscus theologus* (ancient theologian) like Moses, who had received his revelation directly from God. The hermetic writings were, in that respect, comparable to Kabbalah, the secret teachings believed to have been transmitted to Moses by God, purely in spoken form, and preserved in an oral tradition by Jewish priests. Ficino's student Giovanni Pico della Mirandola unified these two occult systems (and he included the Orphic mysteries in his synthesis as well).[36]

Philological research from the seventeenth century to the present has shown that the hermetic writings can be attributed to various authors from the first and second centuries AD. They do not form a cohesive system of thought, apart from their shared Neoplatonic outlook. There are also Stoic and Gnostic influences. The great bulk of this corpus of texts is mystical in orientation, but there are also "practical" passages describing the magic art of making daemons and angels do one's bidding. These magical passages in the Corpus Hermeticum are small in number, however, and could not, on their own, have steered natural philosophy toward magic and experimentation. But the historian Frances Yates has argued that many alchemical and magical writings not belonging to the true Corpus Hermeticum were also associated with Hermes Trismegistus in the Middle Ages, and their reputation rose with that of Ficino's translation.[37]

Ficino's own use of hermetic principles, and Neoplatonic philosophy in general, was highly practical. For him, mysticism was not only an end in itself—the spiritual union of the soul with higher things—but also what we might call a method of obtaining knowledge. According to this view, the soul does what the intellect cannot, namely, rise to the Neoplatonic World Soul. And it is the World Soul (*anima mundi*) on which the order of the material world depends. This World Soul, which encompasses a multitude of forms, controls the corresponding material manifestations in our world through seminal reasons (*logoi spermatikoi*)[38] in its own spiritual domain (*spiritus mundi*). Each of these reasons is associated with particular stars or constellations. The Neoplatonists, including Ficino, also interpreted the seminal reasons as daemons, each with a personality of its own. That made them compatible with Christian cosmology. This brings us to the magical aspect of Ficino's thinking; by communicating with daemons, magicians could focus the energy of the World Soul on parts of the mundane world in order to treat particular body parts or cure disorders such as melancholy. This is what interested Ficino as a physician. In practice, a physician working on hermetic principles did much the same sorts of things as a modern-day medical doctor, prescribing pills, lotions, and so forth. Yet the purported effectiveness of the medicines recommended by Ficino was based on correspondences between the human body and heavenly bodies. Ficino thus became the first person to combine alchemy and astrology systematically into a system of natural philosophy.[39]

This assumption of similarity between the cosmos and the human body is known as the microcosm-macrocosm thesis, and enjoyed widespread support during the Renaissance. Medicinal plants and substances such as unicorn horn were thought to derive their healing powers from the influence of the planets and, in this sense, to serve as intermediaries between the microcosm and macrocosm. Sometimes powers of this kind could be imparted to such objects by the soul of a gifted individual. The magical effects that resulted from all this were described by Ficino as "miraculous" (*mirabilis*). The leading exponents of Renaissance magic in the generations after Ficino's owed at least as much to Ficino himself as to Hermes.

Cornelius Agrippa von Nettesheim was the author of a somewhat skeptical book about vanity, *De vanitate* (1531) and of *De occulta philosophia* (1531), which in its day was considered the most comprehensive summary of occult wisdom. *De occulta philosophia* shows no trace of skeptical reserve. Agrippa's strong emphasis on magic earned him his later reputation as both a dangerous sorcerer and a charlatan, but during his lifetime he was under the protection of benefactors in many European countries, including Emperor Maximilian and his daughter Margaret of Austria. He spent long periods in England and Italy, and shorter ones in France and Geneva, as well as Antwerp, where his books were printed. Surprisingly, his large network of contacts included Erasmus, who sought Agrippa's advice on occasion.[40] Erasmus himself was not terribly interested in the occult (only enough to make fun of it), though his circle of friends also included the physician Johannes Reuchlin, a kabbalist and an inspiration for Agrippa.[41]

Agrippa's systematization of magic relied heavily on Ficino's work. He identified three types of magic: natural, celestial, and ceremonial. The celestial type brought him very close to accepting the existence of daemons. But natural magic held the most important place in Agrippa's system. Like Ficino, Agrippa wished to use magic to bring about miraculous effects, by which he apparently meant not just medical effects, but the control of nature in general.[42] Nevertheless, unlike his contemporary Georgius Agricola, Agrippa was opposed to mining, which he saw as the reprehensible mutilation of nature.[43]

Paracelsus was the great advocate of the microcosm-macrocosm thesis, and responsible for its most systematic expression. His real name was Theophrastus Bombastus von Hohenheim. The son of a physician in the Stuttgart area, he too became a medical practitioner, though it is not entirely clear whether he had a medical degree. In any case, he was usually seen as a surgeon, well below physicians on the social ladder. Paracelsus struggled doggedly to improve the status of surgeons and their methods. He claimed to have cured many patients given up by physicians with official credentials, and his achievements were acknowledged by such luminaries as Erasmus.[44]

Paracelsus combined his attacks on the medical elite with slurs against Aristotle, Galen, and almost every other authority of the ancient world. He presented his own practices as an alternative to conventional medical theories and tended to take a much more active and interventional approach to the human body than was usual among the physicians of his day, using not only surgical techniques but also medicines that he himself had developed. He prescribed a specific medicine for each condition, departing from the usual practice of treating a multitude of complaints with the same herbal remedies. Paracelsus made some use of herbs in his preparations, but even more of chemical substances such as "quicksilver."[45] He was thus the inventor of iatrochemistry, the application of chemistry for medical purposes.

Paracelsus's philosophical outlook was strongly influenced by mysticism and alchemy. His mysticism led him to seek "union" with his objects of study—ailing bodies, chemicals, plants, or whatever they might be.[46] This union took place not in the reasoning mind of the researcher, along logical or deductive lines, but through his "astral body," which connected him with the stars in accordance with microcosm-macrocosm theory. This process was not easy, but required great effort: the object of study had to be investigated, laid open. (Patients' bodies, for instance, were opened up in a literal sense.) In a time of increasing friction between traditional and modern methods in scholarship and the arts, Paracelsus was generally seen as a progressive force. In 1552, Estienne Pasquier, a historian and member of the Bordeaux regional parliament, referred to Paracelsus, Copernicus, and Ramus as the three men who were turning tradition upside down.[47]

Paracelsus's influence on both experimental science and medical reform was profound and lasting. In the mid-seventeenth century, he had a large group of followers among medical practitioners and chemists in many European countries. His contributions to science included his hypothesis that saltpeter was present in the air, which sparked a great deal of research in England around 1600.[48] Paracelsus's own work was too difficult and dark for most readers, but expositions by other writers, such as Oswald Croll in the field of chemistry and the Danish author Petrus Severinus for medicine, attracted a large readership. Bacon admired Severinus's book.[49]

Gerolamo Cardano was another outspoken advocate of the microcosm-macrocosm thesis, which he expounded in a number of works, including *De subtilitate rerum* (*The Subtlety of Things*, 1551), one of two hundred books that he claimed to have written himself.[50] Cardano's interest was in the rare and occult wonders of nature. He wished, first of all, to determine whether these wonders had actually occurred, and then to attempt to *explain* them. Although he attributed some phenomena to planetary influences, as Ficino had, he reduced other marvels to chance convergences of various kinds of natural causes.[51] In practice, that led to a kind of demystification of nature, though it would be wrong to see Cardano as a proto-Enlightenment thinker. It is more probable that he saw no fundamental difference between the two kinds of explanations. Cardano's interest in miraculous occurrences reflected his position as an amateur scholar with aristocratic tastes.

Soon after Andreas Vesalius published *De humani corporis fabrica* (*On the Fabric of the Human Body*, 1543), it found its way into Cardano's hands. He accepted Vesalius's new perspective on anatomy with enthusiasm and applied it as best he could to his medical practice. He was one of the first to value Hippocrates more highly than Galen, the almost universally accepted medical authority.[52]

Despite Cardano's superb understanding of mathematics, he did not use math in his experiments, nor in describing the miracles that were the object of his studies. In this respect, he was typical of the entire sixteenth-century magical tradition. It was not until the seventeenth century that occult properties were analyzed in mathematical terms—primarily by Newton. Even as late as Bacon's day, the experimental approach that was inspired by magic and the occult took a different path than the one advocated by Galileo.

Cardano's somewhat younger contemporary Giambattista della Porta was of noble birth. His book *Magia naturalis* (*Natural Magic*) was first published in 1558 (a greatly expanded edition followed in 1589) and was translated into many national languages. In his home city of Naples he founded a learned society, the Academia Secretorum Naturae. Aspiring members had to show that they had discovered a "secret of nature or art." Della Porta claimed that he had personally discovered more than two thousand such secrets, and on special occasions he would demonstrate one. His emblem was a lynx, an animal known for its penetrating vision. Soon afterward, a group of Roman scholars founded the Accademia dei Lincei, whose members later included Galileo. Della Porta joined the Accademia in 1610, before Galileo did; this shows how respected he was as a man of science.[53]

In 1580, della Porta had to appear before the Inquisition. His academy was shut down, and he was forbidden to publish his books in Italian. No restrictions were placed on the Latin editions; evidently, the authorities' main concern was to make sure that della Porta's secrets did not come into the hands of a wider public. In fact, della Porta himself believed that some knowledge should be kept from the masses.[54]

Natural Magic raised the art of discerning the hidden structure of nature to great heights. Like Paracelsus and Cardano before him, della Porta "read" the external characteristics of plants as clues to their inner character. This was a kind of detective work, like the medical practice of diagnosis.[55] According to this method, fruits shaped like a human heart were a good remedy for heart diseases; walnuts were good for the brain; and leaves shaped like the moon were within the lunar sphere of influence.[56] Della Porta also wrote about his own experiments, including one with a magnet, one in which he produced optical illusions with special lenses, and one involving animal breeding. He had a strong command of mathematics (again, like Cardano), which he put to use in his work on optics. *Natural Magic* was widely read and aroused great interest in the experimental method. Both Bacon and Descartes read it, the latter during a stay in La Flèche.[57] Bacon's *Sylva sylvarum, or a Natural History* was organized in the same way as *Natural Magic*.[58]

Jan Baptista van Helmont was a physician with a practice in Vilvoorde, near Brussels. Though he had received his medical degree in Leuven, he strongly objected to the traditionalism of the medical faculty there. After leaving the university, he devoted himself to chemistry, conducting experiments in isolation for some fifteen years. He became a very skilled experimenter, isolating many different gases in an age when Aristotelian physicists knew of nothing but "air." For instance, he heated coal in a closed vessel, releasing a gas that was neither air nor water vapor. In fact, it was Van Helmont who coined the word "gas," which he almost certainly derived from the Paracelsian term "chaos." Another noteworthy achievement of his was to arrive at the principle now known as "mass balance," by measuring the weight of substances before and after chemical reactions.[59]

Van Helmont believed that by means of "fire" (chemistry), he could reach the essence (or *semina*) of things and thus approach what was hidden or occult. The historian Walter Pagel has called Van Helmont's worldview "baroque," because of the distinctive way in which he combined unbridled mystical and religious speculation with rigorous observation, and non-mechanistic thinking with advanced experimental techniques and quantification.

Despite Van Helmont's isolation, he became very famous during his lifetime. Mersenne wrote him three times to ask for advice. He was interrogated by the Spanish Inquisition and placed under house arrest, but rehabilitated two years after his death.[60]

Van Helmont's medical and alchemical writings, particularly *Dageraat, oft nieuwe opkomst der geneeskonst* (*Daybreak, or the New Rise of the Medical Art*; 1659), were posthumously published by his son Franciscus Mercurius and widely distributed.[61] A large proportion of English physicians studied in Leiden and were exposed to his work there. Around 1650, most physicians were Helmontian (or Paracelsian, which often amounted to the same thing). The influential Walter Charleton translated Van Helmont into English.[62]

Georgius Agricola, the author of a treatise on metallurgy (*De re metallica*, 1556) was scornful of alchemy. In that sense, he is out of place in this series of magicians. His book is more closely related to the technological tradition of the engineering profession, and is illustrated with precise and detailed drawings of machines for use in mining.[63] Agricola was certainly a humanist, and a friend of Erasmus and Melanchthon. He undertook diplomatic missions for the Holy Roman Emperors Charles V and Ferdinand I. As a member of the cultural elite, he hoped to elevate the ignoble craft of metallurgy to the status of natural philosophy. The alchemists, whose working methods he believed were unscientific, were little more to him than a millstone around his neck. Yet Agricola was not an Enlightenment scholar before his time. His use of the term *chymia* (chemistry) rather than *alchymia* can be seen as part of a humanist project to ward off Arabic influences and return to the original classical forms. His view of daemons was peculiar, and apparently based on pre-Christian folklore; he accepted that there were daemons in mines, but saw them as good-natured gnomes who could help the miners.[64]

The physician William Gilbert was found in the company of mathematicians and instrument makers at least as often as in that of alchemists, yet his worldview was thoroughly occult and magical. Gilbert was the author of *De magnete* (1600), which described numerous ways of using magnets for compass needles, as well as a variety of difficulties. Through much experimentation, he assembled a convincing case that the earth itself is a large magnet. This led him to conclude that the earth has a soul and is capable of moving of its own volition, and he used this to argue in favor of the Copernican system.[65]

Gilbert collaborated with "applied" mathematicians, a fact which draws attention to an interesting convergence between the mathematical and alchemical traditions in England. Practical mathematics was viewed with contempt in early seventeenth-century England, as it had been in sixteenth-century Italy.[66] To the dismay of theoretical mathematicians at universities, this diminished their own professional status; their field was "scarce looked upon as Accademical Studies, but rather Mechanical; as the business of Traders, Merchants, Seamen, Carpenters, Surveyors of Lands, or the like."[67] Some of the more artisanal mathematicians later joined the Royal Society; Robert Hooke might be considered an example of this type. Such men tended to be, at least in part, inventors whose chief concern was to profit from their inventions, rather than to share information freely with the scientific community, and Boyle condemned their conduct in almost the same terms in which Bacon had once criticized the alchemists.[68]

Occult Knowledge, Magic, and Science

It was not only by actually performing experiments that individuals like della Porta helped to establish the experimental style of science. There are important conceptual ties between experimental science and the occult philosophies of the Renaissance, ties so strong that without those occult philosophies, the new sciences could not have come into being. Superficially, the Renaissance interest in the occult seems to have been no more than a passing phase. If you browse a bookstore's occult section today, you will find books that draw on the writings of the Renaissance philosophers discussed above, but which have very little to do with modern science. Present-day readers of the genre even consider that an asset. But Renaissance philosophers regarded themselves as model scientists.

Admittedly, some aspects of Renaissance science were discredited as early as 1600. The doctrine of sympathies, which was connected to the microcosm-macrocosm thesis, found few new supporters among scientists from the early seventeenth century onward. Robert Fludd, a late representative of this view in its pure Renaissance form, was hardly taken seriously by his contemporaries. But the work of Robert Hooke, one of the leading experimentalists in the Royal Society, sometimes clearly reflects a mind-set based on sympathies and antipathies.[69]

Francis Bacon harshly criticized magic, but chiefly for its secretive nature, which was repugnant to him. He warmly approved of other elements of the magical tradition—and not just the experimental method. Bacon's terminology betrays his considerable debt to alchemy.[70] Like the alchemists, he was obsessed with the idea of gaining power over nature and extending the human lifespan.[71] He cited the name of Hermes Trismegistus in a new context, as we will see. In any case, there was no clean break between Renaissance science and the Scientific Revolution, as most historians, including even Frances Yates, believed until recently.

Occult and Manifest Properties

The term *occult* is a major source of misunderstandings. The Latin word literally means "hidden," but also has the connotation of "incomprehensible." A look at the word's etymology may shed light on a few points.

From the Middle Ages until well into the seventeenth century, the Aristotelian philosophy of the scholastics used the paired terms "manifest" and "occult" to classify sensory experience. That which was directly accessible to observation was manifest. The classical sciences of astronomy, statics, optics, and music theory were grounded in observations of manifest qualities. But other qualities of things, such as the healing properties of an herb, could only be ascertained through their effects; they were hidden, occult. For the Aristotelians, this meant that experiential knowledge of occult qualities was the most one could hope for; furthermore, this knowledge was unreliable. As everybody knows, medicines do not always work. It was therefore impossible to build certain knowledge, *scientia*, on occult qualities. Experience was a category that belonged to the domain of craftspeople.[72]

Occult qualities appeared to be incompatible with the expectations people had about the normal behavior of the four elements—earth, air, fire, and water—on which the Aristotelian theory of perception was based. Augustine's classic example was quicklime (calcium oxide), which crackles noisily and gives off heat when immersed in cold water, but does nothing when it comes into contact with flammable oil. Augustine could not "observe" what caused these phenomena, and the word that he used to describe how he obtained his knowledge of the behavior of quicklime, namely "experience," was the same word used for the experience of a miracle in the Christian tradition.[73]

The list of "wonders of nature" like this one was fairly long and, by current standards, remarkably heterogeneous. They included Mount Etna, which burns without ever being consumed; theriac, an herbal mixture used to treat snake bites; mandrake, a plant whose thick root resembled the human form;[74] the salamander, an animal that could survive exposure to fire, mentioned by Augustine and observed firsthand by the goldsmith Benvenuto Cellini as late as 1505;[75] mercury, a metal that was liquid at room temperature; asbestos, which could catch fire but was unchanged after the fire went out; the magnet, whose workings did not follow from the qualities of iron; and the shelduck, which grew on trees in Ireland.[76] Many people did not hesitate to add wonders of technological invention to this list.

Yet natural philosophers in the Middle Ages and Renaissance had little interest in strange, rare, and miraculous phenomena. Roger Bacon was in some ways an exception, but even he tried to trace occult properties to natural causes as much as possible.[77] Around 1600, when Johannes Kepler defended physicians' experience with medicinal herbs as a reliable source of knowledge and stood up for experienced astrologers (though he added that astrology was sometimes misguided), he was expressing a view that was foreign to the intellectual circles of natural philosophers, but had taken shape over the course of centuries among physicians and alchemists and their aristocratic patrons.[78]

Experience

The nature of observation was also changing. From the sixteenth century onward, it was considered extremely important for students of nature to observe rare and unusual phenomena. Aristotle had always based his theories on his observations of "ordinary" things; these were the raw materials from which one could derive generalizations by induction and intuition. While Aristotle could always appeal to what everyone had seen, or could see, it required a degree of practice to see and appreciate rare or wondrous things. It was necessary to become an expert. At the same time, generalizations came to be seen in a bad light. Bacon actually warned against them, arguing that it was better simply to collect observations.[79]

Around the same time, a similar change took place in the classical sciences. For Galileo, too, "experience" no longer meant what it had to Aristotle. Instead, he came to see observation as "constructed" by mathematics; the possibility of quantification and later mathematical analysis was the criterion that determined which observations were meaningful and which were not. When observations are intended to support or disprove a theory, their constructed nature becomes still more apparent.

Making observations thus became a job for experts. Examples include observations of sunspots (which had to be shown not to be phenomena in the earth's atmosphere) and of mountains on the moon.[80] It is ironic that during Galileo's lifetime he was most famous for discovering a "wonder"—namely, the mountains and craters on the moon—and not for his more abstract work in physics and mathematics. In 1615, the Jesuit Giuseppe Biancani (Josephus Blancanus) made a distinction between "phenomena," which could be witnessed by anyone, and "observations," which took hard work (for instance, the use of scientific instruments) to produce.[81] These developments in science happily coincided with the tastes of rulers such as the Grand Duke of Tuscany, Galileo's patron and employer. At the supper table, Galileo might be expected to explain the nature of sunspots to a company of interested nobles—perhaps in debate with an Aristotelian. But an argument about why ice floated in water could also be amusing.[82]

Hidden Causes, Hidden Knowledge

Thomas Aquinas had postulated that no animals could exist below the threshold of human perception; otherwise, how could Adam have given all the animals their names? Moreover, if people could not observe certain things, it was obviously not God's intention that they do so, nor did He wish them to seek knowledge about those things.[83] This brings us to an aspect of the occult that combines the senses of "hidden," "unknowable," and "forbidden."

Aquinas may in part have been inspired by a passage in the Vulgate, the Latin translation of the Hebrew Bible completed by Jerome in the year 405. In it, the apostle Paul admonishes the faithful with the words *noli altum sapere* (do not desire to know higher things). For many centuries, this motto was interpreted as a warning not to pry into the secrets of God, Nature, and the authorities.[84] Augustine characterized curiosity as a form of temptation, and the desire to probe the unprofitable secrets of nature as a perversion.[85]

This idea of forbidden knowledge may have mingled with a tradition of secret knowledge rooted in the Arabic world, which was familiar to Westerners mainly through a work erroneously attributed to Aristotle that bore the Latin title *Secretum secretorum* (*Secret of Secrets*). Its contents were believed to be instructions from Aristotle to Alexander the Great, who was said to have carried it with him during his Persian campaign. The core of the book consists of secret advice on statecraft, but later editions were embellished with encyclopedic information about medicine, magic, alchemy, and astrology. Roger Bacon referred directly to this book when he argued that many of nature's secrets should be kept out of the hands of the common people.[86]

The thirteenth and fourteenth centuries saw the emergences of a new genre, devoted to *secreta*, *problemata*, and *experimenta*. The third category was made up largely of medical recipes arrived at through experience.[87] The *secreta* included magical procedures, and their authors were almost always anonymous. But of course, it was impossible to write a book about magical secrets while actually keeping them secret, especially after the invention of printing. In the field of statecraft, Niccolò Machiavelli's *Il principe* (*The Prince*) clearly explained to all the world how rulers obtain and maintain power.[88] Nevertheless, most Renaissance magicians believed that knowledge of magic should be restricted to a small circle of adepts. As mentioned before, Francis Bacon disagreed, and he accused Agrippa, Cardano, and della Porta of using magic purely for their own greater glory. He demanded openness and selfless cooperation in the name of progress when it came to technical knowledge (which in his view included knowledge of magic).[89]

Bacon was writing in the early years of the seventeenth century. A century earlier, Erasmus had proved that the phrase *noli altum sapere* is a mistranslation of the original Greek, which says, "Do not be haughty."[90] But at the Council of Trent, the church decided to continue using the Vulgate, in honor of the tradition that this translation represented. The problem of the mistranslation therefore persisted.

The Vague Border between Natural and Technical Wonders

Ever since the thirteenth century, royal collectors had made no distinction between natural and technical curiosities. This stance is illustrated by the nautilus shells and ostrich eggs mounted in wrought silver that are found in many collections. In conceptual terms, magicians and alchemists from

Ficino onward saw technical and natural wonders as more or less the same. Natural wonders resulted from the improbable confluence of a variety of causal factors. The products of a magician or alchemist differed only in that the maker had molded the circumstances to his will. What was coincidental in one case was intentional in the other (and possibly brought about with the aid of a daemon). Natural wonders were a kind of proto-technology.[91] Bacon's point of view was that the wonders of nature would inspire humans to create wonders of technology.[92] In doing so, craftsmen, engineers, and experimenters would put nature under pressure, subjecting it to their will—in other words, compelling it to do what it would never do on its own (or only very infrequently). Bacon employed a vast arsenal of metaphors, mainly legal and sexual, to lend force to his argument.

Natural Magic, Demonology, and Wonders

In Neoplatonic philosophy, daemons are intermediaries between God and the world. We should not think of them of little devils, but as abstract "intelligences," or formative principles, transmitting the Forms to the material world and thus giving things their "forms." Almost all Neoplatonic authors emphasized that daemons had nothing to do with witches or malicious spirits. If daemons "did" anything at all, or were asked or forced to do anything by a magician, then it would not, in any case, be supernatural, because daemons too were subject to natural law. If daemons could do things that ordinary mortals could not, it was because their senses were more powerful, according to Augustine and the entire tradition that followed him.[93]

But not all students of nature were wedded to every detail of Neoplatonic philosophy. Consequently, the activity of daemons became a catch-all for unusual phenomena for which natural philosophy could find no explanation. In the sixteenth and seventeenth centuries, numerous belief systems, both Christian and pre-Christian, had a place for devils and witches. These systems were often at odds with one another.[94] It was the period of witch trials, in which many well-known scholars participated.

These demonologists established criteria for the presence of genuine witchcraft. They defined four categories—true demonic activity, apparent demonic activity, true non-demonic activity, and apparent non-demonic activity—and the procedures that they developed for classifying individual cases were not so different from modern scientific methods.[95]

Collecting Facts and Natural History

Cardano was a typical example of a Renaissance physician who accumulated knowledge through practice, often in the form of isolated facts, which he then catalogued. He was especially interested in the unusual and bizarre, and described his own life as a "collection of events."[96]

In Cardano's day, almost anyone of any social importance was a collector. From 1556 to 1560, Hubert Goltzius, an engraver from the southern Netherlands, traveled through Germany, Austria, Switzerland, Italy, France, and the Netherlands. Wherever he went, he introduced himself to the local collectors, who varied from kings and cardinals to physicians, poets, and military officers. By the end of his travels, he had visited 968 collections. Most of them contained antiquities, in the humanist spirit of the Renaissance, but also just about anything exotic or outlandish. That included many objects of natural history.[97]

For many Renaissance collectors, all these facts were interrelated in a web of meanings. Every entity in the natural world and every artifact referred to other entities and other artifacts. An object could thus represent, or emblematize, the entire world.[98] Genres such as the *materia medica* (works on pharmacology) and the *secreta* books formed an integral part of natural history with a significance that went far beyond the medical.[99]

Bacon was a fervent reader of works on natural history and argued that forming a collection like the ones in natural history cabinets should be an activity of the state: "a goodly great cabinet, wherein whatsoever the hand of man by esquisite art or engine hath made rare in stuff, form or motion; whatsoever singularity, chance and the shuffle of things hath produced; whatsoever Nature has wrought in things that want life and may be kept; shall be sorted and included."[100]

Bacon felt no affinity, however, with the emblematic view of the world.[101] He had plans for radically reforming the cabinets by removing antiquities and anything that reeked of superstition. Bacon believed that natural history could and should be an inexhaustible source of inspiration for experimental scientists.[102] In *New Atlantis*, he advocated the establishment of various types of collections, botanical gardens, arboretums, and menageries where della Porta–style experimentation could take place: breeding, cross-fertilization, anatomical research, and so forth. [103]

In 1668, a member of the Italian Accademia del Cimento visited the collection of Royal Society in London, which was kept in the home of its curator Robert Hooke. The visitor was unimpressed. Having expected to view a collection that attested to cultured tastes, like the ones he was familiar with in Italy, he instead found only a storehouse of objects selected purely for their direct scientific utility. The German traveler Zacharias von Uffenbach had a similar experience when visiting the Ashmolean Museum at Oxford University. To his horror, the collection was open to the general public, and he had to rub shoulders with the common folk. The public orientation of Baconian science evidently repulsed him.[104]

The activity of collecting facts was also criticized. Descartes could not see the use of natural history, and Galileo made snide remarks about the poor taste of those who collected curiosities.[105] Pascal wrote that curiosity was nothing more than vanity; this was in part a dismissal of the *curieux*, as collectors were called.[106] This opinion was only to be expected from scientist-philosophers who believed in the clear and unambiguous deductive order of nature. Natural history as Bacon envisioned it was not entirely inductive, however.

Francis Bacon

The experimental philosophy of Francis Bacon includes many elements borrowed from diverse traditions. He fiercely attacked the dominant philosophical traditions, especially scholasticism and deductivism more generally, and regarded scientific hypotheses as abhorrent fabrications. He also rejected mathematics, associating its use with deductive argumentation in general and hypotheses in particular.[107] As we have seen, he was critical of magic and alchemy, but did not reject them entirely. He embedded the elements that he adopted in a new framework: the utopian dream of utterly reforming science by placing it under the patronage and supervision of the state and using it for the benefit of the public. Bacon believed that this would give rise to many useful projects for humanity and ultimately make it possible to subject nature entirely to human will. The English Dissenters, who had separated from the Church of England, felt inspired by Bacon's vision, and many of them firmly believed that Oliver Cromwell's Puritan Revolution would turn it into a reality.[108]

The main task of Baconian science was to collect facts, possibly but not necessarily through experimentation. The sorts of facts that Bacon had in mind related in large part to occult qualities. These facts would then be organized into tables, which would make it possible to draw conclusions. Was this a form of inductive reasoning? That interpretation has often been advanced, but it does not entirely do justice to Bacon's intentions.

Statesman and Jurist

Bacon never made it entirely plain what sort of knowledge was supposed to result from his inductive method. If this was ordinary induction, it raised the question of whether his method was simply complementary to the theoretical sciences. After all, even Aristotle had approved of induction (provided it was guided by intuition) as a way of arriving at first principles. Yet this cannot have been Bacon's view, given his hostility to theoretical science in general and Aristotelian science in particular. He had no qualms about using the whole range of Skeptical arguments in his attack on scholasticism.[109] So what was it that Bacon sought? What did his inductive method really amount to?

The answers lie in Bacon's background as a legal scholar and statesman. His method did not grow out of a philosophical background, though we have grown accustomed to describing him as a philosopher. Before taking up science as his main activity, Bacon was a senior official of the English government, and in that capacity, he proposed reforms of the legal system.

Francis Bacon's father, Sir Nicholas Bacon had been Lord Keeper of the Great Seal for the first twenty years of the long reign of Queen Elizabeth I.[110] Sir Nicholas had been one of the driving forces behind Elizabeth's reform of the English state, her "common weal" policy aimed at promoting the public interest. Much later, in 1617, when Francis Bacon himself was named to the position of Lord Keeper, he noted that all the highest officials and legal scholars in England at that time were the sons of jurists. Bacon believed that this was no coincidence, and he proposed that in judicial appointments, preference should be given to those whose fathers had also served their country as judges.[111]

In religious terms, Bacon was a Puritan; that is, a Protestant for whom the Church of England was not Protestant enough.[112] But when the Puritans began to radicalize, he turned against the movement's sectarian tendencies. Remaining staunchly loyal to the Crown, Bacon severed his ties with his patron the Earl of Essex, a Puritan leader, even before Essex fell out of favor with Queen Elizabeth. Under King James I, Bacon rose to the summit of the legal profession, becoming Solicitor-General, Attorney-General, and finally, in 1618, Lord Chancellor. The King elevated him to the peerage as Lord Verulam.[113]

Bacon's great legal project in the 1580s and 1590s was the reform and improvement of the English judiciary. One aspect of this project was that judges were asked to provide reasons for their decisions, in order to foster greater respect for precedent. The present-day English legal system, largely based on case law, goes back to this reform from the days of Elizabeth I.[114] Bacon and his contemporaries did not see this change as modernization, however, but as a return to historical sources in a humanist vein; specifically, a return to the origins of English law.

English law fundamentally deviated (and still does) from the civil law system in the rest of Europe, which is based on the Roman tradition of a codex, a written body of law.[115] In English law, in contrast, the original tradition of oral transmission is accorded greater weight than the later written record (statute law). This oral tradition is known as common law, a system of legal rules and practices established in England by William the Conqueror and the rulers who succeeded him, mainly in the eleventh and twelfth centuries. In those days, kings would travel through the country dispensing justice, but there are no written records of their decisions. Starting in the late twelfth century, rulers showed an increasing tendency to delegate the administration of justice to the courts. Clerks were appointed to these royal courts to keep records of every case: what type of investigation had taken place, what facts had been ascertained, and what conclusions the judges had reached. It is important to keep in mind that in the twelfth century there was hardly any distinction between governing and administering justice. The expression "to hold court" had both meanings, and the courts were one of the instruments through which the king ruled the country.

Sixteenth-century legal scholars proceeded on the assumption that the common law of the kings of the four previous centuries was a real entity, but hidden, so to speak, because it had not been recorded in writing. By studying decisions in court cases, they reasoned, it was possible to catch a glimpse of the true common law. And by doing their best to combine past judgments into a coherent whole, they could rediscover the common law (at least in part) and apply it to new cases. Legal scholars also believed that the common law was comprehensive, with a legal rule to cover every case, and that its overall structure was rational.[116]

In English law, the term "rule" is used as a synonym for "maxim," "principle," "axiom," and "postulate." In other words, these terms are interchangeable. Continental legal scholars saw this as a barbaric abuse of terms that were well-defined and distinct from one another. English legal rules were not postulates or first principles in an Aristotelian sense, and continental jurists concluded from this that English law had no valid claim to the title of *scientia*, as their own systems did. For the English themselves, it was sufficient that the hidden rules of the common law formed a coherent system and did not conflict, and they simply assumed these things to be the case.

Reformer of Science

Bacon launched his campaign of scientific reform in 1605 with the publication of *The Advancement of Learning*. A long series of books followed, such as *The Masculine Birth of Time* (1608), *Novum organum* and *Instauratio magna*, which both appeared in 1620, and the unfinished, posthumously published *New Atlantis* (1627), the utopian tale of an institution called Salomon's House in which scientists played a key role. Bacon's *Sylva sylvarum, or a Natural History* contained a large number of "experiments" and facts, which in his eyes were the building blocks of a new science. It was a mammoth project, beyond the capacity of any one person, and Bacon devoted himself to it for many years.

His vision of science was modeled almost precisely after the reform of common law. Like the rules of common law, Bacon wrote, the laws ordained by God and established in nature had to be discovered by scientists.[117] This was the purpose of the inductive method. According to this vision, the set of natural laws should be interpreted as a coherent rational system, rather than a deductive one organized on the basis of first principles. There are limits to induction in Bacon's system; the inductive method should not produce excessively abstract generalizations that would make us forget the need to perform experiments.[118]

The central legal term in Bacon's approach to scientific research is "facts." Facts are not easy to come by. In a court case, the accused has his own view of the facts and the prosecutor must convince the judge and jury of the superiority of his version. Courts had developed procedures for determining the facts accurately. Bacon made it clear on various occasions that he saw a connection between proper experimental procedure and the procedure in a legal case. First, as much information as possible had to be collected; this explains Bacon's interest in natural history. Every piece of information had to be correct, was "under oath," as Bacon put it. He compared experiments to witness examinations, the primary purpose of which is to gather new information. Causation and guilt are determined at a later stage. Dubious "testimony" should not be covered up, Bacon wrote, but exposed for what it is. This first stage led to the establishment of the facts by the court or researcher.[119]

In Bacon's day, officers of the court were sometimes permitted to "vex," or apply pressure, when examining witnesses, for instance, when unmasking Catholic conspiracies. In 1612, England's witch trials were also in progress. In his judicial practice, Bacon was very conservative in his use of torture but did resort to it on occasion. As he observed, testimony obtained under duress was not admissible as evidence but could provide new information. The historian Carolyn Merchant has argued that the sexual imagery in Bacon's description of the methods for interrogating nature and subjecting it to one's will by mechanical means was informed by witch trials. "Nature must be taken by the forelock," Bacon wrote.[120] Though he may have drawn a line between vexation and torture, in practice this line could be very thin indeed.[121] He also used more affectionate metaphors for the study of nature, such as marriage between the male researcher and feminine nature, and he wrote that nature had to be approached gently. But in his conceptual world, marriage could be brought about forcefully through abduction, and witnesses could be compelled, gently or brutally, to tell all.[122]

Once the facts were established, analogous cases could be investigated. In legal practice, a set of historical cases could be cited to establish a general axiom or law. The prosecutor would then argue that this law had to apply in the case at hand. The defense could present a different set of historical cases and infer a different law. The judge and jury had a second type of experiment at their disposal, the *instantiae crucis* (crucial instances), to suggest which law was applicable.[123]

By long tradition, going back to the classical authors Cicero and Quintilian, this form of legal deliberation and inference was known as induction. This is probably what Bacon had in mind when he used the term "induction," rather than the definition familiar to us from the field of logic. This implies that Baconian science is not purely empirical and theoretical, but similar to the practice of law, in that the goal is always to arrive at a theoretical explanation, a law of nature. But this strain of science is not committed to any single type of theoretical explanation; instead, there are many laws of nature, which are assumed not to conflict. One example of Bacon's inductive method in science was his proposal to connect the phenomenon, or "case," of heat with that of magnetism. He suspected that the underlying cause in both instances was a certain type of motion of material particles.[124]

As Bacon saw it, God had ordained laws of nature that could be discovered, just as the common law could be discovered. In *New Atlantis*, he argued that experimental scientists should enjoy the same status as judges, and even proposed that they wear similar paraphernalia, such as robes.

Bacon was highly optimistic about what could be achieved through this approach, suggesting that even the Fall of Man could be reversed. The meaning of the Fall, in his opinion, was that Adam and Eve had placed human law above God's, but now humanity could return to God's natural laws.[125] And as for the king, envisioned as the leader of this scientific enterprise, Bacon predicted that he would be no less than the new Hermes Trismegistus![126]

British Baconianism

Bacon's work was read widely, both in England and in the Dutch Republic. Immediately after his death in 1626, a collection of his previously unpublished writings appeared. The first Baconians, such as Samuel Hartlib, Benjamin Worsley, and Gabriel Plattes, were chemists and men of business. Plattes was the author of a book about metallurgy, which dealt with transmutation into gold. Worsley spent a great deal of time on developing a new procedure for manufacturing saltpeter so that England could reduce its dependence on imports. Saltpeter was used both in the production of gunpowder and in the metal industry and was in great demand during the

English Civil War. Worsley's Paracelsian principles ultimately failed to produce the desired result, but he had taken the first step toward a large-scale program of innovation.[127]

Hartlib's projects included a plan to reorganize London's Gresham College into a modern university. He called for the abolition of the chairs of theology and rhetoric and proposed that the professors of physics, geometry, music, and astronomy teach experimental science and its social applications. New chairs were to be created in fields such as chemical technology.[128]

Another project of Hartlib's was the establishment of an Office of Address on the model of the *Bureau d'Adresse* in Paris, which the physician Théophraste Renaudot had founded in 1628 for the benefit of job seekers. It was a central register of vacancies, a kind of employment agency. Renaudot also produced statistics on related matters, as well as the widely read periodical *Prix courants des marchandises* (*Current Commodity Prices*).[129] His bureau was an entirely new type of institution. The Office of Address was intended to perform the same functions, and more. One section of it, the Office of Communications, was supposed to put Bacon's ideas about science into practice, promoting and disseminating ingenious new technologies. Robert Boyle worked with Hartlib to bring this about, and the republican Council of State awarded Hartlib a stipend in support of his activities and appointed him Agent for the Advancement of Universal Learning, a resoundingly Baconian title. Unfortunately, nothing concrete ever came of these plans, and after the restoration of the monarchy they were no longer within the realm of possibility. The project did lead to publications, however. *Ireland's Natural History* (1632) was the first study to use information collected by questionnaire. The project moved science and technology to the top of the national agenda as an engine of economic development.[130]

In drawing a connection between the promotion of technological innovation and economic progress, Hartlib and his circle were following Bacon's lead. For both Bacon and his followers, the most direct competitors were the Dutch. For instance, Bacon was closely involved with England's patent system and had pushed for a ban on the export of lead ore, because the Dutch were buying up all the English ore at low prices and extracting lead and silver from it in their own country. Bacon wanted to put a stop to this, and hoped that an export ban would allow the British to develop the necessary expertise themselves.[131]

The English experimental scientists who founded the Royal Society in 1660 called themselves Baconians, and even though their organization was a very incomplete realization of Bacon's plans, it was Baconian in its experimental orientation and its drive to collect scientific facts and observations. John Wilkins took up the plan of cataloging "All Things and Notions."

Yet the English Baconians (with the exception of Newton) were much less averse to the use of hypotheses in science than Bacon himself had been.[132]

The French Académie Royale des Sciences, founded six years later, is not generally seen as Baconian because of its exclusive nature. Only a small scientific elite was invited to join. These French academicians tended to have an excellent command of mathematics, which many members of the Royal Society were lacking. The Académie also secured the patronage of the state, unlike its English counterpart; the restored King Charles II had granted a charter to the Royal Society but otherwise kept it at arm's length. Louis XIV awarded the members of the Académie a generous annual salary, many times greater than that of a university professor of natural philosophy. Christiaan Huygens, who was essentially the first scientific director of the Académie, made practical utility central to its program, "in accordance with Bacon's plan."[133]

Occult Qualities and the Experimental Program

One noteworthy feature of Bacon's analogy between magnetism and heat was that it linked manifest and occult qualities, which natural philosophers had treated as distinct phenomena. Magnetism was a textbook example of the occult, while heat had played a prominent role in natural philosophy at least since Aristotle's *Physics*.

After Bacon, occult phenomena were no longer esoteric experiences known only to a handful of insiders. They became ordinary subjects of the sciences, though at first mainly of the "new" sciences. The discussion of Newton in the next chapter will illustrate that over time, occult qualities also found their way into the subject matter of the "classical" sciences.

Two general points can be made about the role of the occult in seventeenth-century science. One is that science offered explanations of occult phenomena. The most daring of explanatory models was probably the mechanical philosophy developed by Isaac Beeckman, Descartes, and Huygens. Mechanism can be described as "explaining away" the occult by sketching a lucid model of the functioning of a system in terms of its basic parts. Huygens compared his work to breaking a secret code.[134] Another well-known example is Descartes's theory of magnetism, which postulated particles of a special kind of matter circulating in vortices.

Until fairly recently, historians of science believed that the occult had ceased to be a relevant category by the seventeenth century.[135] But right up to the start of the eighteenth century, reputable scientists went on using fairly traditional occult explanations. One example is Robert Boyle, author of *The Sceptical Chymist*, who claimed that his scientific approach repudiated theoretical explanation. While the truth was somewhat more complex,

Boyle was certainly no advocate of mechanism. His atomism was derived, not from Gassendi or Descartes, but from an earlier alchemical tradition.[136] Boyle speculated that saline substances might be produced by "some internal thing, that is analogous to a Seminal Principle," a near-orthodox affirmation of Neoplatonism in the spirit of Ficino. In another passage, about the properties of air, Boyle wrote of a "vital substance" in the air that was irreducible to the mechanical properties of matter. He believed that this vital factor might be of an "astral or some other exotic nature."[137] This idea seems not only Ficinian but more specifically Helmontian, assuming the inherent activity of matter. Boyle sang Van Helmont's praises on many occasions, replicating and confirming the results of his experiment with a willow in a pot (which increased in weight tremendously on a diet of nothing but air and water).[138] Boyle was walking a fine line. He had to defend himself against accusations by Thomas Hobbes that Boyle's philosophy was "vulgar" because it assumed that matter moved of its own accord.[139]

There is a second point to be made about the occult, and about rare and wondrous phenomena in general. Empirically inclined English scientists of the seventeenth century still regarded curiosities as a source of inspiration crying out to be studied, experimentally if possible. Boyle is the most conspicuous example of this tendency. Once, as he was about to go to bed, a frightened kitchen maid accosted him and told him that a veal shank was giving off light. He had the large piece of meat hung on a hook in his room and spent all night studying the greenish glow that it produced. This phenomenon, which was linked to decomposition, was also observed in a rotten fish placed in Boyle's air pump in total darkness on a moonless night, with Boyle and his assistants dressed in black.[140]

Another issue about which Boyle displayed his theoretical open-mindedness was the existence of ghosts. In the 1660s, the nation was gripped by tales of the Drummer of Tedworth, a poltergeist, and the Demon of Mascon. Eminent Anglican clergymen wrote firsthand accounts of their supernatural experiences, roughly according to the protocol that Boyle had developed for his experiments, and with Boyle's approval.

Bacon and Boyle were very similar in their appreciation of exceptional and unusual phenomena. In instructions that Boyle wrote for fellow members of the Royal Society who were planning to travel abroad, he urged them to keep an eye out for "unusual and remarkable" symptoms of diseases and "any thing that is peculiar" about animals.[141]

The National and Religious Context

Boyle consistently assumed that he was dealing with natural phenomena, even when those phenomena were poltergeists. It does not seem unreasonable to suggest that this is a national character trait of the English—not a belief in poltergeists, but a level-headed determination to investigate their existence empirically. The historian Peter Dear has argued that the attitudes of Boyle and his colleagues should be viewed in the religious context of their day. Boyle was a Protestant, like most of the English, whether they belonged to the Church of England or were "Puritans." His Protestant denomination believed that while God had the power to perform miracles, the last time He had done so was during the life of Jesus. Boyle therefore felt safe in assuming that he would not encounter true miracles. He believed that special and even unique natural occurrences could reveal important principles about the workings of nature, and therefore merited investigation by natural philosophers.

In contrast, the Catholic Church held that miracles remained possible as signs from God (though after the Council of Trent it moderated its belief in the omnipresence of miracles, which had been a product of popular superstition). The church established strict procedures for validating miracles, procedures that are still in force today. Two of the main criteria are that a miracle must be a unique event and that it must depart from the regularities of natural law. This implies that the church had a deductive view of natural science, and saw no obstacles to maintaining this view. It does indeed appear to be the case that most scientists in the Catholic country of France still adopted a deductive style of thinking when conducting experiments.

In one famous experiment, Blaise Pascal showed that at a high elevation—specifically, on top of the volcano Puy de Dôme—the air pressure is lower than at sea level. Rather than adopting Boyle's dry tone and emphasizing the exact time, place, and circumstances, Pascal's prose gives readers the sense that they are witnessing the experiment themselves, or even performing it in their imaginations. Pascal was clearly dissatisfied with Boyle's contention that even unique events can provide valid scientific evidence as long as they are carefully recorded and substantiated. Instead, he harked back to Aristotle's view that experience can in principle be shared with everybody else, and that the justificatory significance of experiences follows from their compatibility with a correct, mechanistic explanation.[142]

This deductive, theoretical approach remains part of French science to this day, at least at elite institutions such as the École Polytechnique.[143] In 1666, the Académie launched a Baconian program of research, but its members retained a high degree of freedom in their investigations. After the death of Jean-Baptiste Colbert, a new minister, Michel-François Le Tellier, Marquis de Louvois, assumed responsibility for the Académie and he sparked a crisis within its ranks when he called for a more practical orientation.[144] In this context, Thomas Kuhn pointed out that a number of distinctively "new" or "Baconian" seventeenth-century sciences were mathematized in the early nineteenth century.[145] Until about 1800, the classical and the new sciences remained separate. But between 1800 and 1830, Laplace, Fourier, and Sadi Carnot developed a deductive framework for thermodynamics and described it in mathematical terms. Poisson and Ampère did the same for electricity and magnetism. Finally, Fresnel accomplished the same feat in the field of optics, which until then had not lent itself to a classical approach. It can hardly be a coincidence that all these scientists were French. Jean-Baptiste Fourier, the most outspoken of them on this issue, postulated that only an internally coherent theory along purely deductive lines could yield mathematical equations capable of expressing the relationships among phenomena.[146] In the same spirit, Pierre-Jean-Marie Flourens, the most prominent member of the Académie Royale des Sciences at that time, declared in 1864 that the logic of Darwin's *Origin of Species* was fundamentally unsound.[147] And indeed, Darwin considered himself a Baconian. His method, as he described it, was to gather a huge mass of factual material and then distill the process of speciation from it by inductive means. Darwin's British contemporaries approved of this style of accumulating probabilities and analogies.[148]

France did, of course, have practitioners of the Baconian sciences, and very good ones at that, but they never managed to become part of the country's scientific elite through admission to prestigious organizations such as the Académie.[149]

Protestantism

Following in the footsteps of the sociologist Robert Merton, historians have tried very hard to find a connection between religion, particularly Protestantism, and the growth of science in the seventeenth century.[150] Various hypotheses about relationships of this kind have been refuted by careful historical research.[151] But it may be possible to draw a link between the practice of the Baconian sciences (as opposed to science in general, or even experimental science in general) and what the philosopher Charles Taylor has described as the basis of Protestant ethics: the affirmation of everyday life.[152] The culture of Protestant countries was strongly oriented toward the new sciences and treated them as no less important than deductively organized sciences.

The Protestant understanding of holiness entailed that a baker, for instance, should simply be a good baker (rather than a holy man withdrawn from society). Bacon advocated the study of "small and mean things," which he believed were the most fundamental for those who sought to understand natural phenomena. Though there was certainly something special about Boyle's rotting veal shank, it was not an object of a particularly refined sort—quite the opposite, in fact. Boyle was a nobleman, but not above hanging putrid meat in his bedchamber so that he could study it. This illustrates the serious attention that Protestants devoted to small, ordinary things. Every individual had to make the best possible use of the capacities and resources at his disposal; that was the way to glorify God.[153]

The English Puritans were highly receptive to the "new philosophy," Baconian science, which they felt was highly compatible with their Protestant faith; just as they had broken with religious tradition and returned to its roots (the text of the Bible), the "new learning" went back to the roots of knowledge about nature: direct observation.[154] Radical Puritans hoped that science could help them to achieve their ambitious, millenarian objectives. More conservative Puritans saw science as a vehicle for personal growth and insight into God's creation.[155] Their views were in harmony with the theory known as natural theology, or physico-theology, and a Dutch writer on this subject, the mathematician and pastor's son Bernard Nieuwentijt, had an avid English readership.[156]

Protestant values included a degree of anti-elitism, an attitude found among Paracelsians. Paracelsus himself had never left the Catholic Church, though he had agreed with Luther about many issues. But from the 1580s onward, his writings were tremendously popular among English Puritans. What especially appealed to them was his contemptuous attitude toward intellectual authorities, particularly Galen and the Peripatetics. They were also pleased with Paracelsus's emphasis on direct experience and practical results.[157] In the first half of the seventeenth century, the Paracelsians and Helmontians among English physicians gained the upper hand over the traditional Galenists. When the plague broke out in London in 1665, they stayed at their posts while the Galen supporters fled.[158]

Dutch Baconianism

In England, the connection between Protestantism and the social prominence of the Baconian sciences has proved to be very close.[159] Comparing case studies in a Baconian manner, we might expect to find a similar connection in the seventeenth-century Dutch Republic.[160] This is not straightforwardly the case, however. When the cultural historian Johan Huizinga argued that Calvinism had no special impact on the rise of Dutch science in the seventeenth century, he meant that it had had no *negative* impact. He did not even consider the possibility of a positive relationship.[161]

No systematic study has yet been conducted of Baconianism in the seventeenth-century Dutch Republic. Most historians have focused on the debate between Cartesians and their Calvinist adversaries. But a preliminary reconnaissance shows that there was great sympathy for Bacon and his style of scientific practice.

Without question, the leading Dutch scientist was Christiaan Huygens, son of the poet and diplomat Constantijn Huygens. Christiaan was a mathematical physicist of the Galilean type, but even more rigorous than Galileo in his use of mathematics. From 1645 to 1647, he studied at Leiden University with Frans van Schooten Jr., a mathematician and admirer of Descartes. At the age of seventeen, Huygens sent Mersenne the proof that a catenary, a cord suspended from two points, does not have the shape of a parabola. Huygens was responsible for many influential experiments, but all of them were strictly theoretical or mathematical in character.[162] In other words, he appears to exemplify the "classical" sciences.

But we should not let that obscure our view of Huygens the Baconian, whose crowning achievement was his pendulum clock, perfected only after many stages of development. Huygens's Baconian program for the French Académie des Sciences was mentioned previously. He brought this program into practice immediately, with a proposal for a major project: the publication of a voluminous natural history of plants, animals, and minerals.[163] He later wrote to Leibniz, "One must reason methodically about observations and gather new ones, more or less according to Bacon's method."[164] This remark is connected to the opinion, which he had expressed earlier, that mathematics should be practiced not for its own sake but to solve problems in the natural sciences.[165] It is worth noting that Huygens never had official ties to the Dutch scientific community. From 1666 to 1685, he was in the employ of the Académie des Sciences in Paris.

We find no immediate signs of Baconianism when we turn to the University of Leiden, established by William the Silent in 1575 after the rebellious Dutch provinces had lost Leuven and its university. Leiden began as an educational institution, and was not intended as a center for scientific research, any more than any other university of the time. Until well into the seventeenth century, most of its professors of natural philosophy were Aristotelians. One exception was Willebrord Snellius (or Snell). In 1621, after much experimentation, Snellius discovered the law named after him, which describes the refraction of light at the boundary between air and glass or water. But his student Franco Burgersdijk, a Leiden professor until his death in 1635, was an Aristotelian and had no interest in experiments. The philologist and historian Gerard Vossius also taught Aristotelian physics as an assistant in Leiden for a time.[166]

After 1650, Leiden had more and more professors of a Cartesian bent. Starting in 1675, Burchard de Volder gave lectures that he illustrated with experiments carried out in the lecture hall. At the same time, he was performing experiments as part of his own research. Yet he attached greater value to the Cartesian method of "clear and distinct perception." Late in his life, he was appointed professor of mathematics.[167] De Volder, who received a visit from Boyle and a copy of the recently published *Principia* from Newton, seems to have wished to incorporate his experimental program into a broadly Cartesian or Spinozan deductive framework.[168]

Many Dutch Cartesians, such as Henri Regius and Henri Renerius, were not as rigidly mechanistic as Descartes himself. The principles that Descartes took to be unquestionable were, for them, a hypothesis to be verified, and this perspective gave them greater scope for experimentation. One of Descartes's central lines of reasoning, his refutation of Skepticism in the *Meditations*, was read by his Dutch admirers as a sufficient guarantee for the reliability of the senses in gathering as many facts as possible. This interpretation would have taken Descartes by surprise,[169] but the Dutch Cartesians had no difficulty combining their eclectic natural philosophy with Baconianism.

The Baconianism becomes more evident if we broaden our focus to include medical practitioners. One of the first professors appointed by Leiden University was Rembert Dodoens, who taught the medical uses of herbs. Dodoens had published a renowned botanical survey, a collection of facts that seemed to anticipate the Baconian method. Even before 1600, Leiden had a botanical garden and an anatomical theater, two outstanding venues for practical instruction. It was Leiden's clinical medicine program, which began to take shape in 1638, that earned the university a reputation for excellence. During Cromwell's regime, as England's scientific productivity somewhat decreased, the Dutch Republic was a model to be admired and envied.[170]

The everyday scientific practice of the Leiden professors of medicine was Baconian; Jan de Wale (Walaeus), for instance, conducted experiments on living animals to learn about the circulation of the blood. Chemistry also held an important place in Leiden. The first chemistry professor was not appointed until 1669, when a chemistry laboratory was set up, but Helmontian medical professors recognized the relevance of chemistry to medicine well before then. One such medical professor was Franciscus Sylvius, who came to Leiden in 1658. He was a follower of both Descartes and Helmont,[171] and seems not to have been bothered by the fact that their perspectives on natural philosophy were irreconcilable. This kind of eclecticism was typical of Baconians. One unique case was the autodidact Antoni van Leeuwenhoek. His relationships with Dutch scientists were strained, but his microscopic investigations received an enthusiastic hearing among

English Baconians, who rewarded him with a membership in the Royal Society.[172]

Through most of the seventeenth century, Dutch Calvinists, unlike their English brethren, largely failed to develop a foundational discourse for the new sciences. The figure of the clergyman-scientist, so prominent in England, was almost entirely absent from the Dutch Republic.[173] This seems especially strange when we consider John Calvin himself. While it would be going too far to describe him as a champion of the sciences, his opinions about natural theology led him to glorify nature as God's creation. He adopted the existing metaphor of the two books of God (scripture and the book of nature) and held that the visual revelation of God in the latter was no less significant than the written revelation in the former. Calvin was especially interested in the many details and components of creation, and when he urged scholars to study them, he was probably thinking mainly of botany and other branches of natural history.[174] These were the same sciences that Bacon later thrust into the spotlight.

Calvin had many followers who embraced this aspect of his theology, especially in France and Switzerland. They included both scientists and ministers of religion. The Swiss zoologist Conrad Gesner drew moral conclusions from his study of nature that seem to come directly from Calvin's writings. A French nobleman, Du Bartas, wrote a didactic poem about God's creation that included elaborate descriptions of nature. Simon Goulart de Senlis, a Walloon preacher who spent some time in Amsterdam after 1600, introduced Du Bartas's poem to the Dutch Republic, where it became quite popular.[175] But we know relatively little about Goulart and the other Dutch Calvinist ministers who propagated the study of nature.

In contrast, we know a great deal about anti-Cartesians. Descartes's mechanical philosophy, which had never really taken root in England, met with resistance from a number of influential Dutch Reformed theologians. Initially, they were motivated mainly by distrust of this new philosophy that attached so little value to practical utility.[176] But soon enough, the conservative Calvinists found a philosophical bulwark of their own. The best-known Dutch debate about these issues took place in 1642, between Descartes and the Utrecht university director and theologian Gijsbert Voetius (Voet). Voetius's opposition to mechanism was uncompromising. Paradoxically, he based his theological arguments not on Calvin but on Aristotelianism, as interpreted by the most conservative Jesuit school.[177] The controversy spread to Leiden, where, in 1647, Adriaan Heereboord, professor of logic, defended Descartes. The trustees of the university admonished him to keep quiet. This should not be attributed to intellectual conservatism, however; their goal was simply to avoid conflict. In 1655, orthodox Calvinists in Leiden launched another attack on Cartesianism. This time Johan de Witt, the Grand Pensionary of Holland, had to intervene to protect the freedom of philosophical opinion. He was only partly successful, and on paper the Cartesians suffered a defeat, though it was one without any serious consequences.[178]

Jonathan Israel has placed this dispute in an Enlightenment context, characterizing it as a clash between conservative religious functionaries and secular scientists.[179] The freethinking Calvinist Johannes Cocceius adopted a "secular" point of view, opposing Voetius and advocating a strict dividing line between philosophy and theology, as well as between church and state. The conflict between Voetius and Cocceius in the 1660s had a substantial political dimension: Voetius was an Orangist, a supporter of the governing House of Orange, while Cocceius belonged to the opposing Republican faction.[180] In 1672, known as the "year of disaster" for the Republic, William III of Orange became stadtholder and Johan de Witt was murdered. University life again became very difficult for the Cartesians, who were accused of having wanted to surrender to the French. At the urging of William III, a Voetian was appointed in Leiden, but when he requested other appointments of this kind, the trustees dug in their heels.[181]

Newton's *Principia mathematica* was at first ignored or summarily dismissed by Dutch natural philosophers; this response was informed by Cartesianism. It was not until 1713, two years after the second edition was published, that a Dutch university scholar, namely, Herman Boerhaave, recognized the book's importance.[182] The last stirrings of loyalty to Descartes gradually died away, and what took their place was a kind of Newtonianism. This Dutch Newtonianism was remarkably similar to utilitarian Baconianism, much more so than English Newtonianism was at this early stage. It is a striking coincidence, to say the least, that between 1687 and 1727 at least nine editions of Bacon's works were published in the Dutch Republic and none in England.[183]

The *Principia* itself was anything but Baconian in outlook. Yet Dutch scientists did not hesitate to approach Newton's gravitational theory as if it were the product of empirical research.[184] It is not unreasonable to describe the physician and chemist Herman Boerhaave and the physicists Willem J. 'sGravesande and Petrus van Musschenbroek as Baconians. For them, experimental work preceded theoretical explanation, and theoretical system-building had no place in science.

The most important propagator of Newtonian physics on the European continent was 'sGravesande, who wrote a physics textbook that was also popular in England. But in a major controversy between Newton and Leibniz about the correct measure of the force of a body in motion (mv or mv^2), 'sGravesande sided with Leibniz.[185] His Baconian view of science may have helped him to determine the best approach on a case-by-case basis. The advisory positions held by 'sGravesande as a hydraulic engineer to the Republic show that he was a practical man.[186] Van Musschenbroek greatly admired the "Newtonian philosophy," but like 'sGravesande, he was more impressed by Newton's method than by his theoretical assumptions; in particular, his force of attraction and repulsion.[187] Boerhaave spoke out against the deductivist ideal in science, relying in part on Calvin's doctrines. He believed that chemists must always base their reasoning on facts, observations, and experiments. Although he advocated a very cautious approach to formulating natural laws, he did have intuitions about the workings of nature and leaned toward vitalism rather than mechanicism.[188]

The appreciation of Newton (albeit after some delay) and the decline in support for Descartes can also be seen in the next generation of Dutch natural scientists, who were more directly inspired by religion than their predecessors. In the domains of religion and natural philosophy, England and the Dutch Republic converged, around 1700, in the physico-theology of Bernard Nieuwentijt.[189] A minister's son and physician in the town of Purmerend, Nieuwentijt was a very popular author in both countries. Physico-theology (better known in modern English as natural theology) taught that God's greatness was confirmed by the perfection of His creation. The Dutch scientists who subscribed to this view included Boerhaave, 'sGravesande, and Van Musschenbroek.[190]

In Boerhaave, in particular, one can see an intimate relationship between Baconianism and Calvinism. Boerhaave's way of life left no doubt that he was a deeply religious man. He was closer to Voetius than to the Cocceians in both his religious and his scientific views, even though his political sympathies lay with the freethinker Spinoza.[191] Spinoza's natural philosophy was repugnant to him, and Cartesianism no better.

The views of Voetius, Nieuwentijt, and Boerhaave confront us with several marked differences between English Puritanism and Dutch Calvinism. The Dutch believed that knowledge of nature could be acquired through large numbers of individual observations, but did not believe in the possibility of discerning the hidden causes of things. As they saw it, the only appropriate response to God's wisdom, as it manifested in nature, was humility. They did not believe, either, that humanity could reestablish its dominion over nature and recreate paradise on earth—at least, the idea struck them as arrogant.[192]

While they did not profess this belief, they did put it into practice. The creation in 1611 of the Beemster polder, the first truly large-scale land reclamation project, inspired a sense of national pride. For contemporary observers, the strictly geometrical layout of the polder was reminiscent of the organization of paradise. Joost van den Vondel, the greatest Dutch poet of the seventeenth century, immortalized the project in verse: "Here laughs the golden age, in gladsome pleasure-gardens."[193] The Dutch Republic never had an Agent for the Advancement of Universal Learning, but it did have the celebrated engineer Jan Leeghwater, who supervised the reclamation of land from numerous lakes, creating not only the Beemster but also the Purmer, the Wormer, the Schermer, the Heerhugowaard, and many other polders.

The province of Holland was commonly referred to as a garden; in metaphorical terms, Holland was a *hortus conclusus*, a walled garden, ringed by fortified cities such as Groningen, Nijmegen, and Bergen op Zoom.[194] This led to the use of a walled garden—in fact, the Garden of Eden—as an icon of Holland. Had the Dutch discreetly reestablished a paradise of their own?

Many questions remain unanswered about the relationship between Dutch Baconianism and Calvinism in its Voetian, Cocceian, and other varieties. In any event, Nieuwentijt's physico-theology did not fall out of a clear blue sky but was anticipated by Calvin himself. The hypothesis that Dutch ministers and natural philosophers alike were sympathetic to Bacon, in part because of their religious beliefs, appears to be supported by the facts. But the relationship between Baconian science and Protestantism need not be attributed entirely to the dogmatic natural philosophy of the Calvinists. It is also inherent in a more general Protestant ethical stance: the affirmation of everyday life. Pending further research, we can provisionally conclude that this virtue became part of the Dutch mentality at a very early stage. In the Netherlands, the Baconian sciences were able to develop from the late sixteenth century onward and disturbed only by occasional skirmishes between the Cartesians and the strict Calvinists who remained loyal to Aristotelianism.

Instruments and Laboratories

Sixteenth-century *virtuosi*, artist-engineers, had mathematical techniques at their disposal, while alchemists had their arsenal of flasks and crucibles. *Virtuosi* had mathematical training, while alchemists might have picked up a few quantitative methods (as Van Helmont had), but were far from being mathematicians. In the seventeenth century, the two traditions began to come together. This was due in part to the development of the analogical style as a mode of scientific thought: analogies inspire experiments. Another factor was a set of newly invented "philosophical instruments," such as the telescope, microscope, thermometer, barometer, and air pump. Polished prisms served as "philosophical" toys before they were put to practical use. It was mainly thanks to instruments like these that Baconians could use mathematics (for instance, to describe the results of their measurements) without feeling compelled to commit to any theoretical "causes." The practitioners of the classical sciences, who had been accustomed to passive measurement instruments, learned to appreciate instruments that actively intervened in the ordinary course of nature, such as the air pump and electrostatic generator.[195] It is not always easy to draw the line between passive and active instruments. Microscopes, for instance, may seem like passive aids to observation, but any microscopist examining a biological sample has undoubtedly sectioned the tissue in ingenious ways and perhaps stained it with dye or adjusted the illumination.[196]

In social terms, the distinction between natural philosophers and artisans remained significant. The aristocrat Huygens may have spent years of his life grinding his own lenses, but he still tried to stop his assistant Nicolaas Hartsoeker from becoming a natural philosopher in his own right. (Hartsoeker later succeeded in spite of Huygens.[197]) Likewise, Newton saw Francis Hauksbee, the curator of experiments at the Royal Society and caretaker of its collection of instruments, as a mere hired hand, paid to carry out Newton's experiments. Hauksbee's predecessor, Robert Hooke, had clashed with Newton several times because of Newton's refusal to take Hooke's "philosophical" ambitions seriously. Newton also had an uneasy relationship with the glassblowers on whom he depended for his equipment, because they adhered to the old craft tradition of keeping some crucial pieces of knowledge secret.[198]

Chemistry, which succeeded alchemy, took science in an even more craft-oriented direction. But even as late as the eighteenth century, there was still a clear social distinction between natural philosophers of a speculative bent and practically minded apothecaries and metallurgists. What both groups aimed to do, each in its own way, was to separate and isolate substances. Accordingly, they all had to deal with an ever-expanding set of chemicals and materials. The "artisans" (some of whom were actually wealthy manufacturers) showed a stronger tendency to use pragmatic criteria in naming "pure" substances, and often, all that was meant by "pure" was that no method had yet been developed to divide these substances into still more basic components. It was this pragmatic working method that ultimately led the natural philosophers to abandon their Paracelsian principles.[199]

The Reform of the Universities

Until well into the nineteenth century, experiments were not performed in university laboratories, for the simple reason that such laboratories did not exist. At one moment in history, universities even had reason to fear for their institutional future. In France, the National Convention abolished all twenty-two French universities in 1792, including the Sorbonne, one of the three longest-established universities in Europe, and the venerable University of Orléans, founded in 1306. Four years later the French, who by then had conquered much of Europe, closed the University of Leuven (est. 1425) in the former Habsburg Netherlands. Cologne (est. 1388) followed in 1798, and then more than ten other German universities. Napoleon later reversed some of the educational reforms made during the French Revolution, but he too did away with several universities.[200]

This assault on the universities took place for sound reasons. For the most part, the rise of the experimental sciences in the sixteenth and seventeenth century and the reformist ideals of the Enlightenment had not penetrated the groves of academe. Many universities were bulwarks of the *ancien régime*, and a few were still directly controlled by the Catholic Church. The Jacobins created new institutions for their most prestigious educational programs: the École Polytechnique for the natural sciences and the École Normale Supérieure for the humanities.[201]

In the eighteenth century, experimental research was conducted at academies and institutions that were independent of the universities, such as the Collège du Roi and Jardin du Roi. These organizations were left untouched by the French Revolution and in 1793 were transformed into the Collège de France and the Muséum national d'Histoire naturelle, respectively. The Muséum later had a major chemist on staff: Joseph Louis Gay-Lussac, who had studied at the École Polytechnique. Still further into the nineteenth century, the physiologist Claude Bernard was affiliated with both the Collège and the Muséum. In his research on the physiological mechanisms of digestion and other topics, Bernard refined the methodology of the controlled experiment. He often used living animals to ensure that the internal workings of the organism would be intact, and he investigated the effects of all sorts of variables one by one, keeping them separate through careful experimental design. His colleagues called him the legislator of the experimental method.[202]

Some amateur scientists needed neither a patron nor an institute to perform experiments. James Prescott Joule was the son of a wealthy brewer in Manchester. He is known for being one of the discoverers of the law of conservation of energy, and particularly for the concept of the mechanical equivalent of heat, which states that mechanical energy can be converted into heat.[203] Joule had never studied at a university. An Oxford or Cambridge education would have been a natural step for him, but he presumably thought of that as an old-fashioned choice, and was apparently no more impressed with the universities in Glasgow and Edinburgh. Instead, he received his scientific education at home from private tutors, including training in mathematics from John Dalton, the originator of the modern theory of the atom. Dalton was the secretary of the Manchester Literary and Philosophical Society, a learned society founded in 1781, most of whose members were physicians, apothecaries, and manufacturers with a predilection for science. Joule also became a member.[204]

He started out in a laboratory that his father provided for him, later moving to the cellar beneath the brewery. In his best-known experiment, Joule used two weights to spin a paddle-wheel in a barrel of water. He measured the temperature of the water at the beginning and end of the experiment with a very sensitive thermometer made by a Manchester instrument maker. He calculated the energy needed to keep the wheel in motion by counting the number of times that the weights were raised above a certain height. It was extraordinarily difficult to obtain precise measurements, and he spent years refining the process. The historian Otto Sibum, who reproduced Joule's experiments, encountered many unexpected surprises. Among other things, the experimenters discovered that if they produced sweat during the process of taking measurements (a difficult thing to avoid when lifting weights), their body heat would disturb the experiment.[205]

Joule's skill in measuring slight differences in the temperature of water had been developed at his father's brewery. The years that preceded Joule's experiments had witnessed a revolution in brewing techniques, as British authorities and consumers had raised their expectations of the quality of beer. The only way to meet those expectations all year round was to control the temperature of the mixture very carefully at each stage in the brewing process. Joule, who performed his scientific experiments in the evenings, was involved in this type of work every day at the brewery.[206]

Charles Darwin is probably the best-known gentleman scientist, and conducted experiments at Down House, his home in Kent. The original purpose of these experiments was, of course, to find supporting evidence for evolutionary theory. But after the publication of *The Origin of Species* in 1859, Darwin also turned to other topics. Plant stems, especially those of climbing plants, seem to determine the direction of their growth through a searching process, as if the tip of the stem housed a brain. In 1880, Darwin published *The Power of Movement in Plants*, which presented the results of twenty years of experimental research on plant physiology.

The book was a success in Britain, but German botanists were considerably more skeptical. Julius Sachs in Würzburg was especially critical of Darwin. Perhaps for the first time in the history of science, Sachs drew a dividing line between professionals like himself and amateurs like Darwin.

He argued that experiments in plant physiology can only be carried out in a well-equipped laboratory with modern instruments (such as the auxanometer in Würzburg, which could measure the growth of a plant stem with greater precision than a simple ruler). Above all, Sachs wrote, such laboratories provided conditions under which experiments could be repeated and were open to scrutiny.[207]

Darwin, who died in 1882, was one of the last of the gentlemen amateurs, in the original sense of the term. But the public image of the gentleman amateur was not as quick to fade. When Emile Zola devoted *Le docteur Pascal* (*Doctor Pascal*, 1893) to the theme of scientific practice, he made his main character a country doctor. Too absent-minded to charge his patients, Pascal Rougon conducts his experiments in heroic poverty.

Sachs worked in a laboratory set up for him by the grateful University of Würzburg in 1869, after he had turned down an offer from the University of Jena. Würzburg was following a trend that had been visible at German universities for about twenty years; university laboratories were the new standard in experimental science. This was a surprising development, given the sad state of the universities just a few decades earlier, around 1800.

The new research universities were the unanticipated result of an earlier reorganization, which had produced Germany's answer to the Napoleonic reforms: the neo-humanist university. The Friedrich-Wilhelm University was founded by Prussian royal decree, in 1809 in Berlin. Important models for this university were Göttingen and Halle. If those two institutions had not developed an Enlightenment-inspired utilitarianism in the late eighteenth century, Germany might well have abolished its universities, as France had. But the neo-humanist reformers, who framed the research ethic of Göttingen and Halle in the ideal of academic freedom and the unity of research and teaching, saw their university as a genuine new institution.[208]

The aristocrat Wilhelm von Humboldt is still renowned as a promoter of the ideal of *Bildung* (formation of character), embodied in the new nineteenth-century organization of Germany's *gymnasia* and universities. Humboldt and the utilitarians had a great deal in common. Both were opposed to the old university system, dominated by theologians, and both were supported by the members of the bourgeoisie (and, in Germany, of the nobility) who had stepped up as the new elite at state level and in the rapidly expanding machinery of the Prussian state. The authorities supported the neo-humanist program, even to the point of acknowledging that Germany's reluctance to reform had contributed to its military defeat by Napoleon; the country's leaders had proven unable to strike out in new directions. Prussia also actively supported industry, but this policy did not affect the universities.

The humanistic study of the classics was the focus of a great deal of ambition. Whereas theologians at most universities used Latin reflexively as the medium of the traditional *disputatio*, the University of Göttingen had invented the seminar, where the classics were the object of philological research and the study of Latin was an objective in its own right. Professors in Göttingen were expected to publish. This research ethic spread from the study of the classics to the fields of history and mathematics, and ultimately to the natural sciences such as chemistry. The relatively small research laboratory of the organic chemist Justus von Liebig at the University of Giessen became a much-emulated model for combining education and research. In 1838, when Liebig called on the Prussians to provide financial support for a large research laboratory, he boasted of the practical applications of chemistry, but otherwise rejected utilitarian criteria. For him, the formation of character was central.[209]

Liebig first made an impact through his science courses for students of medicine and pharmacy. He had a small laboratory with ten to fifteen student researchers until 1840, when the number started to rise, eventually reaching about fifty. Liebig had a keen sense of what research topics would attract international interest, partly thanks to advice from his former teacher Joseph Louis Gay-Lussac in Paris. He helped his students to approach these topics, keeping in mind their individual levels of ability. Most of them did not embark on research careers, but became pharmacists.[210] In 1860, when Prussia finally acceded to Liebig's request and invested one million German marks in chemical research laboratories in Berlin and Bonn, it was motivated by the desire to compete with other German states for top students. German industry had no real demand for university-trained chemists until after 1870. By that time there were qualified chemical researchers available, just when the German dye industry needed them most.

Until the 1880s, Americans who wanted doctorates would travel to Germany. All attempts to entice them to pursue their doctoral studies in their own country failed until 1876, when the first American research university, Johns Hopkins, was founded. Almost simultaneously, elite colleges such as Harvard and Yale began offering graduate programs, and the number of doctoral dissertations grew rapidly. In 1870, virtually no PhD degrees were awarded in the United States, but by 1900 that number had increased to 250, and by 1915 it had reached 650, about half of which were in the natural sciences.[211]

How did Johns Hopkins and the other graduate schools manage this feat? Earlier initiatives had been based on vain hopes of demand for research skills in the business world. Johns Hopkins, in contrast, acted at just the right moment to take advantage of a rapid increase in enrollment at American colleges, which brought about a collegiate reform movement that sought to improve the qualifications of college teachers. For college graduates, who had mostly gone into business until about 1850, secondary schools became the main job market. For the newly minted PhDs, just about the only job market was the colleges. The American educational sector, like Germany's, became a self-reinforcing system in which the rise of research both resulted from and contributed to an increase in educational levels.

Since the beginnings of experimental science, researchers who actively intervened in nature have brought about billions of phenomena, most of which do not take place spontaneously and can only be witnessed under laboratory conditions.[212] The experimental style of science, more than any other, has made science a fundamentally open-ended activity. Descartes was the last great natural philosopher to believe that science would ever be "complete," if not during his lifetime then soon after. But as early as the seventeenth century, experimentalists were generating new phenomena too quickly for theoreticians to keep up with them, and they have gone on doing so ever since. The experimental way of thinking and acting has also spread to areas in which intervention in nature is not really possible, such as field biology. "Naturalists," who collect specimens and think in evolutionary terms, have nonetheless developed quasi-experimental procedures.[213]

Karl Popper and other twentieth-century philosophers of science saw experimentation purely as the handmaiden of theoretical science, serving purely to test hypotheses. Ian Hacking and Alistair Crombie rightly insist that experiment has its own "logic" of exploration, equal in significance to the hypothetical style and other styles of science.

The *virtuoso* and the alchemist have persisted as distinct archetypes, and to some extent they are still with us today.

What all forms of experimentation have in common is their character as explorations of nature. Modern biochemical experiments, for instance, can often be performed in a single day. This makes it possible to try a new approach without a great deal of delay, explore paths that might turn out to be dead ends, and so on. No experiment stands alone; each one refers to a series of other experiments, which biochemists sometimes call their "experimental system."[214] Similarly, physicists often encounter "resistance" from their materials and experimental set-ups, and in many cases, this makes the development of their research more improvisatory than outsiders might think.[215]

The experimental style is connected with other styles in all sorts of ways. The results of experiments can be classified; like Bacon, one can develop a taxonomy by comparing cases in an inductive manner. This method can be used to generate new concepts.[216] Once again, we see that in the traditional relationship between the experimental and hypothetical-analogical styles, there is more involved in experimentation than just testing theories. Like sets of data gathered from nature, experimental results can be analyzed statistically, and the development of the statistical style in science brought about further changes in the experimental style. In scientific fields where statistics plays a role, the experimental method that has developed since about 1930 includes protocols and principles of good practice that Boyle could only have dreamed of. Clearly, the experimental style is constantly in development. But it remains its own arbiter, refusing to subordinate itself to other styles of science.

7

The Hypothetical Style

Analogies between Nature and Technology

The true and the made are interchangeable.
—Giambattista Vico

What, then, is truth? A movable army of metaphors, metonyms, anthropomorphisms, in a word: a totality of human relations.
—Friedrich Nietzsche

IN THE MIDDLE AGES, there was a well-known theological argument that man will never be able to understand nature, because nature was created by God, and man cannot fathom God's purposes. But in the domain of technology, Nicholas of Cusa (Cusanus) wrote in 1450, man is a "second god."[1] In other words, what man makes himself, he can understand completely.

Cusanus's argument shed a different light on Aristotle's view that there was a sharp distinction between nature and technology, providing a conceptual basis for using "unnatural" technology to understand unforced nature. Yet there was a price to pay: this understanding was analogical, and the deductive framework sketched by Aristotle left no room for analogy, at least not in science. Artistotle had envisioned a hierarchical universe with causal chains extending from the Unmoved Mover at one end to the observable world at the other, while the explanatory link made by analogy is utterly independent of hierarchical structure.

From the early Middle Ages onward, philosophers saw the clock as a scale model of the cosmos and as a model for animal locomotion. These analogies were hinted at by Thomas Aquinas, and Jean Buridan and Nicholas Oresme fleshed them out in greater detail a century later.[2] What intrigued these thinkers most was that once a clock is set in motion, it goes on moving of its own accord. God, they reasoned, could have set the celestial spheres in motion in the same manner. In 1370, during Oresme's life, King Charles V had a clock installed in his palace in Paris. This clock was intended to serve as a standard for all the other clocks in the city. Its novel feature was that every hour of the day was equally long. Before that time, there had been fixed numbers of hours between sunrise and sunset in summer and winter; accordingly, the hours had varied in length.[3]

In 1348, Giovanni de' Dondi, a physician and professor of medicine in Padua, developed an astrarium based on the Ptolemaic system, which was driven by clockwork.[4] (Long before, Archimedes had built a planetarium, which was plundered by the Romans when they conquered Archimedes's home city of Syracuse. It remained in Rome for a couple of centuries before it disappeared.[5])

An astronomical clock built in Strasbourg in 1352 inspired scientists for centuries afterward. It was a monumental clock whose attractions included moving automata, one of which (a rooster) is still on display in a local museum. Between 1571 and 1574, the clock was replaced by a new and larger one.[6] This only reinforced the clockwork image of the universe. Several scientists made repeated reference to the Strasbourg mechanism: Rheticus mentioned it in his defense of Copernicus, and Kepler and Boyle discussed it as well.[7] Those who saw the clock, Boyle wrote, felt reassured that God was the maker and protector of the world.[8]

Scientists saw the clock metaphor as an invitation to learn more about nature by recreating it. A mechanical model, it seemed, was a hypothesis about how nature works. Nevertheless, we should not assume that this line of thought foreshadowed the hypothetical-analogical method of modern science, especially when we look back on the Middle Ages. Medieval thinkers would not have seen analogies as a viable philosophical method for gaining new insights about nature. Scholastic philosophers and theologians made a sharp distinction between literal truth (the domain of science, or natural philosophy) and allegory, in which analogical thinking could take place freely and served a spiritual and theological purpose, especially in the area of salvation history. Plato's dialogue *Timaeus*, which draws quite a number of analogical connections, was read in this way. Medieval literature was full of analogies and allegories that drew readers' attention to the unity of creation and God's plans for the world. One example is the link between the red rose and the blood of Christ, a link made even stronger by Christ's crown of thorns and the thorns of the rose bush.[9] The controlled movement of a clock (or an hourglass) was associated with the allegorical figure of Temperance.[10] Undoubtedly, the aforementioned similarities between the cosmos and clockwork were also placed in the service of a spiritual concept.

In the Renaissance, however, the two levels of literal truth and allegory began to mix. Connections were drawn without constraint, both between natural phenomena and between nature and culture, through intellectual systems such as the doctrine of sympathies, the microcosm-macrocosm thesis, and emblematics. Ficino's new way of reading the *Timaeus* played a role in this shift. The theological aspect of analogy and allegory did not disappear, but it gradually mingled with the natural-philosophy dimension. This was most clearly visible in the writings of Paracelsus. Cesare Ripa's well-known *Iconologia*, the first edition of which appeared in 1593, depicted a person of a "sanguine" temperament with a number of symbolic attributes, including roses.[11] Apparently, in addition to their religious meaning, roses could also refer to the physical aspects of blood. Natural history was another field in which factual and emblematic associations were interwoven.

Around 1650, the allegorical-emblematic style of looking at nature lost its intellectual appeal, but this did not herald the end of the analogical mode of thought. Though analogy was abandoned in natural history, it remained in use in other fields of science, contrary to what many historians have supposed. One might even speculate that the use of analogy from 1600 onward was a highly watered-down version of Renaissance thinking.

The new analogical style focused primarily on the resemblances between nature and technology and may have owed something to the long-term impact of the medieval technological analogies discussed previously. The closing of the divide between nature and technology, particularly by Francis Bacon and the magical tradition that preceded him, undoubtedly helped to make analogies between nature and technology seem more and more plausible. In the seventeenth century, Marin Mersenne and René Descartes were the two great natural philosophers of the hypothetical-analogical method. Mersenne advocated a skeptical version: technology was comprehensible because it was produced by man, while nature was incomprehensible because God surpassed our understanding. Descartes, in contrast, supported a more optimistic version of this method, and was hopeful that analogical reasoning could reveal the truth about nature.

In the modern hypothetical method, when the word "truth" is applied to an analogical comparison, it is always understood to be in scare quotes. An analogy remains a figure of speech, whether or not the comparison is drawn from the technological domain. It is instructive to compare this to figurative literary language. Aristotle made no objection to the use of analogical comparison in poetry, and modern theories of metaphor still build on Aristotle's insights. Literary theorists have observed that the content of a metaphor can never be made fully explicit, because the two terms of the comparison each belong to a separate domain of reality. But this is precisely what gives a metaphor its power; it can inspire new ideas, which can then lead to new discoveries. Metaphor, or analogy, thus has enormous potential as a heuristic method. The first to make conscious use of analogy was the astronomer Johannes Kepler, in his account of the working of the eye (discussed in the next section).

One might question whether a hypothesis based on a technological analogy can ever be true, no matter what philosophical views on truth a person might have. After all, technology changes, and scientists have a natural tendency to base their theories on the latest technologies of their day. A comparison between the human mind and a telephone exchange evokes different associations if we think of a manual switchboard from 1940 than it does if we think of an automatic digital exchange from 1990.[12] In general terms, we can conclude that scientific analogies reflect the cultural and social preoccupations of their day.

The Eye and the Camera Obscura

The ancient Greeks regarded the eye as a kind of lantern, which made objects visible by illuminating them, as it were. Alhazen was the first to describe the eye as a passive receiver, but like the Greeks, he believed that the image of the external world was formed in the *crystallinus*, a small, sensitive organ within the eye. Andreas Vesalius, in *De Humanis Corporis Fabrica* (1543), noted that the *crystallinus* was shaped "just like a lens," but was unable to say anything about how the eye worked. In 1583, Felix Platter (often Latinized as Plater) identified the retina as the visual receiver. Like Vesalius, Platter was an anatomist. Although his wording was suggestive— "light is sent into the dark room of the eye"—he too failed to discover how visual images are actually formed. When the answer came, it was not from an anatomist but from a mathematician. Yet despite mathematicians' insight into geometry and optics, they did not solve the problem for quite some time.

In 1545 Gemma Frisius, in Louvain, gave an accurate description of the working of a camera obscura, which since then has been an indispensable observation instrument in astronomy. In 1554, Francesco Maurolico performed optical analyses of both the camera obscura and the eye, though without drawing a connection between the two. These analyses remained unpublished, however, and though mathematician and astronomer Christopher Clavius was probably aware of the manuscript, Maurolico's work had no influence. Publication did not take place until 1611, after Kepler had made his discovery. Likewise, Leonardo's speculations in the *Codex Atlanticus* had no influence on his contemporaries because they did not find their way into print.[13]

Kepler's interest in the operation of the eye was originally provoked by problems of astronomical observation. These problems cast doubt on the reliability of the human senses, but Kepler was unwilling to yield any ground to this type of pessimism. With his analogy between the eye and a camera obscura, Kepler dodged a number of important questions, such as what visual sensation really is and how it comes about, instead concentrating solely on the mechanical analogy and treating the eye as a picture-taking machine. His actual discovery was that the retina is the screen onto which visual images are projected: "The retina is painted with the colored rays of visible things."[14] He simply accepted the fact that the image projected onto the retina is upside-down. It was this, more than anything else, that made his contemporaries deeply suspicious of his theory, but Kepler shrugged and made no attempt to discover what happened to images after they were projected onto the retina.

Kepler conducted his most significant research on the eye in 1600 and published the results shortly afterward. In 1625, his theory was confirmed experimentally by Christopher Scheiner in Rome. Scheiner prepared a human eye, scraped off the sclera (the tough outer coat) to reveal the retina, and demonstrated that the inverted image of a candle flame placed in front of the eye was, in fact, displayed there.[15] Kepler himself was mainly interested in the development and improvement of optical instruments such as the telescope, and his work provided the impetus for new types of spectacles, not only for near-sightedness and far-sightedness, but also for such conditions as astigmatism.

Metaphors for the Heart and the Blood

William Harvey (1578-1651) discovered that blood circulates in a closed system and identified the heart as the pump that propels the blood through the arteries and veins. This story could be told in much the same way as that of Kepler's optical discovery, and was told that way until fairly recently, with one difference of emphasis: in Harvey's case, the experimental aspect outweighed the metaphorical. He demonstrated his discoveries through experiments on living animals, by cutting off the flow of blood through the veins, for instance. His experimental evidence—in combination with anatomical features such as the structure of the heart valves, which are strikingly reminiscent of the valves of a pump—lent additional plausibility to his comparison between a pump and the heart. Another mechanical comparison made by Harvey was between the heart as it fills with blood and a leather bag filling with liquid. In his most famous work, *De Motu Cordis et Sanguinis in Animalibus* (*On the Motion of the Heart and Blood in Animals*, 1628), Harvey mentioned the pump analogy only in passing, but did give a series of examples of cyclical processes in nature.[16]

From a modern-day perspective, the analogy with the pump is an extraordinary case of the mechanistic mode of thought in biology, and Harvey has been revered at least as much for his method of discovery as for the discovery itself.[17] In the process, the term "mechanistic" (or "mechanical") has become heavily laden with significance. Mechanistic reasoning is said to rely on strict causal connections that can be demonstrated experimentally, dispensing with all the "unscientific" ballast from the period extending from the ancient world through the Renaissance: Aristotle's teleological view, occult powers, animism, vitalism, and so on. Used this way, the term "mechanistic" stands for nothing less than the entire modern scientific method, as it is understood to this day. Harvey is often credited as the originator of "mechanical" thinking in this general sense.

Yet more recent historical research has seriously complicated this picture. Harvey was a staunch Aristotelian who emphatically disavowed the Baconian views held by many of his contemporaries. He even applied Aristotle's theory of impetus to the propulsion of the blood through the heart. Vitalism informed his work in countless ways, and moreover, the role of experiment in his overall working method was ambiguous, to say the least. Mechanistic analogies played a role in Harvey's thinking, but alongside or even subordinate to analogies of a very different nature.

None of this detracts from the importance of Harvey's achievement. Even considering that his work grew out of an established tradition, the discovery of the circulation of the blood in a closed system made all established views obsolete. From Galen's day to Harvey's, it had been assumed that after blood left the heart it was consumed, so to speak, by the tissues, and that the liver was constantly generating new blood. Furthermore, no clear distinction had been made between the two types of circulation (pulmonary, between the heart and the lungs, and systemic, between the heart and the rest of the body), and there had been a great deal of confusion about whether the septum between the right and left sides of the heart was or was not permeable. Even after Vesalius had demonstrated its impermeability, many people had been convinced that blood must be able to pass through this barrier, even if that required a miracle. Michael Servetus had discovered pulmonary circulation, however, and had published his findings in 1553, the year that he was executed for heresy in Geneva at John Calvin's instigation. Andrea Cesalpino had drawn attention to the constant motion of the blood through the veins into the heart and out of the heart again through the arteries.[18] Finally, in 1602, Hieronymus Fabricius had demonstrated the existence of valves in the veins, though he had not sought to explain their presence by ascribing a function to them. Harvey pondered this discovery for many years, and it eventually sparked the insight that led to his own breakthrough. Fabricius was an anatomist at the University of Padua and had supervised Harvey's doctoral studies there.[19] Padua had been the center of European medical research ever since Vesalius had become a professor there, and Servetus had studied there too.

Harvey himself saw the circulation of the blood in a closed system as his most important discovery, but it took several decades before his findings were generally accepted.[20] Descartes criticized Harvey's ideas because they were not mechanistic enough for his taste. How could the heart expand and contract spontaneously? Descartes's own hypothesis was that there was a fire in the heart that made it expand and served as a kind of engine.[21] The first person to confirm Harvey's results experimentally was Johannes Walaeus at the University of Leiden, whose work greatly contributed to the acceptance of Harvey's ideas.[22]

Harvey published *De Motu* in 1628, when he was already fifty years old. In the first half, he described the working of the heart; in the second half, he presented his discovery of the circulation of the blood in a closed system. For a long time, Harvey's readers saw this as a rhetorical masterstroke, but historians have ultimately reached the conclusion that Harvey wrote the two parts at different times.[23]

The first section was begun while he was still in Padua, while the second was not written until after 1625. It is important to keep in mind when Harvey's books were written, because it gives us a way of thinking about the consistency of his oeuvre as a whole. The question of consistency is most relevant to two much later publications, *De Circulatione Sanguinis* (*On the Circulation of the Blood*, 1649) and *De Generatione Animalium* (*On the Generation of Animals*, 1651). In these books, the heart hardly plays any meaningful role. It is the blood itself that circulates through the body, on its own initiative and under its own power, to perform its life-giving work. In 1928, Eric Nordenskiöld analyzed this as the conservatism of an old man, a reversion to pre-scientific thinking.[24] Later historians have instead emphasized the continuity of Harvey's work and the persistence of vitalistic themes in his thought alongside his more scientific achievements.[25] Nevertheless, there do seem to be real shifts of emphasis over time in Harvey's interpretations of his findings, and if we seek the origins of those shifts, we find a diverse thicket of factors that contributed to his discoveries.

Harvey always underlined the importance of his observations and his experiments. He had a good reputation, and figures such as Robert Boyle came to witness his anatomical exercises. In reporting his discovery of the circulation of the blood, Harvey stressed an empirical consideration; according to his calculations, the volume of blood pumped away by the heart in an hour exceeded the volume of milk produced by a cow in a day.[26] The implication was clear: Galen's idea that the liver was constantly making new blood was obviously a physiological impossibility. Harvey's tendency to assign the blood an increasingly important role, relative to the heart, was also based on his observations. In his embryological work, he had seen that the circulation of the blood begins before the heart is formed.[27]

But alongside observations, analogies played too prominent a role in Harvey's work to be overlooked. In his earlier work on the heart (in the relevant section of *De Motu*), he used a metaphor that overshadows the pump analogy: the heart as a sun and a king. The context was a three-way correspondence between the cosmos, the country, and the body. Harvey also described the heart as a fountain and a hearth.[28]

Was this only rhetoric intended to charm his patron? Harvey, court physician to King Charles I, was married to the daughter of the previous queen's physician. In 1628, when *De Motu* was published, the king's position was secure. Harvey pushes his comparison between the heart and a monarch too far for it to be dismissed as a vacuous display of fealty. For instance, Harvey claims that the heart is in charge of distributing nutrients throughout the body by means of the blood, just as the king oversees the distribution of food to his subjects. Harvey was an ardent royalist, and by all biographical indications, he remained so his entire life—even after 1649, when Charles I was deposed and publicly beheaded.

Nevertheless, there is evidence that as early as 1628, Harvey was uncertain whether the heart or the blood wielded supreme authority. In any case, he believed the blood was crucially important and regarded it, among other things, as the material embodiment of the soul, Aristotle's *anima*.[29] In theory, the fact that the blood circulated though the body could be considered independently of whether the heart or the blood provided the motive force. But from the very start, Harvey drew a close connection between the circulation of the blood and his vitalistic beliefs, which also strongly resembled the beliefs of leading vitalists of his day, such as Johannes Baptista van Helmont. In his reflections on the motion of the blood, Harvey referred to Aristotle's remarks on the cycle of rain and cloud formation, another cyclical system in nature. Then, in his 1649 treatise, he described the blood itself as the "fountain of life," with its own inherent heat that he apparently believed could be transmitted independently of the heart. The blood was responsible for its own propulsion, while the heart was demoted to a reservoir of blood.[30]

Could Harvey have become a republican at the end of his life and incorporated this political view into his work? Christopher Hill has put forward the controversial suggestion that Harvey's 1649 book could be seen as an attempt to obtain a new patron. This idea cannot be called far-fetched, though it finds little support in Harvey's biography. For several of Harvey's contemporaries, such as Robert Boyle and Walter Charleton, a link can be drawn between their vitalist views and their moderate reformism in religion and politics.[31] And if Hill's suggestion is not true of Harvey, it is true of Harvey's intellectual admirers.

Two years after Harvey's death in 1653, the publisher and translator of the English edition of Harvey's work called him the "seditious Citizen of the Physicall Common-Wealth."[32] This description, evidently approving in tone, referred to the republican form of government instituted by Cromwell. It is as if blood were made up of free republican citizens. The literary historian John Rogers also sees Harvey's vitalistic view of blood as connected to republicanism, but on a more general conceptual plane; he points out that the terms "individual," "agency," and "organization," which became part of the English language around this time, have their roots in vitalism.

In 1668, two members of the Académie des Sciences hypothesized that sap circulates in plants in the same way as blood in animals. This general idea led to a number of more concrete claims; they suggested that plants had two kinds of vessels, analogous to veins and arteries, and consequently two kinds of sap as well, and that the roots of plants made sap, just as the liver made blood. In their work on this process, Perrault and Mariotte also postulated that plants had a kind of controlling organ. This idea came from the French physiologist Jean Riolan, who only partly agreed with Harvey and believed that blood itself was a nutrient.[33]

What developed from these ideas was an experimental program in which Perrault and Mariotte were engaged for at least twenty years with the participation of other members of the Académie such as Christiaan Huygens and Joseph Pitton de Tournefort. This led to a number of discoveries; for instance, that sap flowed not only from the roots upward but also in the opposite direction, from top to bottom, and that upward- and downward-flowing sap differed in color and consistency. But it was very difficult to demonstrate that these different types of sap were transported by different systems of vessels. For instance, none of the vessels were found to contain valves. In the course of their research, they discovered that not only roots but also leaves absorb water. This led them to search for a similar process in animals; the absorption of fat through the skin was one potentially analogous process.

The greatest difficulty was that a pump playing the same role as the heart could not be found in plants. But there were also animals without hearts, as Harvey had shown. To cope with this fact, they developed new hypotheses about the causal mechanism responsible for the upward-moving sap: capillary action, a chemical process, or air pressure. Huygens supported this last hypothesis, but Perrault disagreed. Tournefort pointed out that there were sponge-like structures around the vessels in which sap could be stored. Most members of the Académie finally resigned themselves to the conclusion that the phenomenon must have multiple, simultaneous causes.[34]

The Mechanical Philosophy of René Descartes

Descartes tried to picture how God would have gone about creating humans. The outcome was an image of the human body as a hydraulic machine, which led Descartes to a wide variety of hypotheses about transmission of impulses from the eye to the brain to the muscles (for example, hand-eye coordination) and about involuntary movements that proved to be of lasting value in the physiology of the human body.[35]

Descartes may have derived this image from existing hydraulic devices. From the mid-sixteenth century onward, hydraulic automata were a source of amusement in the gardens of Florence; bronze angels lifted trumpets to their mouths, and nymphs rowed back and forth across a pond.[36] This fashion spread to other places in Europe. Around 1614, Descartes spent some time in the Paris region, and it is quite probable that he went for frequent strolls in the nearby gardens of Saint Germain-en-Laye. These royal gardens had been laid out by Italian landscape architects (and described by the Huguenot engineer Salomon de Caus), and there were a number of hydraulic devices installed there, including singing birds.[37]

During his stay in Amsterdam, Descartes paid regular visits to the butcher to watch animals being slaughtered. He took pieces home with him to dissect, a practice which he kept up for about ten years. Though he made no anatomical discoveries, he said he was content just to gain a better grasp of physiology.[38] His ultimate objective was to find mechanistic explanations for all biological processes. From the moment his book *L'Homme* (*Man*) was published in 1667 until the end of the eighteenth century, it remained one of his most widely read works.[39] For instance, Descartes describes the working of memory as the formation of an image (in the pineal gland) when the original object is not present.

Descartes's emphasis on automatic movements in physiology, such as hand-eye coordination and the pain reflex, resulted from his aim of strictly separating living matter from anything that was "mental" in nature. Renaissance naturalists such as Bernardo Telesio believed that the mind and body could overlap fluidly, or perhaps were different manifestations of one and the same thing. Descartes was determined to see his mechanistic view prevail.[40]

Descartes's speculations were fruitful, but he too often assumed that they must be true, even in cases where he had missed the mark. Descartes hoped all his life that his method would be a route to a true deductive science that could offer certainty. Christiaan Huygens and other Cartesians were the first to acknowledge the speculative nature of Descartes's physics, and therefore its hypothetical character.[41]

Descartes's analogies typically have several levels. First, they have a firm metaphysical basis in his view of the nature of matter: matter is passive, which is to say that it has no inherent activity. Insofar as matter is in motion, that motion has an external cause, because when God created the world, He endowed it with a certain amount of motion. This fixed amount is subject to a principle of conservation and never decreases or increases. For a long time, historians saw this as the beginning of the new, modern science. Consider, for example, the title of Dijksterhuis's great historical work *The Mechanization of the World Picture*. But today, it is thought that Descartes was fairly isolated in this view. Most seventeenth-century natural philosophers of a mechanistic bent were much less stringent than Descartes about this conservation principle, and by the end of the century, it was seen as utterly outdated.

The metaphysical basis of Descartes's mechanical philosophy had a second aspect, which was endorsed by more of his contemporaries: the analogy between microscopic and macroscopic reality, which Descartes borrowed from the Dutch natural philosopher Isaac Beeckman.[42] Processes of change in the visible world, at the macroscopic level, can be interpreted as the effects of mechanical processes at the microscopic level—that is, at the level of atoms.[43] This is an important connection between the mechanistic view and atomism, but it is also important to keep in mind that there was no consensus whatsoever about the nature of atoms. Atomism could therefore engage in multiple alliances—in particular, it sometimes went together with vitalism, which held that matter was imbued with its own inherent activity.

Another aspect of Descartes's mechanical philosophy was his belief that, at the microscopic level, the universe was composed of fluid masses, currents of particles. This is the background to his renowned vortex theory. Fluid masses, whether static or in motion, could be used to clarify an array of mechanical operations. Hydrostatics and hydrodynamics thus formed the foundations of Descartes's physiology and physics.

Descartes did almost no work in the field of chemistry, but the Cartesian program, of course, entailed that chemical processes had to be reducible to mechanical processes at the microscopic level. As a result, there was a great deal of speculation about the shape of atoms, in which various ideas drawn from Lucretius were adapted to fit into the mechanistic view. Descartes explained why oil adheres to other substances more easily than water by reference to the shape of the oil particles, which he said must have very long protuberances that make them cling to things more easily.

Most seventeenth-century natural philosophers, however, were not in the least convinced that mechanistic explanation was the only kind that was valid in chemistry. Most of them, following Robert Boyle's lead, assumed that a combination of mechanistic and vitalistic principles were at work in patterns predetermined by God. In this approach, the chance nature of the processes described by Lucretius did not have troubling implications (such as the nonexistence of God).[44]

Many elements of the mechanical worldview can be explained by analogy to machines. But this is not the whole story. Dijksterhuis has noted that some meanings of the term "mechanical" have more to do with mechanics as a branch of physics. Mechanics in this sense involves a much more abstract notion of causality than the operation of specific machines. Given the potential for confusion, Dijksterhuis felt it was regrettable that the term "kinetics" had not come into wider use.[45]

Two reservations should be made, however. One is that the concept of mechanism derives its meaning from the program that it entails for natural sciences other than kinetics. It is through its application to these other sciences that the metaphorical character of mechanism becomes apparent.

The other reservation is that when Descartes worked on specific applications of his mechanical philosophy, the underlying model was not usually kinetics, but fluid mechanics. One example is his model of light as a current of fluid in *Le monde* (*The World*, 1634), in which he claimed that this approach provided a physical explanation for his geometric treatment of light.

Light is one of the three types of matter identified by Descartes; he calls it the *materia prima*. In his system, this *materia prima* is the sole constituent of the sun and the stars. The *materia secunda*, later known as ether, is the substance that fills the spaces between planets. This is a very subtle material, though not as subtle as light. The third material is ordinary earthly matter. These types of matter do not differ "in quality," but only in size.[46] Particles of *materia secunda* can penetrate earthly matter just as easily, Descartes wrote, as lead shot pours through a net full of apples. Descartes described the propagation of light as analogous to the propagation of pressure in a column of liquid: motion on one side of the column immediately leads to motion on the other side.[47]

Descartes's examples seem very concrete, but appearances are deceiving. They are analogies, intended to make the structure of the world comprehensible on the most general level possible, rather than empirical examples like Bacon's. "Not wishing to overlook any of that which is most general on this Earth," Descartes wrote.[48] When presenting his physical theories, Descartes constantly appealed to the imagination of the reader to stay with him as he constructed a new world: "Permit your thoughts to leave this World for a while and watch as I bring another, wholly new one into being before you in the realms of imagination."[49] Descartes then went on to say that such a world not only could have been created by God, but also corresponded very precisely to our own.

Descartes made a similar appeal to the imagination in presenting his vortex theory of the solar system. This theory was first published in *Principia Philosophiae* in 1644 (the French translation, *Principes de la philosophie*, appeared in 1647), but Descartes had suggested it earlier, in *Le monde*. A vortex is a whirl of matter around a central point. Descartes saw the solar system as a whirl of *materia secunda* particles around the sun, dragging the planets along as they revolved. He evoked the image of eddies in a river: the water moves in circles and objects floating in the water circle along with it. These circles are never perfect, as one can easily observe. "One can just as easily imagine," Descartes wrote, "that all the same things are happening to the planets, and that is all that is required to explain their behavior."[50] And just as in a river a large whirlpool may have a smaller eddy nearby, the moon has its own smaller vortex around the earth.

To complete his picture of the solar system, Descartes needed to make additional assumptions, such as differences in the density of the *materia secunda* that made it possible for each planet to remain in a stable orbit, and still other factors to explain the velocity of each planet.[51] Numerous times, he invoked the "naturalness" of circular motion.[52] However, he did not quantify the many aspects of vortex theory, a step that would have made it possible to investigate the theory's consistency in greater depth. This work was carried out by later Cartesians such as Christiaan Huygens.

Despite the limitations of Descartes's vortex theory, its achievements were tremendous. It was consistent with strictly mechanistic principles, such as the inertia of matter—that is, the fact that matter has no intrinsic motive force. Furthermore, in a cosmos filled with something like a fluid, there is no vacuum, an implication that was important to Descartes for metaphysical reasons. In a theological sense, it was crucial that the earth "did not move," at least relative to the celestial bodies surrounding it. Descartes seems to have truly believed that this approach could reconcile Copernicanism and Catholicism.[53] In passing, we see that Descartes applied vortex theory not only to the problem of why planets remain in stable orbits around the sun, but also to that of why objects on earth fall downward.

Except for Kepler, no one else had ever drawn such a strong connection between these two issues before, although neither Kepler nor Descartes formally united them as Newton later did.

Descartes's vortex theory even offered a mechanistic explanation of magnetism, though that phenomenon was the most resistant to this treatment. Kepler had posited a force of attraction, an unacceptable assumption for a mechanical philosopher. Descartes's point of departure was the well-known phenomenon that iron filings on a piece of paper with two magnets underneath it form vortex-like patterns. He suggested that the filings are deposited in patterns of this kind by invisible currents of a special kind of *materia tertia*. The currents form because magnets have tiny pores, or channels, through which they can pass, whereas other materials do not have such channels.

Descartes also presented a mechanical alternative to Kepler's theory of gravitation. Kepler, inspired by the astrological tradition, had assumed that when a stone falls downward, the earth is exerting a force of attraction like the force that magnetically charged particles exert on each other. Kepler also saw magnetic forces at work in the pull exerted on the earth by the sun.[54]

A different theory of gravitation was developed in 1644 by the French mathematician Gilles Personne de Roberval, a professor at the Collège Royal (now called the Collège de France). Roberval explained the stability of planetary orbits by reference to a hydraulic theory somewhat like Descartes's, but for bodies falling to the earth, Roberval assumed a force of attraction that operated over short distances.[55]

In 1646, Descartes mocked this explanation in a letter to Mersenne, commenting that Roberval supposed matter was intelligent and "knew" when other matter was nearby.[56] In contrast, according to vortex theory, earthly matter was pushed to the ground by rapidly whirling *materia secunda*. This theory also explained why things remained stable on a turning globe and were not thrown off the surface. Again, the argument was based on observations of eddies and whirlpools in water. Descartes described an "experiment" (which it is unclear whether he actually performed) involving a rotating vat filled with small lead balls and larger pieces of stone. The lead balls gradually "push" the pieces of stone toward the center of the vat.[57]

In 1669, Huygens presented an experiment to the Académie Royale des Sciences that seemed to confirm Descartes's theory, but also to offer a crucial correction. He put powdered wax into a cylindrical dish filled with water, which was then rotated. The particles of wax were flung toward the outer edge of the dish, but no sooner did the motion cease than they returned to the center. Huygens concluded that this phenomenon was caused by the greater friction of the particles of wax against the bottom of the dish. This led him to draw an analogy. For an earthly body to undergo the acceleration of gravity that had been observed, the *materia secunda* had to be rotating much faster than the earth itself: seventeen times as fast, according to Huygens's calculations.[58]

Soon after Huygens's presentation, his theory was attacked by Roberval, another founding member of the Académie. Once again, Roberval confronted Huygens with the possibility that matter exerted a force of attraction. He also criticized Huygens for accepting the circular motion of the particles as natural. Roberval argued that it was merely the product of other underlying events. Huygens was easily able to dismiss the possibility of an attractive force by appealing to mechanical principles, as Descartes had before him. Roberval's second criticism was not as easy to refute, however. According to Huygens's biographer, the physicist C. D. Andriesse, Huygens could have known that Roberval was right, and in fact, he did know that. In his defense, Huygens remarked that he accepted circular motion only as a fact, and not as a principle.[59]

For a mathematical physicist of Huygens's stature, this was a highly unusual appeal to experiential fact. But it is important to recognize that circular motion had been regarded as natural since Aristotle's day by thinkers including Copernicus and, later, Galileo, Descartes, and Thomas Hobbes, although those last three also subscribed to a seemingly incompatible principle of inertia.[60] Furthermore, in defending his vortex theory, Descartes had used analogies from experience, and Huygens may have thought of himself as doing the same thing. It was later said that Huygens was motivated by loyalty to Descartes, but nothing could be further from the truth. Huygens's loyalty was to mechanical principles and their compatibility with experimental results, which admittedly did not demonstrate the truth of those principles, but did provide strong evidence for them.[61]

After weeks of deliberation, the other members of the Académie threw their support behind Huygens, affirming that vortex theory was correct. Huygens went on believing in his revised vortex theory, even after reading Isaac Newton's *Principia Mathematica*. Vortex theory also enjoyed the support of the French scientific elite a full century longer. In 1730, Jean Bernoulli derived Kepler's third law from a vortex hypothesis, a feat that Newton had declared impossible. The Académie rewarded him with a prize.[62]

Newton's *Principia* and the End of the Traditional Mechanical View

In 1687, Isaac Newton's *Principia Mathematica* was published. Its most striking contribution was its mathematical derivation of the law of gravitation. Newton defined gravitation as the mutual force of attraction between two bodies, which is proportional to the product of their masses and inversely proportional to the square of the distance between them. The law applies both to two celestial bodies, such as the sun and the Earth, and to objects on Earth's surface.

Newton's derivation of the law of gravitation was immediately accepted in England, but on the European continent it was highly controversial. Huygens and Leibniz rejected it, as did Burchard de Volder and his students. Others ignored Newton's work.[63] Mechanists reacted much the same way as Descartes had to Kepler's theory, complaining that Newton's force of gravitational attraction was occult and inadmissible, in that it involved action at a distance. According to the mechanical worldview, for one body to have a causal effect on another, the two bodies had to come into contact, either directly or through a medium. Newton's law of gravitation clearly violated this principle.

Gravitation truly is *actio in distans*, action at a distance. Newton defended his theory against the mechanists' criticisms in two ways: in part by declaring his innocence, and in part by acknowledging that a mechanistic explanation would be more satisfactory. There was a third possible response that he never gave: namely, admitting that when he designed his theory of gravitation, an occult *actio in distans* was what he had envisioned all along. Even so, historians are becoming more and more convinced that this is what actually happened.

When Newton presented his derivation of the law of gravitation, he took an unusual approach.[64] He started with three laws of motion that he himself had formulated (now known as Newton's laws), as well as Kepler's second and third laws, relating to planetary orbits (which Newton regarded as mere observational regularities). From these premises, you can mathematically derive Kepler's first law (which describes the elliptical form of the planetary orbits) and the law of gravitation. This derivation is *deductive* in the mathematical sense. In a philosophical sense, however, it is *inductive*, since Newton regarded Kepler's second and third laws as "phenomena," that is, empirical generalizations from observation.[65]

Newton could also have proceeded in the opposite direction. If you start from Newton's three laws of motion plus his law of gravitation, you can derive Kepler's three laws. This derivation is also deductive, and this mode of presentation would have had a more traditional deductive structure, in the sense that observational regularities would be derived from theoretical laws of physics. Those theoretical laws would then have a hypothetical status, however. Newton had two objections to this method. First, it would be open to the objection that other theoretical principles might also produce the same observational results (Kepler's laws). Second—and this was the critical point in the debate—it would assume the truth of the law of gravitation and therefore of action at a distance.

Fortunately for Newton, the mathematical derivation worked either way.[66] This made it possible for him to maintain that he had not made any prior assumptions about the physical nature of the phenomenon of gravitation. In Newton's words, "Hypotheses non fingo" (I do not fabricate hypotheses).[67] Newton labored to create the impression that the law of gravitation had simply emerged from his calculations through no fault of his own.

One might defend Newton by saying that the non-necessity of making assumptions about the structure of physical reality seems to be inherent to the mathematical method itself. Later generations of physicists were eager to interpret Newton's writings in this way, especially in the first half of the twentieth century. But this perspective only became plausible after nineteenth-century physicists abandoned, out of necessity, the assumptions about physical reality that had seemed intrinsic to their mathematical models, a rather painful process that we will return to later. In contrast, the mathematicians who came before Newton, from Copernicus to Galileo and after, had always insisted on their right to interpret their mathematical manipulations of reality in concrete terms.

Newton was consistent to the extent that in developing the general concept of force used in his laws of motion, he remained noncommittal about what a "force" actually is. Descartes had made no reference to forces in his laws of motion, limiting himself to spatial extension and motion as the only two relevant properties of matter. Undeniably, he went on to refer informally to the "force of a moving body," which may, for example, be expressed when it collides with another body (a concept now referred to as momentum). In the context of his mechanistic beliefs, however, Descartes's main concern was to make certain that force could not be interpreted as an inherent property of matter.[68]

Newton defined force as "action exerted upon a body, in order to change its state, either of rest or of uniform motion in a right line."[69] He then formulated his well-known second law, which is now familiar to us in the form $F = ma$, or, "Force is equal to the product of mass and acceleration." Newton was, of course, aware that forces that meet his definition may have various physical causes: muscular exertion, pressure, centrifugal force, and so on. But he restricted his attention to force in its logical guise—that is, force as it could be defined solely on the basis of its observable effects. "For I here design only to give a mathematical notion of those forces, without considering their physical causes or seats." It was emphatically *not* his intention to "attribute forces, in a true and physical sense, to certain centres . . . when at any time I happen to speak of centres . . . as endued with attractive powers."[70]

Newton had also covered himself in advance. In 1692, he had written to one of his supporters, Richard Bentley, "You sometimes speak of gravity as essential and inherent to matter. Pray, do not ascribe that notion to me; for the cause of gravity is what I do not pretend to know, and therefore would take more time to consider it."[71]

Both before and after completing the *Principia*, Newton toyed with the idea of a mechanistic explanation of gravity, but he never fleshed out his wide-ranging ideas on the subject in any detail. In 1675, for instance, he suggested that the sun fed on particles of ether, which functioned as fuel. The suction that this created, he speculated, might keep the planets in place.[72] In 1717, he presented a new theory of ether, which he saw as an adequate response to his critics. He later explained gravitation by reference to the repulsive effect of particles of ether. Attraction by means of repulsion—it seemed his investigations had left him none the wiser.[73]

In his *Opticks* (1706), Newton explained his thinking in more detail, writing that particles of light have "certain powers, virtues, or forces, by which they act at a distance." He then extended this generalization to include particles involved in chemical reactions, and used it to explain capillary action (the behavior of fluids in very thin tubes). This was enough to disturb the mechanists deeply, even though Newton was talking about tiny particles and not about the sun and the earth. He specified two "active principles": gravity and whatever caused "fermentation," which included any spontaneously occurring chemical reaction, generally in the context of biological processes. And while Descartes had assumed that the purely mechanical law of conservation of impulse would keep the world in motion forever, Newton speculated that without an active principle, the universe would ultimately perish.[74] "And what that Principle is is a mystery to me," Newton wrote.[75]

He did not entirely rule out—at least, not in these passages or in his other publications—that a mechanistic explanation for the active principles might one day be discovered. But Newton also left behind other, unpublished writings. The historians Richard Westfall and Betty Jo Dobbs have determined from Newton's notes on his alchemical experiments that he did, at one point, become convinced of the reality of his occult *actio in distans*. They derive this conclusion in part from Newton's increase in alchemical activity in the period 1679–1680 and his conviction that chemical substances acted on one another at a distance. Consider Newton's account of the intense heat and motion generated when vitriol (sulphuric acid) and metallic powder are mixed: "the rushing together of the particles with violence could not happen unless the particles begin to approach one another before they touch one another . . ."[76] This, they argue, must have given Newton confidence that if active principles are at work between tiny particles, they are also at work between celestial bodies. The preceding quote comes from a conclusion that Newton originally intended to include in the *Principia* but later withdrew. The ideas contained in this conclusion would later be fleshed out in query 31 of the *Opticks*, but at this earlier stage Newton did not acknowledge any possibility of explaining the active principles by mechanistic means.

Newton practiced alchemy very intensively for almost thirty years. It would be no exaggeration to say that he took more thorough notes on his alchemical experiments than on his optical research.[77] But even if it is probable that these experiments gave Newton sufficient confidence to accept *actio in distans* despite its mysterious nature, other intellectual sources besides alchemy can be identified for his concept of active principles. In the generations prior to Newton's, many English natural philosophers in the mechanist camp had not been very consistent in their devotion to mechanism, and had invoked active principles regularly. It has even been suggested that Newton only took up alchemy in order to find experimental evidence that such principles existed.

In the first half of the seventeenth century, England was a melting pot of numerous intellectual traditions, and Newton was influenced by two in particular. The first was Neoplatonism, inspired by the fifteenth-century Italian scholar Marsilio Ficino. This school of thought had branched out into the field of alchemy and, in some cases, had distinctly "hermetic" features. One of its representatives was Henry More, a leading member of the Cambridge Platonists. More was primarily a theologian, and the explicit objective of his reconciliation with Cartesian mechanism and Neoplatonism was to avoid the shoals of atheism. Newton is known to have studied More's writings in depth.

Another school of thought to influence Newton was that of the French philosopher Pierre Gassendi. Entirely ensconced in the humanist tradition, Gassendi had translated the surviving Greek texts relating to Epicurus into Latin and furnished them with commentary, which was strongly colored by the ideas of Paracelsus and Petrus Severinus. Adopting from Epicurus and Lucretius the idea that atoms move by themselves, Gassendi developed a vitalistic atomism in which atoms were considered centers of force. In England, Gassendi's followers included Boyle and the physician Walter Charleton, both of whom were sympathetic to the ideas of Johannes Baptista van

Helmont. Charleton merged these elements into a vitalist perspective on matter in which material particles known as atoms were inherently active and, in a sense, alive.[78]

Newton speculated in a vitalist vein on numerous occasions. In 1669, he theorized that a small fraction of the matter in the universe was imbued with activity, a kind of *flamma vitalis* (vital flame).[79] This was a well-known term from the iatrochemical tradition, often used in connection with fermentation. Newton had fermentation phenomena in mind when he wrote his *Hypothesis of Light* (1675), in which he describes a "vital aerial spirit." And in 1705, he wrote, "We cannot say that all nature is not alive."[80]

Historians of science have offered various explanations of Newton's inconsistent remarks about the physical character of forces of nature. He was constantly in search of a reasonable balance between mechanical principles and vitalist beliefs. It has been noted that he expressed himself more cautiously when his audience included theologians. This should not, however, lead us to conclude that Newton, as a scientist, was opposed to theology. On the contrary, surprisingly enough, Newton not only spent more time on alchemy than on optics and mechanics put together, but in fact spent more time writing theological tracts than on all his scientific interests put together—including alchemy. But it was precisely his theological beliefs that required him to proceed cautiously. Newton secretly subscribed to the heretical doctrine of Arianism, which denied the divinity of Christ. If discovered, he would have risked severe persecution. As a fellow of Trinity College and a professor of mathematics at Cambridge University since 1669, Newton had sworn an oath of allegiance to the Church of England. In fact, one of the conditions of his fellowship was that he become a member of the Anglican clergy within seven years. This could have been Newton's undoing, but unexpectedly, the king granted him a dispensation from this obligation. The reason for this dispensation has never become clear, but it must have come as a great relief.[81]

Richard Westfall has argued that Newton's theology was shaped by his scientific beliefs, and not the other way around. This could be true but is difficult to verify. One of Newton's strongest religious beliefs was in God's absolute power over nature through the rules that he had made. Through his experiments, Newton hoped to demonstrate the operation of these rules. The term "hypothesis," as used and understood in the seventeenth century, was both too weak and too strong for his purposes: too weak because it suggested that the hypothesis might in time be replaced by a better one, while Newton's aim was to reveal the workings of nature (more precisely, of God in nature) once and for all; too strong, because it suggested that there was an underlying causal explanation. For Newton, "force" itself was enough of an explanation, as long as it was clearly demonstrated. But if absolutely necessary, force could be explained as God's continual activity in nature, and not as the action of matter itself (as the vitalists believed) or the intercession of the World Soul (as the Neoplatonists said), no matter how much these ideas may have inspired him.[82]

A hint of Christian voluntarism can be detected here: God's rules are as they are, end of story. Considering the eclecticism of Newton's thought (his mixture of theology, mechanism, vitalism, Neoplatonism, and even Stoicism), he clearly fits perfectly into the English tradition in which the long-lived influence of Helmont counterbalanced Descartes.[83]

The historian Otto Mayr has emphasized a political image that was influential in Newton's day: namely, the balance of power. Oliver Cromwell and the poet John Milton had embraced the balance of power as an anti-absolutist principle during Cromwell's republic, and the subsequent stadtholder-king William III had made this principle central to his foreign policies. Even during Newton's lifetime, his followers made use, on several occasions, of the image of a balance between attraction and repulsion that held the universe together.[84] It is this, more than Newton's unimpeachable mathematical proofs, that explains why English natural philosophers without exception embraced the *Principia* immediately, shedding no tears for classical (that is, Cartesian) notions of mechanism. They revered Newton as the designer of what would later be called classical mechanics.

In a fairly typical historical pattern, Newton's followers developed a more coherent natural philosophy than their master ever had. The leading members of this group, Richard Bentley and Samuel Clarke, were chiefly active as Anglican theologians and clergymen. The two men used the Boyle Lectures, which were established in 1692 and funded by an endowment from Robert Boyle, as a means of spreading Newtonianism throughout England. Bentley, the first Boyle lecturer, gave talks in the style of sermons at the London church of St. Mary-le-Bow. Bentley's and Clarke's activities enjoyed Newton's wholehearted approval.

In a lingering dispute between Newton and Gottfried Wilhelm Leibniz (about who should be considered the inventor of differential and integral calculus), which began around 1700, Clarke acted as Newton's defender. The debate went beyond calculus, turning to natural philosophy in 1715, when Leibniz raised objections to Newton's views on the nature of gravitation. Leibniz would not accept *actio in distans* since he associated force with living things. Leibniz regarded the quantity now known as kinetic energy (mv^2) as far more significant than Descartes's mv. To Leibniz, mv was *vis mortua* (dead force), while mv² was *vis viva* (living force). Force was a characteristic of bodies, of matter.

Clarke and the other Newtonians, however, adhered to the original mechanistic principle that matter was "brute and stupid." Gravitational force, and force in general, were manifestations of God's action in nature. Clarke's God was an absolute monarch; everything that occurred in nature was subject to His supervision and control and, where necessary, His intervention. It was the idea of divine intervention that seemed most absurd to Leibniz. From time to time, Newton's God had to wind the watch he had made (that is, the cosmos), or even repair it. Clarke, however, charged that Leibniz's vision of the universe as a perfect clock made God superfluous and therefore encouraged atheism.[85]

By raising the specter of atheism, Clarke brought a dimension to the debate that went far beyond science and theology. He argued that whoever did away with the need for God also, by implication, did away with the need for the king. And anyone who believed that the king was superfluous might just rise up and depose him one day. In other words, Clarke warned that Leibniz's views would lead to anarchy.

Was this a figure of speech? Presumably not, considering that this debate took place at a time when the English monarchy was in a rather perilous state. In 1688, the stadtholder of the Netherlands, William III (the country's military and de facto political leader, though not its formal sovereign), had overthrown the king of England and, with the support of many members of Parliament, assumed the English throne and the governorship of the Church of England. In 1714, when William III's sister-in-law Anne died without issue, the English Crown was inherited by a German duke, who happened to be the son of Leibniz's former patron in Hanover. But this son, Georg Ludwig, who became King George I, did not include Leibniz in his retinue, and in fact favored Newton, who was honored as the greatest English scientist by far. George, like his immediate predecessors, was politically affiliated with the Whig political party and the so-called Low Church—the tolerant, Protestant-oriented faction within the Church of England. Clarke and Bentley belonged to this faction and were both closely associated with George's court. In other words, Clarke supported the constitutional monarchy of the House of Hanover, rather than the tyrannical absolutism of James II, the last king in the Stuart dynasty. In a skillful balancing art, Clarke brought his theological and political opinions into harmony, partly by emphasizing the king's absolute freedom of will and partly by observing that the king freely chose justice rather than force.[86]

But the conservative Tories had never entirely acknowledged the legitimacy of the new dynasty, and the Low Church bishops were under considerable pressure from their conservative High Church brethren, as Clarke discovered when he was persecuted for holding the same heretical views as Newton. The affair ended with a whimper, but the chastened Whigs and Low Church faction chose to play it safe by publicly opposing radical views in politics, religion, and science. One of their main targets was the attribution of independent thought or activity to matter. The real enemy was not Leibniz, who was far away in Germany, but John Toland and other followers of Spinoza. A book published by Toland in 1696 had been denounced as an atheistic work and publicly burned in Dublin; ever since then, many people had seen Toland as a threat to the nation.[87]

Jean Desaguliers was another member of Newton's inner circle, but his work as a propagandist of Newtonianism spread a different message. He was a Huguenot, born in France, who had grown up in England. In 1713, he became the Royal Society's Curator of Experiments, a position long held by Robert Hooke before him. Desaguliers translated a textbook of Newtonian physics by the Dutch philosopher Willem Jacob 'sGravesande from Latin into English. 'sGravesande's approach greatly appealed to Desaguliers's taste.[88] Desaguliers also wrote a textbook of his own, which was fairly technical in the sense that it moved from more general Newtonian views about atoms and gravitation to practical applications such as the motion of levers, weights, and pulleys. Ultimately, this practical mathematical and physical framework encompassed the entire world. For example, 'sGravesande's definition of a river was "water that flows because of its own weight, in an open channel."[89]

Desaguliers achieved his greatest fame by giving courses in coffee houses in London and clubs in the provinces for audiences with no previous knowledge of mathematics. He vividly illustrated Newton's three laws of motion, made mincemeat of Descartes's vortex theory, and demonstrated a Newtonian universe using pulleys and levers.[90] In this "popular Newtonianism," the whole universe was described in terms of an equilibrium between attraction and repulsion. Desaguliers had no qualms about applying this philosophy to the political sphere. At the coronation of George II in 1728, he was permitted to recite an allegorical poem containing the following lines: "The limited Monarchy, whereby our Liberties, Rights, and Privileges are so well secured to us . . . makes us sensible, that Attraction is now as universal in the Political, as in the Philosophical World." He went on: "By Newton's help, 'tis evidently seen / Attraction governs all the World Machine."[91]

But much more than a state ideologist, Desaguliers was an engineer. The practical orientation of his lectures was intended to achieve practical effects, promoting the application of scientific knowledge to technical projects such as canal-building, harbor improvements, advances in transportation, and industrialization in general. Desaguliers set an example followed by many others, and by midcentury, lectures like his were being given all over England. In other words, he and his kindred spirits paved the way for their country's Industrial Revolution. Desaguliers was fascinated with early steam engines and conveyed that fascination to others. According to the historian Margaret Jacob, he can be seen as the intellectual forefather of the civil engineering profession.[92]

Desaguliers also lectured in the Netherlands in both Latin and French, usually for audiences with a stronger mathematical background than his English listeners.[93] Throughout the eighteenth century, the Dutch took an active interest in natural science, forming scientific associations such as Teylers in Haarlem and a society of learned women in Middelburg with forty members.[94]

Even so, the practical application of Newtonianism never really caught on in the Netherlands. The consequences are well known: the Netherlands industrialized much later than England, and later than present-day Belgium, which was then under Austrian rule. Was there an "ideological" factor at play, rather than simple nostalgia for the commercial supremacy of the Dutch Republic in its Golden Age, the seventeenth century? Jacob has suggested that in the Netherlands, the use of new technology in industrialization was associated with the ideas of the Patriots, supporters of popular rule who were critical of the Dutch stadtholder and absolutism. A Dutch importer of English steam engines, J. van Liender, wrote to James Watt in 1788 that his machines were objects of loathing because they were linked to the Patriot movement.[95]

Neo-Mechanism: Field Theories

The eighteenth-century history of physics is typically seen as the onward march of Newtonian physics and, more specifically, Newton's theory of gravitation, which won more and more support, even among the French. In 1735 and 1736, the astronomers Pierre Bouguer and Pierre-Louis Moreau de Maupertuis made measurements in Peru and Lapland, respectively, which showed that Earth is slightly flattened at the poles. This phenomenon (though not the exact degree of flattening) was consistent with Newton's predictions and not with the views of Cartesian physicists.[96] The mathematician Alexis Clairaut predicted the return of Halley's comet (which takes approximately seventy-five years to complete its orbit around the sun) one month before it reappeared in 1759. The *philosophe* Voltaire also became an influential promoter of Newton's ideas in France, and the mathematician

Jean d'Alembert later took up the cause, accusing the remaining supporters of Descartes's theories of forming a sect. Finally—and this brings us almost to the end of the century—astronomer Pierre-Simon Laplace dealt with a major difficulty in the Newtonian picture of the solar system. Newton had known that planetary orbits can display slight eccentricities, and he imagined that God was intervening to keep them in place. As we saw above, this suggestion had been mocked by Leibniz. Laplace demonstrated that these slight eccentricities ultimately cancel each other out, so that the solar system is stable in the long term. Strange irregularities in the orbits of Jupiter and Saturn, for instance, proved to be part of a 929-year cycle. Likewise, the three moons of Jupiter were found to keep each other in perpetual balance.[97]

The only task remaining to eighteenth-century thinkers was to transform Newton into an Enlightenment figure. In this new age, God was no longer necessary (a notion that would in fact have horrified Newton). Halley's comet would keep revolving around the sun and not, as Newton had believed, plunge into it, causing the end of the world. Newton had seen this prediction of his as empirical confirmation of his analysis of the biblical prophecies in the book of Daniel and the Revelations of St. John.[98] But aside from this typical seventeenth-century Christian "superstition," Newtonian physics was a monumental scientific achievement that stood largely unaltered, at least until Einstein developed his theory of special relativity in 1905. Or was it?

If we examine the late eighteenth-century formulation of Newtonian gravitational theory in more detail, we find, beneath the superficial picture that it presents to us, a very different type of mathematics than Newton had ever employed. This was more than a shift from one abstract method of calculation to another. The new mathematics had developed out of fluid dynamics, a field that had not yet existed during Newton's life.

Fluid dynamics was developed primarily by three Swiss mathematicians: the aforementioned Johann (or Jean) Bernoulli, his son Daniel Bernoulli, born in Groningen while his father was a professor there, and Leonhard Euler, a student of Johann Bernoulli's. None of the three was deeply committed to Newton, either personally or intellectually. Jean had assumed a somewhat dubious intermediary role in the dispute between Newton and Leibniz about the discovery of calculus.[99] At heart, he remained a Cartesian all his life, as shown by his continuing efforts to develop a vortex theory of gravitation.[100]

Daniel Bernoulli's *Hydrodynamica* dates from 1733. He took a very different approach to fluid mechanics than had Newton, who had seen fluids as a collection of separate moving parts. Daniel Bernoulli saw fluid as a continuous medium, and this assumption of continuity made his ideas about causal mechanisms in matter (like the ideas of the other Swiss scholars) much more closely related to Leibniz's than to Newton's. A few years later, Johann formulated general differential equations that employed Daniel's insights.

For many years, Daniel Bernoulli assumed the existence of a vortex-type ether, but unlike earlier scientists, he never sought to use this assumption to explain planetary orbits. Intellectually, he probably had a greater affinity with Newton than did Euler, who rejected action at a distance (though he accepted many other aspects of Newton's legacy). In 1740, Daniel made the leap to accepting Newtonianism almost without reservation.[101]

Euler did not follow suit. Daniel Bernoulli, d'Alembert, and the other French Newtonians had essentially given up on finding an underlying causal explanation of gravitation, and they derided the ongoing search for such an explanation as mere "metaphysics." Euler, however, continued searching for a mechanistic account. In 1760, Euler posited that there were two camps of physicists: the attractionists (a group that included all English physicists) and the impulsionists, who explained gravitational effects as the result of bombardment by particles of ether. Both camps, Euler believed, could agree that matter behaves *as if* a gravitational force is at work.[102] Euler did not win over any other scientists to his point of view, however.

Still, Euler was influential in the long run because of his mathematical work. In 1750—with the help of the differential and integral calculus, which he developed further (working with Leibniz's version rather than Newton's)—Euler reorganized the entire framework of Newton's mechanics so fundamentally that Newton's laws, as we know them today, were in fact formulated by Euler.[103]

Finally, in 1750, Euler took an interest in fluid mechanics, taking the results achieved by the Bernoullis and greatly simplifying them. He described a fluid in motion in the most abstract possible terms, as a "field": a space in which each point is characterized by a quantity that is a function of its space-time coordinates. The concrete meaning of these quantities depends on the specific physical theory to which one subscribes; in essence, Euler allowed individual researchers to "fill this in" as they chose.[104]

One of the striking features of Euler's work is that he did not perform a single experiment to obtain his results. Instead, he developed the "idea" of a fluid in purely conceptual terms by making the fewest possible physical assumptions.[105] Then, in 1766, he applied his theory of fluid mechanics to the problem of the sun, Earth, and the moon (regarded as a single system) in relation to the terrestrial tides. (Newton had not been able to deal with interactions involving more than two bodies.) Later applications of Euler's theory included the turbine and the propulsion of ships and aircraft, areas that were of course well suited to an experimental approach.

But even after gravitation was furnished with a new, metaphorical basis—the image of a fluid in motion—there was an important difference between a gravitational field and the "field" formed by an actual moving fluid. In the latter case, while we can infer the existence of a current from the behavior of reeds in the water, there are other ways of determining the existence of the water, other criteria for its reality. A gravitational field, in contrast, has no detectable features other than its effects on masses located within the field. It is solely through these effects that we can ascertain its existence. Throughout the nineteenth century, this situation remained a thorn in the side of many scientists, such as Michael Faraday, a gifted experimenter and one of the discoverers of the connection between electricity and magnetism.[106]

But it was another British scientist, William Thomson, who profoundly influenced the questions that physicists would investigate for the rest of the nineteenth century. Thomson is best known by another name. In 1891, late in his life, he was given the title of Baron in recognition of his contributions to science and industry, and since then, he has been known as Lord Kelvin. But it was half a century earlier, at the age of twenty-two, that Thomson became a professor of natural philosophy (in other words, physics) in Glasgow. In 1841 and 1845, he showed that there was a mathematical analogy between heat flow, electrostatic fields, and Faraday's electrical and magnetic lines of force. In 1856, influenced by Faraday's discovery, he proposed a connection between the medium of light, the ether, and Faraday's lines. This strengthened his conviction that there must be some underlying physical reality at work.[107] A few years later, he and one of his colleagues set to work on a textbook, the *Treatise on Natural Philosophy*, which was intended to replace Newton's *Principia*. This book, which took "energy" rather than Newton's idea of "force" as its fundamental concept, did indeed lay out an influential program of research.[108]

Thomson had a strong tendency to think in analogies. The steam engine was a major source of inspiration for him, especially in its industrial uses, as were observations of everyday phenomena such as the lines made by milk as it mixed with his coffee in a Paris hotel. From the time he published the *Treatise*, however, "fluid" became the master analogy, used to address a host of problems, including gravitation. Thomson also made increasing reference to Euler's work. Later generations of physicists complained that Thomson had been "broad, but not deep." But Thomson's axiom was: imaginable = measurable = practical.[109]

The British physicist James Clerk Maxwell was more comfortable than

Thomson with abstraction. He built on Faraday's work and Thomson's 1856 proposals by means of mechanical analogies and models, which culminated in an electromagnetic field theory. This theory rose to international prominence in the late nineteenth century, in part through the work of the Dutch scientist Hendrik Lorentz. But Lorentz had no interest in providing mechanistic foundations that would reveal a physical "reality" underlying the whole range of phenomena at issue. Still, until the early twentieth century, such a model remained the fond wish of some scientists, such as the mathematician Carl Adam Bjerknes.[110]

Newton's mechanics was the crowning achievement of the Scientific Revolution. Of course, the Scientific Revolution is a construct developed in a later period, but it is certainly true that Newton's *Principia* made a strong impression on his contemporaries, whether they were with him or against him. Yet ironically enough, because gravitation was a form of action at a distance, it lacked "certainty." It had to be accepted as a phenomenon, but no necessary cause could be identified. *Scientia* thus lost ground to *opinio*, the domain of uncertain knowledge.[111] But is this not intrinsic to the analogical style of science?

In scientific practice from the seventeenth century onward, the analogical mode of thought served as a cover for the barely concealed tension between an "as-if" mentality ("it's just a hypothesis") and the certainty that other thinkers, particularly Cartesians, found in mechanical analogies. Mechanism was not the only form of analogical thinking, either during or after the Scientific Revolution. Atomism, in its diverse forms, was another. Gassendi, Descartes, Newton, and Leibniz were all atomists, but each in his own way. For generations, historians of science saw in mechanism and atomism the essential components of the Scientific Revolution's new worldview. In this chapter, however, we have remained as neutral as possible about the truth of that worldview. It was not the worldview as such that interested us, but the way in which it was applied to particular scientific issues to yield new insights.

For the past two hundred years, analogy has played an ever-larger role in scientific thought, a tendency that has sometimes displeased more doctrinaire researchers in the deductive or experimental tradition. In the eighteenth century, Albrecht von Haller complained that analogy had become a substitute for experiment.[112] Around 1900, philosophers of science sought to expunge metaphor from scientific practice, and even from language in general. It hardly need be said that they were unsuccessful.

In the first half of the nineteenth century, a variety of French and English physicists and biologists embraced the use of models as a scientific method. They saw this as a form of induction; in other words, they probably believed that empirical research into one phenomenon can suggest an analogy to another, more familiar phenomenon, an inductive step. But they undoubtedly saw the next stage, working out the details of the hypothesis, as deductive in nature.[113]

It should be noted that the deductive style was never entirely subsumed within the hypothetical-analogical style. Jean-Baptiste Fourier is one example of a physicist who adopted a purely deductive style, illustrated in his *Theorie analytique de la chaleur* (1822). He presented his theory of heat as entirely autonomous, with no conceptual ties to mechanics. Most of Fourier's contemporaries took a different view, however. The difference between them is illustrated by their diametrically opposed ideas about the status of mathematics. For scientists like Poisson and Hamilton, mathematics was no more than a blind set of tools, a symbolic language, while Fourier saw in mathematics the structure of nature itself.[114]

Technology was always the primary source of inspiration for the hypothetical-analogical style. By the mid-twentieth century, one of the cornerstones of control theory, cybernetics, played a foundational role in countless sciences, from organizational studies to ecology to human physiology. The complex structure of ecosystems was described using the model of a self-regulating mechanism. By expressing this model mathematically, scientists could simulate the behavior of ecosystems on a computer. Later, however, mechanical models in ecology lost a great deal of their appeal.[115]

Soon after 1953, a cybernetic view of the cell took shape. This involved all sorts of terms and images derived from natural and artificial language, such as "genetic code," "information," "computer program," "copying," and "reading" (in the synthesis of proteins, RNA molecules are "read"). These somewhat immaterial metaphors supplanted previously popular images such as the "lock and key," which referred specifically to the three-dimensional structure of large proteins. Until 1953, most scientists believed that proteins were the carriers of inherited traits, even though there had been strong evidence (or so it is claimed) since 1944 that DNA molecules were responsible. As late as 1947, Jacques Monod at the Institut Pasteur in Paris described the cell as an ecological system in which numerous specially shaped protein molecules interacted with one another. Seen from this perspective, life was "complex, fluid, and contingent." But when Francis Crick and James Watson clarified the structure of DNA in 1953, Monod's view almost instantaneously became old-fashioned and obsolete. The DNA molecule's double helix allowed it to be replicated simply and efficiently, a fact which was quickly apparent to the scientific community. Once DNA was assigned a central role, the way was clear for a new image: the linear transmission of information from DNA to RNA to protein. This image squared nicely with a principle that had been formulated in 1940, the "one gene-one enzyme hypothesis."[116] Then, in 1958, François Jacob and Jacques Monod presented a cybernetic model in which the DNA sequence that codes for a particular protein can be "turned on and off" by regulatory mechanisms in the cell (mechanisms which, in this vision, are ultimately controlled by DNA). The concerted action of coding DNA (which carries genetic information) and regulatory DNA was to have been the theme for a book that Monod never completed: *Enzyme Cybernetics*.[117]

The metaphor of the "genetic code" became enormously successful, much more so than enzyme cybernetics. What would we be able to read in the Book of Nature once the code was cracked? A series of commands: under the control of genes, life would unfold as if by command. It was a language that was in step with the Cold War.

Cybernetics is now seen as too crude a regulatory mechanism for use in cell biology, and the term "code," too, is used with much greater care. Comparisons between cells and computers have remained fairly popular, partly owing to the continued development of computer modeling, which has led to the emergence of entire new fields of research such as computational physics and computational chemistry.[118] But many researchers in molecular biology now view computing analogies with extreme suspicion. The one gene–one enzyme hypothesis has been discarded. In a more recently developed view, cellular processes control DNA rather than vice versa. DNA is now seen as the stored memory of the cell, to be consulted when necessary. In this view, RNA plays a more central role. But it should be kept in mind that RNA is not a single, large molecule but rather a collection of smaller molecules.[119] We are thus witnessing a return to a more "ecological" or "social" view of the unity of the cell.

Many other fields of science have also made use of metaphors drawn from social life. Modern conceptions of the workings of the immune system, which protects the body against intruders, are reminiscent of the Harveyan conception of blood as a free republic of citizens. This immunological metaphor focuses on the improvised cooperation and spontaneous coordination between different types of white blood cells.[120]

One very special case is the theory of quantum mechanics formulated in the 1920s, with its principle of indeterminacy. The classical atom behaved as if it were a familiar particle, subject to the principles of mechanics. But it is not clear whether the new fundamental particles of quantum physics are "particles" at all, since they have features such as continuous, closed orbits. In 1924, Niels Bohr and Hendrik Kramers published an article claiming that the law of conservation of energy did not apply to interactions between these particles. Apparently, the mechanistic relationship between cause and effect no longer held sway. The historian Paul Forman has made a convincing argument that this form of noncausal thinking was inspired by cultural views that were on the rise during the Weimar Republic, especially the very popular book *Der Untergang des Abendlandes* (*The Decline of the West*) by Oswald Spengler.[121] According to Spengler, thinking in terms of cause and effect leads to the perspective of death. Those who instead wished to "experience the world as becoming" or "lift the stiff mask of causality" could learn to see fate as the meaning of history: "The idea of fate presupposes life experience instead of scientific experience, the force of vision instead of calculation, depth instead of spirit."[122] Spengler was the apostle of an outlook on life in which as little heed as possible was paid to all sorts of so-called necessities. Will power and the ability to act were more important. Prominent physicists such as Max Planck were, by their own account, drawn to this outlook.

Metaphors and analogies derive their plausibility from a kind of intellectual appeal, which is more often than not influenced in part by considerations of "social" desirability. There are no fixed rules for the use of metaphors in science, nor are there any discernible, enduring patterns in their use. People who believed in a hierarchy of the natural sciences, with mathematical physics at the summit, followed by chemistry, and then biology, and finally psychology and the social sciences, have suggested that ideas and concepts might "trickle down" from the top to the bottom. But historians of science have seen that physics, too, has adopted ideas and metaphors from other, "lesser" domains of scientific knowledge.

8

The Taxonomic Style

Thee, too, Lucerne, the crumbling furrows then
Receive, and millet's annual care returns.
—Virgil

Similarity is an institution.
—Mary Douglas

OF ALL THE STYLES of science, the taxonomic is the least respected. A taxonomy is an arrangement of facts, things, or entities on the basis of a comparative method. It brings order, but only a provisional order, because it lacks a prior theoretical basis—or at least, this has been the dominant belief for the past hundred years. If any foundation is available for a taxonomic system, then according to the usual line of argument, it is not a product of the taxonomy itself; in chemistry, for instance, the periodic table of elements is a system that we could not truly "understand" until the advent of modern atomic theory. Are taxonomies chiefly a practical tool for dealing with the world?

In a classic 1903 anthropological study of the "primitive" mind, the French sociologists Emile Durkheim and Marcel Mauss discussed the speculative nature of the classification methods used by so-called primitive peoples. In those categories, they discerned systems of conceptual relationships among people and things, a kind of natural philosophy analogous to scientific thought.[1] Since then, the investigation of ways of classifying the world has been a widely used method in cultural anthropology.[2]

Later scholars, however, have questioned the sharp distinction that Durkheim and Mauss claimed to have found between theoretical knowledge and practical utility.³ If we look at Western societies through an anthropological lens, we find taxonomies everywhere. Companies use taxonomies to organize their employees, social security systems put labels on applicants for benefits, and hospitals categorize patients and causes of death. Mortgage lenders and supermarkets group their customers by income, or at least by postal code. If you consult the annual reports of the British Office for National Statistics, you will find many tables subjecting everything under the sun to some form of classification.⁴ In all of these cases, the classification systems reflect both practical concerns and theoretical assumptions about the nature of certain populations.⁵ Unfortunately, the relationship between styles of science and "primitive" thinking has not been examined as thoroughly with regard to styles other than the taxonomic.⁶

If we focus on the Western tradition, it would be shortsighted to consider the taxonomic style somehow inferior to other scientific styles. The periodic system, for instance, can boast significant achievements. Long before atomic theory was developed, the periodic table helped scientists to make predictions about the existence of previously unknown elements, thanks to the "logic" of its unfilled cells. This system was developed by Dmitri Mendeleev in 1869 to do what all taxonomies do—namely, bring order to a chaos of facts, in this case the properties and potential uses of the chemical elements. Mendeleev had an educational objective in mind: making the great mass of factual knowledge about chemistry manageable for his students. He ignored his contemporaries' theories about the atom, considering them irrelevant to his purpose.⁷

Elementary particle physics has been jokingly described as a menagerie because it postulates such a large number of particles—quarks, muons, electrons, and positrons. Physicists hope and expect they will one day arrive at a deductive theory allowing them to reduce this number. Meanwhile, they have developed a supposedly provisional "taxonomy" of particles, arranged according to their properties (such as the presence or absence of spin or charm). Just like the periodic system, the classification proposed by Murray Gell-Mann and Yuval Ne'eman in the early 1960s predicted missing particles that were later found to exist.⁸

Yet in the history of science, taxonomy has not always been a step toward a deductive system; the taxonomies of natural history, systems for organizing plants and animals, were ultimately provided with a foundation by the theory of evolution. This was a new type of theory, which did not become a full-fledged style of science with its own distinct status until the nineteenth century. In the eighteenth century, no one suspected the existence of the theory of evolution or felt the need for a foundational theory. The greatest taxonomist of the eighteenth century, the Swede Carl Linnaeus, believed that his classification system reflected, or at least approached, nothing less than God's plan for creation. Linnaeus was honored by his contemporaries and the next generation in a way that no later taxonomist ever would be—and that in itself is sufficient reason to call the eighteenth century the age of taxonomy.

This shows that a taxonomic system does not necessarily lead to what we would consider a more ambitious theory, such as atomic theory or the theory of evolution. The converse is also true; the theory of evolution, for instance, does not entail a single correct taxonomy for the animal kingdom. The fact that, in this domain, at least four competing classification systems still exist today forms an argument for the enduring autonomy of taxonomic thinking as a style of science.[9]

In other words, the taxonomic style has not disappeared from the scientific world. On the contrary, it crops up in unexpected places, such as the social sciences, where researchers often use what they call the comparative method.[10] A comparative method is a means of building a taxonomy, nothing more and nothing less. Various sciences, ranging from the earth sciences to neurological research and including chemistry, are confronted with vast numbers of measurements as a result of new techniques. To meet this challenge, scientists have to use large-scale computer programs, and to organize all the data, those programs require taxonomies.[11] The taxonomic style may have a very bright future indeed.

Collections and Emblematics

The chief purpose of any taxonomy is to cope with an abundance of data; in that respect, taxonomy is like statistics, which approached this problem in an entirely different way from the nineteenth century onward. First, however, someone must collect all that data. Taxonomic thinking was fueled by collections, particularly those assembled between the sixteenth and eighteenth centuries. You can collect just about anything. There are even collections of collections; think of a guide to British museums. Around 1650, collections started becoming more specialized, in contrast to sixteenth-century collections, which tended to be encyclopedic.[12] The botanic garden in Pisa had a gallery displaying Mexican "idols," concave and convex mirrors, miniature Flemish landscape paintings, and the "horn of a unicorn."[13] The garden itself, like those in Padua, Leiden, and Paris, took the form of an encyclopedia of plants. Bernard Paludanus, who became Enkhuizen's town physician in 1585, had a cabinet of curiosities that contained herbs, poisons, fossils, rocks, coins, and clothing.[14] This collection was not the prelude to a more "mature" form of collecting.

A 1655 book by the Flemish physician Samuel Quiccheberg can be read as a program for amassing a collection that could serve as a *theatrum sapientiae*, or theater of wisdom. In the Renaissance, both scholars and rulers saw the peacock and the chameleon as more than just representatives of poultry birds and reptiles, respectively. Each one stood at the center of a web of meanings, affinities, and similarities spanning a large portion of the cosmos and the realm of human endeavor. In short, they served as *emblems*.[15] There was no denying the importance of practicalities such as the medical properties of certain parts of the peacock. Yet the peacock, the chameleon, the hedgehog, and many other creatures also had moral meanings. To make this web of meaning visible, natural history collections had to cast a very wide net, as did books that surveyed the field. *Ornithologia*, a study of birds published by Ulisse Aldrovandi between 1599 and 1603, consisted of three massive folio volumes and included thirty-one pages about the peacock. As an encyclopedic compendium of all conceivable literary references and emblematic associations, it was unequaled.[16] Likewise, botanic gardens were not only the largest collections of plants believed possible but were also seen as re-creations of paradise, the Garden of Eden, which was said to have held all the world's plants within its walls.[17]

In the mid-seventeenth century, the emblematic significance of natural phenomena faded into the background. In 1657, the Polish physician Jan Jonston published a book in Amsterdam that seemed at first to be yet another encyclopedic survey of natural history. Yet he devoted just two pages to the peacock. All the references to emblems, proverbs, myths, and legends had vanished, and medicinal uses of the peacock were included only if they had been satisfactorily demonstrated.[18] The rest of the entry described the bird's external features, and it was just such external features of animals and plants that provided the sole foundation for the great taxonomic systems of John Ray, Joseph Pitton de Tournefort, and Carl Linnaeus.

The historian William Ashworth has hypothesized that the disappearance of emblematics was related to the discovery of New World plants and animals. Not one classical author had ever mentioned the anteater or the armadillo. It was impossible to weave a network of sympathies and antipathies around them, interpret them astrologically, or anything else of the kind. Jonston, drawing on both descriptions of North American and Brazilian plants and Aldrovandi's work, may have aimed for consistency in his entries and left out the emblematics in sheer desperation.[19]

Some uncertainties remain, however. We know that taxonomy was not seriously extended to the higher taxa (groups of organisms) until after 1650. Aldrovandi was not concerned with classification systems, apart from the usual rough-and-ready ones, such as the distinction between birds, quadrupeds, and fish. His limited attempts at classification were based on the "virtues" of plants (essentially, their medical properties).[20] His contemporaries Rembert Dodoens (or Dodonaeus). and Matthias de L'Obel (or Lobelius, or de Lobel) in contrast, did introduce systems of classification in their herbals (as botanical treatises were called), however imperfect these systems may later have seemed. It is also striking that Dodoens and L'Obel limited their work on natural history to plants, and did not write about animals or minerals, thus displaying a degree of specialization at an early date. Their descriptions of individual plants are also considerably more restrained than Aldrovandi's. For instance, the thirtieth book of Dodoens's *Cruydeboeck* (Herbal), which dealt with evergreen trees and shrubs, included an ode to the laurel because of its connection with Apollo, but did not discuss this connection in the actual entry on the laurel. It seems safe to say that, for Dodoens, the mythological aspects of this plant were mere literary frills with none of the importance Aldrovandi had attributed to them.[21]

Dodoens also made it very clear how he felt about the theory "of the Marks or Impressions of things," according to which one could, for example, wear bulbous *Scrophularia* roots around one's neck to treat lumps that were similar in appearance. He did not name the contemporaries who held this view, such as Giambattista della Porta, but he did give dozens of examples, rejecting them all as "untrue and ludicrous."[22]

In any event, the conditions for the development of taxonomic systems emerged during the Renaissance. Chief among them, if we focus on the plant kingdom, was the explosion in the number of known plants. In 1542, the German botanist Leonhard Fuchs described 500 plants. Fuchs arranged his entries in alphabetical order, but this method soon became inadequate. By 1583, Andrea Cesalpino listed 1,500 plants, and Gaspard Bauhin's books, published around 1620, included some 6,000 plants. Finally, in 1683, John Ray identified more than 18,000 species of plants.[23]

The growth in the number of known plants, and the accumulation of the expertise necessary to tell them apart, took place within certain specific practices and intellectual contexts. We will begin by examining these practices and contexts and then move on to the historical development of taxonomy. The emphasis will be on botany, but other objects of classification will also be considered where appropriate.

Collecting Plants

In June 1554, Francesco Calzolari and Ulisse Aldrovandi set off with a few other enthusiasts on an expedition to Mount Baldo, in northern Italy, near Verona. They spent a few weeks there, collecting medicinal herbs in the open fields. Calzolari was an apothecary in Verona who later set up a museum in his home that contained unusual objects relating to natural history, such as the horn of a unicorn, and stuffed animals, such as a chameleon and a bird of paradise. Aldrovandi, who later became a professor at the University of Bologna, also established a home museum, which became Italy's largest. His many treasures included a small dragon found near Florence in 1572 on the same day that Pope Gregory XIII was elected, a confluence of events that brought its owner great renown. Aldrovandi was a physician and a humanist whose contemporaries regarded him as the "present-day Aristotle" and the "modern Pliny."[24]

Though these two men had much in common, there were also great differences between them. Calzolari had not received a university education, and his Latin was rudimentary. In social terms, Aldrovandi was by far his superior, but in the field, among the plants and flowers, the roles were reversed, because Calzolari's training as an apothecary gave him the greatest practical expertise in medicinal plants. In this respect, their joint expedition marked an historic shift. As the leading apothecary of his day, Calzolari gained new recognition for his profession. This recognition also extended to what apothecaries knew; it was unprecedented for a humanist like Aldrovandi to take an interest in an apothecary's knowledge and his methods of acquiring that knowledge. Before this shift, in 1520, a Viennese physician had published a book warning ambitious young physicians about apothecaries and their foolish, deceitful ways, with a list of tips on placing orders with them. He did not advise his readers to venture into the field.[25]

Aldrovandi's trip to Mount Baldo with Calzolari was not the first time the physician had looked beyond his books. In the 1540s, he had studied with Luca Ghini, the first professor of natural history in Bologna and Pisa. Ghini is known for devising a method of drying plants and thereby making it possible to amass collections known as *herbaria*—not an earth-shaking advance, but one that presumably had great practical significance. Ghini was also the first to make fieldwork mandatory for medical students, and in Aldrovandi's student days it was not at all unusual for professors and students to go searching for plants in the countryside together.[26] The 1554 expedition was exceptional, however, and the participants looked back on it as a particularly valuable experience. Calzolari later led several other expeditions to Mount Baldo, which always included inquisitive physicians.

Aldrovandi was a multifaceted scholar who built up a collection of more than 14,500 dried plants, wrote the three-volume *Ornithologia*, and took copious notes on other topics in natural history, which were posthumously published as eight thick folios. His collection served as the focal point of his professional network: scholars came from far and wide to see it. Paludanus signed his guest book, and L'Obel and Bauhin attended his lectures and undoubtedly saw his collection.[27]

Yet above all, Aldrovandi was a humanist, meaning that he had a thorough grounding in the classics. In the field of botany, the main classical authors were Theophrastus (the disciple and successor of Aristotle at the Lyceum in Athens), Dioscorides, and Pliny the Elder. Dioscorides, a Greek author, was active around AD 50 and wrote *De Materia Medica*, a lengthy work describing some six hundred plants. Pliny, a Roman contemporary of Dioscorides, gained both fame and notoriety as the author of *Naturalis Historia*, a thirty-seven-volume collection of some twenty thousand facts about all conceivable topics. Among Renaissance humanists, opinion was divided about the significance of Pliny's work. He had supporters, but other scholars looked down on him for his poor Greek and disorderly presentation. It has been said that Pliny inspired some botanists to go out and study plants for themselves, so that they could correct errors introduced by poor copyists of Pliny's manuscripts, as well as Pliny's own mistakes.

By restoring classical works to their original form, Renaissance botanists hoped to revive the practical information they contained. Their masterwork was the reconstruction of a very special medicine known as theriac, used as an antidote to snake venom, a cure for the plague, and a treatment for many other ills. Good theriac fetched high prices.

Galen had provided a recipe for theriac that included sixty-three different herbs and the flesh of female vipers. In the Middle Ages, some of these herbs had been impossible to find, and others could not even be identified. The only theriac available was therefore of poor quality and hardly even deserving of the name. It took a great deal of effort to reconstruct this medicine.

Initially, Pier Andrea Mattioli was the greatest authority on the subject. Mattioli, the author of a critical edition of Dioscorides and a humanist of the old school, never went into the field, but had plants delivered to him. He respected Calzolari, but wrote that the director of Padua's botanic garden, who, like Calzolari, had no university education, could not tell the difference between basil and lettuce. The poor man was subsequently taunted out of office.[28]

Aldrovandi continued Mattioli's work. He had learned that many of the herbs in the recipe for theriac had not died out, though they did not grow in Italy. Thanks to his large network of correspondents, he had many of these plants sent to him. Still, not all the ingredients were available. Through careful reasoning, Aldrovandi found substitutes for a few of them. He even tested the effectiveness of these substitutes experimentally, sometimes on laboratory animals.

In 1561, Calzolari wrote to Aldrovandi that he had followed the latter's instructions for making theriac and wanted to make sure he had Aldrovandi's approval. Since the trip to Mount Baldo, their roles had reversed. Now that physicians had more experience of botanical fieldwork, they insisted on supervising the work of the apothecaries. By the end of the century, this supervisory role had been enshrined in the law of almost all the Italian states.[29]

This system ensured that almost all botanical scholars were physicians. That remained true throughout the Renaissance and until well into the eighteenth century. Tournefort and Linnaeus were physicians; the latter earned his medical doctorate from the University of Harderwijk (after just one week there) and gave lectures at the faculty of medicine in Uppsala.[30] Even in the nineteenth century, after botany became an academic subject in its own right, most of the people who took an interest in it were still physicians. In Britain, many nineteenth-century "field clubs" had physicians on their boards; this reflected the fact that botany was a compulsory course for medical students until late in the century. Botanic gardens also originated in the cultivation of plants for medical purposes; the University of Oxford Botanic Garden, founded in 1621, was used in medical education for centuries, and the Chelsea Physic Garden, founded in 1673 by the Society of Apothecaries, was, as its name implies, a garden of medicinal plants.[31]

Why were these physician-botanists willing to invest so liberally in the advancement of botanical knowledge, which doubtless benefited their profession but also took them far beyond its boundaries? In the Renaissance, botany was seen as a new science, with all the pros and cons that this entailed. Universities offered only very limited scope for pursuing the discipline, since they were for the most part very conservative institutions. The Sorbonne did what it could to prevent the Jardin du Roi from ever being established.[32] Medical faculties at a small number of universities, particularly the University of Montpellier and the new universities in Basel and Leiden, were more open to botany. Yet for the most part, the botanist-physicians had to attract interest from other parties, and this changed the context in which botany was practiced.

In 1539, Guillaume Rondelet became a professor in Montpellier, responsible for teaching anatomy and botany.[33] Rondelet did not leave anatomical dissection to the surgeons, but did it himself, following the example of Vesalius in Padua. He thus represented the two innovative branches of the medical field, and perhaps surprisingly, his colleagues rewarded him for it by appointing him Regius Professor in 1545. Only a handful of professors were granted this position (endowed by the King of France) and the generous salary that went with it. Rondelet went to Italy in 1549 and visited Ghini and Aldrovandi. After this visit, he introduced botanical field trips as part of the medical curriculum.

Rondelet owes his fame primarily to his disciples. Clusius, L'Obel, Felix Platter, and Jean Bauhin all studied with him in the 1550s. Gaspard Bauhin, Jean's brother, who was born too late to study with Rondelet, became a professor in Basel in the same two fields as Rondelet: anatomy and botany. He later became Platter's successor in an even more prestigious chair in medicine.[34]

All in all, however, opportunities at universities were very limited, and ambitious physician-botanists were forced to look elsewhere. The greatest of them all, Aldrovandi, was not the best-paid professor at the University of Bologna, despite his formidable reputation. Throughout his life, he sought a patron of the highest rank, not only for the status this would confer but also to secure funding for scientific expeditions. He was largely unsuccessful, however, unlike Galileo and a number of Aldrovandi's fellow botanists.[35]

For many physicians, a reputation as a learned botanist was one way of obtaining royal sponsorship. Mattioli managed to become the physician to Emperor Ferdinand I, and Andrea Cesalpino was the pope's personal "physician-in-ordinary." Matthias de L'Obel served as the physician to William of Orange until the latter was assassinated in Delft; L'Obel then went looking for a patron in England and became a botanist at the court of King James I. Although he had not repudiated Catholicism, Dodoens refused to become the personal physician to Philip II of Spain, but he did accept an invitation from Maximilian II. In 1582, however, he left the service of Rudolf II, a devout Catholic, to accept a professorship in Leiden.[36]

Carolus Clusius, another great botanist from the Netherlands and a good friend of Dodoens and L'Obel, was the odd man out among all these physicians, and he deserves special consideration. His career is a dramatic and adventurous illustration of how botanists defined their role as a new variety of scholar and researcher.

Clusius was born in 1526 under the name Charles de L'Ecluse, in Arras, a border town that was then part of the Habsburg Netherlands (and known as Atrecht). His father was a minor noble.[37] He began his studies at the Collegium Trilingue (Trilingual College)[38] at the University of Louvain, with which Erasmus had also been affiliated. Vesalius, too, studied there. Clusius completed his education in the humanities with Philipp Melanchthon in Wittenberg. From 1551 to 1554, he studied medicine in Montpellier, where he lived in the home of Rondelet. He left Montpellier without a medical degree, however, and in fact never obtained one, instead working solely as a botanist from that point on.[39]

Clusius did not become financially independent until 1573, when he came into his inheritance and finally obtained a high-ranking position. The German emperor Maximilian II asked Clusius to become the director of his new herb garden. When Clusius hesitated, telling Maximilian that it was beneath his aristocratic dignity to risk being mistaken for a common gardener, the emperor added his name to the book of nobles and gave him the title of prefect of the imperial medical garden.[40] He thus became a courtier equivalent in status to a natural philosopher; Galileo later held a similar position at the court of the Medici.

In the years prior to his imperial appointment, Clusius lived from hand to mouth. He held jobs as a private tutor, and in 1564 became the mentor of one of the sons of Anton Fugger, a well-known banker whom he accompanied on a business trip to Spain and Portugal. Clusius travelled the Iberian Peninsula for more than a year, bringing back some two hundred dried plants that had never been described before. He had also acquired a work by Garcia de Orta entitled *Coloquios dos simples e drogas he cousas medicinais da India* (*Conversations on the simples, drugs and materia medica of India*), published just a couple of years earlier in Goa, Portugal's colony on the Indian subcontinent, which included descriptions (in Portuguese) of plants from that region.[41] He visited England with another of his pupils.[42]

Clusius also did translation and editing of short botanical works for the Antwerp printer Christoffel Plantijn. In collaboration with the author, he translated Dodoens's *Cruydeboeck* (the 1554 edition) from Dutch into French. Plantijn presumably paid for this. After his return to the Low Countries, Clusius continued his translation work, producing Latin versions of Garcia de Orta's book and a Spanish botanical treatise by Nicolás Monardes that dealt with the New World.[43] All these books were printed by Plantijn, as was Clusius's own botanical catalogue, or flora, of Spain, which was published in 1576.

Clusius's imperial appointment brought new opportunities for botanical travel from his new home in Vienna, and he visited Silesia, Bohemia, Hungary, and the eastern Alps. The result was a flora of Austria, which was published in 1583. Part of this flora was devoted to new bulbous plants that had been brought to Austria from Turkey not long before: the iris, the crocus, the hyacinth and the tulip. In Vienna, Clusius became proficient at cultivating tulips, which he later brought with him to Leiden. He was not the first to describe the tulip, and certainly not the first to introduce the flower to Western Europe, but his collection of bulbs included the most interesting varieties, which spread quickly due to his large network.[44]

In 1592, Leiden University appointed Clusius as the prefect of the Hortus Medicus (a medicinal plant garden) and a professor of botanical medicines. It may appear as if he had come full circle, back to the field of medicine, but the truth is more complex. The position of prefect of the garden had first been offered to Paludanus because of his well-known natural history collection. Paludanus refused, however, saying his wife did not wish to leave Enkhuizen. Clusius was offered more attractive terms; while Paludanus would have worked alongside a professor of medicine and botany (Pieter Paauw), and probably feared that "alongside" would turn out to mean "under,"[45] Clusius was not confronted with any such arrangement. Between 1594 and 1598, without anyone peering over his shoulder, Clusius created Leiden's botanic garden with the help of his assistant, the apothecary Theodorus Clutius.[46]

Patronage, especially of the royal variety, played an important role in weakening the strong link between botany and medical utility. While princely benefactors were very interested in utility, in a general sense,[47] they were also attracted to the prestige that physician-botanists could bring. Clusius even managed to succeed without a medical degree, though he had fewer options to fall back on in hard times than his fellow botanists. His appointment in Leiden shows that universities could also serve as modern patrons. Apparently, for Leiden University, the prestige of a new science like botany took priority over traditional disciplinary considerations.

More "bourgeois" developments also came into play, especially for printers like Plantijn. Botanical works, which included hundreds of woodcut illustrations, were not cheap.[48] Their production was only economically feasible if they could be sold throughout Europe, and this is why they appeared in so many editions, both in Latin for scholars and in a variety of vernacular languages. Printers could save money by using the same woodcuts in each edition.[49] We should not conclude from all this that by 1600 botany was already being practiced for its own sake. Long after that year, university medical curricula remained the main context for botanical studies.[50]

Yet in a growing number of other domains, botany was managing to prove its usefulness. As a result, royal bureaucracies supervised botanical work even more closely in order to ensure continuity. This was especially true in eighteenth-century France.[51]

Botany and Agriculture

The tulip was an ornamental plant, and as the number of bourgeois households with ornamental gardens increased, it soon became economically significant. Some botanists researched the uses of plants in the household, particularly the kitchen. The spice trade was of obvious importance to the economy, and in the seventeenth century not only Orta and Monardes but also Hendrik van Reede tot Drakenstein and Georg Rumphius, both of whom worked for the Dutch East India Company, contributed to the field of colonial botany.[52]

Botanical knowledge was also used to improve agriculture. Many crops were spread from one continent to another, a few noteworthy examples being the potato (which was brought from America to Europe, thanks to Clusius),[53] the coffee plant (which went from Africa to Java, from there to Amsterdam's botanic garden, and ultimately to Suriname and Brazil), and cassava, which was taken from Brazil to Mauritius to feed the population of slaves on that originally uninhabited island.[54] One odd episode with a characteristic Renaissance flavor was the rediscovery of a crop that had been tremendously important in the ancient world but was forgotten during the Middle Ages. Lucerne (*Medicago sativa* is its present scientific name), a plant in the *Leguminosae* family, is one of the first crops ever domesticated. Compared to many similar plants, such as vetch, fenugreek, lupin, and sainfoin, it makes superior horse feed and fertilizer. The Romans grew it in rotation with grain in places where grain alone would have depleted the soil. Since ancient times, lucerne had been regarded as a powerful medicine for sick animals. It did, however, require more precise irrigation than related fodder crops, such as sainfoin and red clover (which, like lucerne, belong to the legume family). The Arabic and Spanish word for lucerne is "alfalfa," the name under which this plant became known in the United States. Today, young alfalfa sprouts are produced for human consumption.

The classical authors Dioscorides and Rutilius Palladius, who wrote about agriculture in the fourth century AD, produced accurate descriptions of lucerne under its old Latin name of *medica*. In the fourteenth century, something very peculiar took place. In translations of Palladius into Florentine dialect, *medica* became *meliga*, as part of the general consonant shift from "d" to "l" that had taken place in the Florentine dialects. Unfortunately, *meliga* was the name of a different plant, namely sorghum, a species of grain adapted to poor and dry soils. In fourteenth-century Italy, sorghum was still being cultivated, while lucerne had been forgotten.[55] It was not until Virgil's *Georgics* and other works on agriculture by classical authors became available that people began to wonder about lucerne.[56]

In 1551, Mattioli was working on his commentary on Dioscorides, and arrived at the entry on *medica*. For help in identifying this plant, he turned to Ghini, who happened to have some valuable information. About ten years earlier, Ghini had been given lucerne seeds by someone who had travelled in Spain, and he had successfully planted them in his own garden. Even so, it was difficult to identify the plant with any certainty. Ghini's plants had blue flowers, but other sources claimed that *medica*'s flowers should be yellow. Dioscorides said nothing about the color of the flowers, and his works were not illustrated. Early medieval manuscripts sometimes had beautiful illustrations in the margins, but when copying the manuscripts, scribes often reproduced the drawings without having seen the actual plants. The outcome is not hard to imagine. And what was the related plant that Dioscorides called *polygala*? Sixteenth-century botanists faced identification problems with which the classical authors could not help them, and they had to draw their own conclusions. Botanists had to distinguish between species of plants and describe them in such a way that their readers could identify them correctly. Leguminous plants such as lucerne were one of the many problematic groups about which botanists sometimes disagreed. Gaspard Bauhin sharply criticized Dodoens's approach to legumes.[57] The many vernacular names in use alongside the Latin names led to even greater confusion.

In 1694, Tournefort clarified the distinction between *Medicago* (lucerne) and *Onobrychis* (sainfoin), the latter of which was Dioscorides's *polygala*.[58] (Tournefort was writing for French speakers, but the French and English vernacular names are identical.) It was essential to name plants properly, because this is what made it possible to identify the right species for each purpose. Lucerne was superior in terms of productivity and nitrogen fixation, but required careful cultivation and regular attention. Sainfoin, in contrast, could be sown on marginal, unproductive land, eventually making it suitable for farming. Lucerne was generally kept separate from other plants, while red and white clover flourished in fields of grass or even grain, where they prevented weeds from gaining a foothold. Some species were grown for their legumes (that is, for their fruit), while others were used for hay. *Melilotus* (melilot or sweet clover) was grown for medicinal purposes, to prevent heart disease.[59] Botanists were not, of course, the only ones who collected information about these properties, but they did make such information widely accessible. They also played a role in the domestication of the above-mentioned plants and others, by taking them from the wild and growing them in their own gardens or, in some cases, in botanical or royal gardens with which they were associated.

Botany was of great economic significance, as the example of clover in seventeenth-century England demonstrates. Sainfoin, red clover, and white clover were not cultivated there until relatively late, after 1650. Sir Richard Weston, an English agricultural expert, was the driving force behind the introduction of these plants, after a period of exile in the Netherlands during which he was able to study farming techniques in Flanders and Brabant. Samuel Hartlib assisted him with this project. The great leap forward in English agricultural productivity sometimes called the Agricultural Revolution was directly related to the introduction of clover.[60]

Collections of Data

Identification and description can be regarded as an indispensable first step toward what might be called a collection of data, which in turn is a step toward a taxonomic system. Conversely, taxonomy has to point the way back to individual plants; its practical purpose is that of a *key*. With an unknown plant in one hand and a guide to the taxonomic system in the other, one should be able to determine the correct name. As we saw in the case of the leguminous species, we must identify plants so that we know whether they are medicinal or malign, useful or harmful. Pre-Linnaean botanists (from the Renaissance onward) treated botanical names as if they had intrinsic value, as if they were capsule descriptions of the essence of the plant.[61]

This was only possible if plants could be identified in a more or less reliable way as distinct species. Given the diversity of the plant kingdom, the many varieties in which plants may be found, and the influence of soil and climate on the appearance of individual plants, drawing reliable distinctions between species was far from easy. For this reason, sixteenth-century botanists established *types*, perhaps without being aware of it themselves: each type was represented by an individual plant that could serve as its typical representative.

Botanical drawing was a powerful instrument in establishing such types. Mattioli (who never went into the field) secured the services of Giorgio Liberale, a student of Ghini's who made excellent drawings from nature. Mattioli's commentary on Dioscorides included 932 high-quality printed illustrations.[62] Sixteenth-century book illustrations were almost always woodcuts, which were much cheaper than copper engravings and lasted a long time without deteriorating. Printers very often used the same woodcuts in herbals by different authors, and some remained in use for more than a century. There were also new woodcuts being produced all the time, however, and Clusius is known to have personally supervised the production of plates for his Spanish flora.[63]

In the seventeenth century, the microscope gave a fresh impetus to the identification of plants, making it possible to examine their parts, their morphologies. Strikingly, it was the mathematician Joachim Jung (also called Jungius)—also a physician and botanist—whose morphological studies distinguished among all relevant parts of plants, investigated their geometric positions relative to one another, and developed a terminology for this purpose, which made it possible to describe any new plant species with great precision. Jung developed this terminology as a step toward a systematic taxonomy. Later, Linnaeus acknowledged Jung as being one of his few "predecessors."[64]

Nehemiah Grew was a practicing physician who had studied at Cambridge and Leiden. He began his research into plant anatomy as an amateur, but later, very exceptionally, he received financial support from the Royal Society. Robert Hooke, the society's curator of experiments, had permitted Grew to use his microscope. Other microscopists included Marcello Malpighi and Rudolph Jacob Camerarius; along with Grew, they were the first to ascribe sexuality to plants. These three scholars, unlike Jung, were not primarily concerned with taxonomic issues. Instead, they were researching the workings of the reproductive system, a mechanistically inspired program with an especially strong connection to the theory of preformation, which held that the fully formed plant was present within the seed.[65]

Classifying Plants

Dodoens openly admitted that it was difficult for him to find a classification system for his *Cruydeboeck*. He decided to use Dioscorides's system rather than that of Theophrastus or Galen. Theophrastus's method was not suitable for a reference work because he compared plants in terms of their individual anatomical parts; information about a single plant was thus distributed over several different places. Dodoens judged Galen's arrangement "in the order of A, B, C" to be equally impractical, because some closely related plants were distant from one another. In contrast, Dodoens wrote, Dioscorides "had done his best to bring together similarities in form and powers." Like Dioscorides, Dodoens adopted the two criteria of outward appearance and medicinal value, but added apologetically that his collection of plants was much larger than his predecessor's, a situation that forced him to order large groups of plants alphabetically.[66] Dodoens was not as explicit about his debt to Theophrastus, from whom he had borrowed the broad distinction between herbs, undershrubs, shrubs, and trees.[67] This system of categories, which reflected the ladder of nature (*scala naturae*) by leading from simpler forms of life to more perfect ones, was in general use by botanists.

Throughout his work, Dodoens brings together similar-seeming groups of plants, but according to a slightly different criterion each time. The section, or "book," on aromatic and ornamental flowers groups roses, violets, and columbine. He made a more meaningful contribution to taxonomy by bringing together bulbous plants: the tulip, the hyacinth, and the onion. He also placed orchids in this category. In the tenth book of the *Cruydeboeck*, which is dedicated to "plants with curled flowers," Dodoens describes the parsley family. In the Latin edition, this family of plants is called *Umbelliferae*, a name it still bears today.[68] Other parts of the book, however, classify plants according to their medical effects. Useful varieties of clover and other leguminous plants were separated from related weeds, such as wild fenugreek.

Dodoens's working method has been described as intuitive. He strove for order without imposing a uniform criterion on nature that would force him to make strange exceptions. His vast erudition made it possible for him to bring similar plant species together, even in sections where he used functional criteria.

Dodoens's Dutch contemporaries used similar methods but emphasized formal resemblances much more strongly, at the expense of medical utility. In organizing his flora of Spain, Clusius was not guided by practical considerations, but considered only the shape of the flowers.[69] L'Obel's system went further than Dodoens's; while Dodoens used just one level of categories for each of his thirty books, L'Obel introduced subcategories.[70] However, functional similarities sometimes reasserted themselves in L'Obel's work; clovers, for instance, were grouped with grasses, apparently because both were used as fodder.

Bauhin's *Pinax theatri botanici* (*Illustrated Exposition of Plants*, 1623) eventually became even more influential than the work of Dodoens and L'Obel. This treatise, also based on the intuitive method, was conceived as a list of synonyms for six thousand types of plants. Bauhin's nomenclature was often, though not always, binomial, using terms from Dioscorides and other classical authors for the "genus." The "species" names were based on more recent experience, including his own. Linnaeus later adopted many of Bauhin's names.[71]

Bauhin's list was not organized alphabetically or by medicinal properties, but on the basis of formal resemblances.[72] For instance, he grouped grasses with rushes, horsetails, and irises. Like L'Obel, he was not always consistent; he had two separate groups for "vegetables" and "aromatic herbs."[73]

After Bauhin's *Pinax*, the remainder of the seventeenth century did not see much progress in system building apart from the discovery of many new species. In *Elemens de botanique* (*Elements of Botany*, 1694), Joseph Pitton de Tournefort left much of Bauhin's work unaltered, describing how you could use Bauhin's system to proceed accurately from a plant genus to an individual species. You had to find the correct genus on your own, however, and this was where Tournefort concentrated his efforts: on distinguishing between higher-level taxonomic categories of plants. Like his successor Linnaeus, Tournefort built on the work of Andrea Cesalpino. The title of Tournefort's work was modeled after Euclid's *Elements of Geometry*, the quintessential example of a deductive system.[74]

Cesalpino, a student of Luca Ghini's, was the only true Aristotelian among the Renaissance botanists. His system was unsuccessful during his lifetime, however. He followed the example of Aristotle's logic of classification, starting from what he considered the most essential parts of the plant. In his day, however, plant anatomy was still too poorly understood for this insight to be useful in practice, and it led Cesalpino to group plants together that everyone intuitively sensed were unrelated.[75]

Though there are countless practical complications, the guiding principle of Aristotle's taxonomy is simple: determine the essential attribute of a class of individuals (the *fundamentum divisionis*) and the number of variations on that attribute, and then divide the class into that number of subclasses. Do the same for each subgroup, and so forth. In theory, this yields structures of the kind familiar to us from tree diagrams or, in terms of another present-day conceptual system, subsets within sets, and still smaller subsets within those subsets. The difficulties arise from the fact that you are supposed to determine in advance, *a priori*, what the most essential criterion is for dividing any given group, while in practice you constantly have to make *ad hoc* decisions about this.[76] Even Cesalpino could not resist the temptation to focus on parts of plants that were striking and distinctive but had no clear function, rather than parts that were less conspicuous but much more essential for the plant.[77]

The above-mentioned *ad hoc* decisions are inevitable in taxonomy because individuals differ from one another in qualitative, rather than quantitative, respects. This makes complete abstraction impossible. The result is a constant tension between one's basic deductive principles and one's experience.[78]

What made Cesalpino the first botanist to apply Aristotle's logic of classification to his field? Did he draw directly on Aristotle's work, or did he have more recent models, that is, Aristotelian classification systems in fields other than botany? To my knowledge, the question has never before been posed. There were classification systems before Cesalpino's, particularly for human knowledge. The dialectician Rodolphus Agricola had developed a system of rhetorical categories, and in 1572 the anti-Aristotelian Petrus Ramus had proposed a method of analysis involving tree-like structures that was deemed useful for case-law and other subjects. The Jesuits developed rules for classifying knowledge and organizing it hierarchically; this involved the classification of scholarly activity into distinct disciplines.[79]

Joachim Jung studied medicine in Padua, and it is assumed that he was exposed to Cesalpino's ideas there. Though he did not produce a detailed classification system, he did work out a rough version with an Aristotelian logical structure, based on morphological categories of his own devising:

I. perfect plants, with both stems and roots
   A. plants with separate stems and roots
   B. plants with structures combining stems and roots
II. imperfect plants, lacking either stems or roots
   A. lacking stems, with the seeds on the backs of the leaves
   B. lacking roots
      1. with separate stems and leaves
      2. with stems only
      3. with structures combining stems and leaves[80]

The generations after Jung were plagued by the question of whether a taxonomic system based on this sort of logical structure could lead to a "natural" system for classifying plants. Whatever criteria one used, it was clear that there could be no natural system unless groups emerged naturally, of their own accord, as seemed to be the case with the *Umbelliferae* and the family of flowering plants known as the *Compositae*. In 1696, John Ray accused Tournefort of promulgating an "artificial" and "essentialist" system. Ideally, scientists would have liked to identify true essences that could form the basis for a natural system, but it was difficult to determine what those essences might be. Tournefort's system was certainly ambitious, but it was also pragmatic; a taxonomy served mainly as a key. There is no harm in a key being artificial, as long as it works.[81]

Tournefort retorted that Ray arbitrarily mixed different sorts of criteria. This criticism cannot have made much impact on Ray, who had no compunctions about working from a plant's outward appearance when necessary, rather than from meticulously defined essential aspects.[82] The historian M. M. Slaughter has drawn a connection between Ray's approach and the ideas of the philosopher John Locke. Earlier in his life, around 1864, Ray had in fact subscribed to the classification principles espoused by Cesalpino. Ray, a Puritan minister and a friend of Boyle's, did not at that point see any reason for Aristotelian essentialism about living nature. With regard to the new physics, he subscribed to corpuscular philosophy,[83] which treated the observable properties of matter as "secondary qualities" that were anything but essential.[84] Yet when Locke used corpuscular theory as the foundation for a general theory of knowledge, Ray accepted his conclusions. According to Locke, there are real essences, but we cannot learn anything about them; our descriptions of the properties of things are linguistic constructs rather than essential aspects of the things themselves. Locke sees no objection to speaking about the properties of things, but argues that it is pointless to arrange them in terms of their essences. This implies that all characteristics of plants are equally suitable, in principle, as bases for a taxonomic system. Locke's anti-essentialism also entails a skeptical attitude toward the higher taxa, which he believed existed only as constructs of the human mind.[85] That was the reason that Ray called Tournefort's system both "artificial" and "essentialist" in the same breath: to give priority to any one characteristic because of its supposedly essential nature was in fact to make an arbitrary choice. The paradox was that Ray himself could not offer any guarantees of the naturalness of his own system. He could only hope, on intuitive grounds, that he had approached the true natural order as closely as possible.

Ray had another reason for skepticism about strictly hierarchical systems: he believed in the *scala naturae* (ladder of nature, or great chain of being), the originally Aristotelian idea that all forms of life were related through small gradations of increasing perfection. According to this notion, no matter how sharp an analytic distinction one might make, somewhere in the natural world a transitional or intermediate stage could be found. From the sixteenth century to the nineteenth, all botanists subscribed to this belief, at least to some extent.[86]

Ray's system was well received not only in Britain, but also in the Netherlands. For instance, Jan Commelin, the director of the Hortus Medicus in Amsterdam (the botanic garden in the Plantage, now called the Hortus Botanicus) adopted Ray's system soon after the publication of the latter's taxonomic work *Methodus Plantarum Nova* (*New Method of Plants*) in 1682.[87]

Carl Linnaeus

In conceptual terms, the differences between Linnaeus and his predecessors Ray and Tournefort were not so great: Linnaeus was a more orthodox Aristotelian and more strongly influenced by Cesalpino and Jung.[88] Yet his system triumphed over all others thanks to its unbending consistency, its ability to absorb newly discovered species with relative ease, and above all the success of its two-part names for specific plants. Even opponents of Linnaeus's system had no choice but to use these new names.

Linnaeus's classification of the higher taxa (above the level of species and genus) was based entirely on their reproductive organs. This was a new approach. In 1694, Rudolph Camerarius, the director of the botanic garden in Tübingen, published experimental evidence that plants only produce seeds if the egg cells in the pistil have been fertilized by pollen from the stamen. Camerarius identified the pistil as the female sexual organ and the stamen as the male one. A few years earlier, Marcello Malpighi in Bologna and the English botanist Nehemiah Grew had expressed suspicions of this kind, and Grew had described the process in sexual terms.[89] Ray, who was in contact with Grew, had immediately acknowledged the existence of plant sexuality. Tournefort, in contrast, still had not accepted the idea at the time of his death in 1708. Tournefort's student Sébastien Vaillant, however, greatly contributed to the acceptance of this idea, in collaboration with Herman Boerhaave in Leiden.[90]

Describing the anatomy of plants in the same terms at that of animals was not in itself unusual. In the previous chapter, we saw that members of the Académie Royale des Sciences tried to demonstrate the circulation of the blood in plants and identify the botanical equivalent of the heart. Malpighi even subscribed to a methodological "principle of uniformity" in his research on plants and animals.[91] Tournefort thought in similar terms, calling the seed the "infant" and the flower the "nursing mother." He did not see the stamen as a sexual organ, however, but as analogous to the kidneys. Pollen, in this view, was likened to renal excretions.[92] Apparently, the sexual analogy was difficult for Tournefourt to stomach. Since the early Middle Ages, it had been assumed that plants, unlike animals, had not been involved in the biblical Fall (with the exception of the date palm, a species with separate male and female trees). Various well-known botanists refused to accept that plants had a sex life until well into the eighteenth century.[93]

It should be added that the sexuality Malpighi attributed to plants in 1675 does not correspond to later ideas about sex in plants and animals. Recall his aforementioned preformationism, the belief that a tiny, fully formed plant was present in each seed. He saw that seed as situated in the base of the pistil, which he referred to as the uterus (now called the ovary). Yet he also saw stamens as a kind of uterus and pollen as "unclean food" excreted by the body, like menstrual blood. Nehemiah Grew shared Malpighi's belief in ovarian excretion of this type, but believed that young stamens, more specifically, were responsible for this activity. He suggested that older stamens changed into testes. Grew drew his own analogy between plants and animals, specifically hermaphroditic snails. We should not imagine that Grew was proposing a miraculous transformation from a female to a male organ. His opinions were in line with the widespread belief that ovaries and testes were anatomical variants of one another.[94]

The idea that female and male sexual organs resembled one another was also present in the work of Vesalius, and is sometimes called the one-sex model. The historian Thomas Laqueur has observed that in the late seventeenth and early eighteenth centuries, the prevailing view of animal and human sexuality developed from a model in which the female was an imperfect version of the male to one in which there were two different kinds of sexual bodies. He and the historian Londa Schiebinger have argued that the idea of sex itself changed as a result. In earlier writings on sexuality, sex characteristics were invariably seen as signs of how individuals related to the cosmic principles of masculinity and femininity.[95] In the new mode of thought, sexuality was a product of the body. This was linked to a new emphasis on the physical difference between masculinity and femininity, which resulted in a two-sex model. It was no coincidence that this intellectual upheaval took place at a time when the class into which an individual was born (as a noble, a merchant, or a farmer) became less definitive of his or her identity. Instead, biology had become destiny.[96]

When Linnaeus published his "sexual system" of taxonomy, *Clavis Systematis Sexualis* (*Key to the Sexual System*, 1735), many scholars were already reasonably open to the idea of plant sexuality, though it was not a generally accepted idea by any means. Even so, Linnaeus's work caused quite a stir, mainly because of its vivid metaphors. Plants celebrated their weddings in bridal costume in the privacy of their petals until the groom's testicles opened and fertilized the ovary. Some species of plants held their weddings in public, and others took part in "clandestine marriages"; this last metaphor was not so far-fetched, since secret marriages were still possible in the first half of the eighteenth century. There was not a great deal of monogamy in the plant kingdom. Some flowers were polyandrous, and others polygamous, in which case Linnaeus spoke of concubines.[97] Numerous scholars felt that Linnaeus's imagery went too far and considered it scandalous that he had named one plant genus *Clitoria*. Still, the outcry died down after four or five years.[98] Schiebinger notes that the source of Linnaeus's metaphors was not sex itself, but sexuality and married love.

Using more or less the same method as Cesalpino, Linnaeus identified the stamen and pistil as the most essential parts of the plant. His first level of classification was based on stamens, and he divided the plant kingdom into twenty-four classes, the first of which had one stamen, the second two, and so forth, up to the eleventh, the *Dodecandria* (literally "twelve-men"; there was no class with eleven stamens). The other classes were for plants with stamens in special positions or arrangements.

Each class was then subdivided into orders (families) depending on the number of pistils, up to and including the *Dodecagynia*, plants with twelve to twenty pistils. The families with more than twenty pistils were simply called *Polygynia*. This made it easy to construct straightforward tables of plant species. In many cases, plants that resembled each other were separated, while plants of different kinds were clustered together. The great French naturalist Buffon, Linnaeus's exact contemporary and only rival of high stature, ridiculed Linnaeus for grouping the mulberry tree with the stinging nettle, the elm with the carrot, and the strawberry with the rose.[99]

Linnaeus frankly admitted that his system for classifying the higher taxa was an artificial one, but he argued that there was no alternative, and most other botanists accepted that, at least until about 1800. Linnaeus would, of course, have preferred a natural system, and he published a "fragment" of such a system, which was based largely on the outward appearance of the plants in question.[100] He also repeatedly pledged to design a natural system one day.[101] In his system for the lower taxa (genus and species), however, he believed he could approach a natural system by strictly arithmetical means, without referring to outward appearance (though he did believe you had to consider outward appearance in the initial stages, "under the table," as it were).[102]

There are two steps in the development of a taxonomic system. We have already discussed the first in detail: the process of logical subdivision, preferably on the basis of a single criterion. This is, in modern usage, the analytic part of the operation, which presupposes a deductively ordered reality. At least, that was how Cesalpino, Jung, and Tournefort saw it.

At the end of the operation, when you arrive at the bottom level of the hierarchy, you group individuals that appear similar. At this point, you are no longer interested in what divides individuals, but in what brings them together, their underlying affinities.[103] In principle, you could then trace back this entire process in reverse order, from bottom to top.

Linnaeus grouped species into genera in this relatively inductive manner. Unlike most botanists, he believed the genera were part of the "reality" of the plant kingdom as God had created it. Later, he even came to the opinion that God had created only the genera, and that the species had emerged from them later.[104] In describing sexual organs, he made use of not only the stamen and pistil, but all the outward features of the plant's reproductive organs, known collectively as the fructification. In his *Genera plantarum*, the first edition of which appeared in 1737, Linnaeus compares these outward features to letters: "These marks [of the fructification] are to us so many vegetable letters, which, if we can read, will teach us the characters of plants; they are written by the hand of God; it should be our study to read them."[105] Linnaeus distinguishes among thirty-eight parts of flowers that were involved in reproduction, including the pistil (of course), the stamen, divided into three components (the filament, anther, and pollen), the various parts of the corolla, and so forth. He rules out a number of properties of plants as criteria, considering them unstable and unreliable: color, odor, flavor, the size of the plant as a whole, the time of flowering, and the place where a plant is found.

Linnaeus describes four different aspects of the thirty-eight "letters": number, shape, proportion, and position (*numerus, figura, proportio,* and *situs*). In his opinion, these were enduring properties, unlike odor and flavor.

They remained intact in a herbarium (a collection of dried plants) and could be inferred from a drawing. Linnaeus also developed a remarkable method of verbally describing the genera. Whereas botanists such as Dodoens, Bauhin, and Tournefort had described each species in half a page or more of complete sentences, this is Linnaeus's description of the stinging-nettle:

Urtica
CAL. Perianthium tetraphyllum: foliolis subrotundis, concavis, obtusis.
COR. Petala nulla
Nectarium in centro floris, urceolatum, integrum, inferne augustius, minimum.
STAM. Filamenta quatuor, subulata, longitudine calycis, patentia, intra singulum folium calycinum singula. Antherae biloculares.[106]

It is as concise as possible and yet describes the plant comprehensively. Even without translating the Latin, we can see more or less how Linnaeus constructed these formulaic descriptions.

On the basis of his thirty-eight parts, he could describe the entire plant kingdom. These same parts were expressed in a unique way in each genus (as we have seen, each part had four distinct aspects to be described). In *Philosophia Botanica* (1751), Linnaeus calculated the total number of conceivable genera: $38 \times 4 \times 38 = 5776$. That far exceeded the number of known genera, and Linnaeus took this as proof that his descriptive system was exhaustive and adequate.[107] In any case, he had more than enough ways to fit newly discovered genera into it.

Linnaeus thought it was possible, in other words, to fit the entire plant kingdom into one huge table of permutations. Ideally, all the cells in that table would be filled, demonstrating the uninterrupted (though finite) continuity of God's creation.[108] It was a kind of enormous *scala naturae*, except that it extended in all directions, like a geographic map.[109] Linnaeus believed that his classification of genera, unlike his exclusively sexual system of classes and families, represented a truly natural order.

Linnaeus's method of classification on the basis of differences and similarities was linked to other developments. In the second half of the seventeenth century, the faculty of medicine in Leiden had developed a new form of education, clinical education, named after a new kind of hospital, the clinic. This educational model spread throughout Europe from 1720 onward, with Boerhaave as its focal point.[110]

Clinics were relatively small, with about twelve beds on average, and patients were carefully selected on the basis of their medical conditions. Earlier hospitals had tended to be much larger, with all sorts of patients jumbled together unsystematically. The philosopher Michel Foucault has called the clinic a "nosological theatre" where medical symptoms were on show. The characters in this drama were not the patients, but the diseases, each of which had its own constellation of symptoms. Students had to learn to read these symptoms correctly and thus arrive at the proper diagnosis. Only then could they predict the further course of the disease and determine the most promising treatment.[111]

In 1748, Emperor Francis I purchased a collection of minerals, stones, and fossils and had them arranged in the library of his palace in Vienna according to their "degree of affinity," in uninterrupted order of superiority. In 1778, when the collection was reorganized, the idea of the *scala naturae* faded into the background somewhat, and instead a systematic organizing principle was adopted. This made it possible to display various forms of affinity, in an arrangement reminiscent of Linnaeus's metaphor of the geographic map.[112] The art historian Debora Meijers has pointed out that around the same period, collections of paintings were reorganized along similar lines.

In the taxonomies of diseases, plants, minerals, and paintings, similarities and differences were at the same level. In biology and medical science after 1800, however, differences were seen as secondary to a deeper identity.[113]

Foucault describes these ways of seeing as fundamentally different knowledge systems, or epistemes. The first, "classical" system was widespread between 1650 and 1800, while the second is still dominant today.[114] These ways of thinking transcend the boundaries of individual sciences; Foucault shows how they influence thinking about language and economics. What is striking about the classical episteme, however, is how precisely it is expressed by Linnaeus's taxonomic method. The new taxonomic methods that emerged after 1800 are much more marginally related to the modern episteme.

Linnaeus considered a collection to be more than just a representation of nature. For him, nature itself was a collection, a museum. In his speculations about how God had created the plants and animals and placed them in the Garden of Eden, he imagined that garden as "a kind of living museum."[115]

Taxonomy and Collections

In 1735, when Linnaeus came to the Netherlands, he found many kinds of natural history collections. The Netherlands was famous for collections of this kind, and they were one of the reasons for Linnaeus's visit. He was not disappointed; in fact, he stayed for three years.[116] There were impressive "living" collections, such as the garden and menagerie of the wealthy banker George Clifford in the Haarlem area, as well as the botanic gardens in Leiden and Amsterdam, which were models for the rest of Europe. Indeed, Linnaeus devoted his first major publication to Clifford's garden. Yet he attached just as much importance to dried plants as to living ones. He showed great interest in the herbarium maintained by Johannes Burman, the new director of the Amsterdam botanic garden, which Burman was using as the basis for a flora of Ceylon.[117] The Amsterdam apothecary Albert Seba also had a renowned natural history collection, though Linnaeus felt its organization was old-fashioned.

Even so, many other collections (mostly cabinets of curiosities and the like) had been abandoned in the decades before Linnaeus came to the Netherlands. Most collections are auctioned off after their owner dies, but there were few buyers around 1715, when the great collector Nicolaes Witsen, a former burgomaster of Amsterdam, complained that there were few "curious amateurs" remaining.[118] The problem was that around 1700 natural philosophers believed in an orderly, homogeneous, and regular universe, while the collections displayed just the opposite qualities.[119]

Toward the end of the eighteenth century, it again became fashionable to collect shells, minerals, fossils, mounted animals, and dried plants. For the period from 1700 to 1790, the art historian Krzystof Pomian has managed to trace 723 Parisian collections with inventories (these collections did not consist entirely of books). At the start of the century, from 1700 to 1720, 15 percent of the collections related to natural history, a figure that increased to 39 percent by period between 1750 and 1790. The market value of the objects rose along with the social status of natural history collectors. In 1715, such collectors were still marginalized, but by midcentury, Diderot had written an article in his *Encyclopédie* praising them for their contributions to science. The collections of his day were usually taxonomies in concrete form.[120]

By this time, doctors and apothecaries were in the minority among collectors; instead, the tone was set by amateurs from the upper classes and professional naturalists. The rare and exotic, which had played such a central role in cabinets of curiosities, were gradually displaced by the relatively ordinary, which was valued because it was seen as scientific or useful.[121]

This new wave of enthusiasm about natural history was also expressed in other ways. Linnaeus himself led day trips into the field, escorting as many as three hundred men and women on each trip.[122] In Verona in the second half of the eighteenth century, new botanical expeditions to Mount Baldo were mounted, but this time, most of the participants were not doctors but amateurs from numerous walks of life, including quite a few women. A Veronese geologist led a group of women into the mountains in search of fossils. Both the initiators of these activities and the participants tended to subscribe to the new ideals of the Enlightenment, one of the most important of which was the expansion of scientific knowledge.[123]

Seba's cabinet of natural curiosities was finally auctioned off in 1752, sixteen years after its owner died at an advanced age. One of the main buyers at this auction was the unknown amateur Arnout Vosmaer. The proceeds of the auction amounted to 24,400 guilders, an enormous sum; Vosmaer paid 2,433 guilders for his acquisitions. A few years later he sold his collection to the stadtholder, William V. That was the beginning of the natural history cabinet of which Vosmaer became the director; over the years, he expanded the collection considerably. Vosmaer's scientific approach was somewhat eclectic: both the organization of the cabinet and his publications on the taxonomy of the animal kingdom were inspired by both Linnaeus and Buffon, in roughly equal measure. This placed Seba's curiosities in a new context.[124]

William V's collection was open to the public, though with many restrictions. In 1795, it was confiscated by the French and taken to Paris, where Georges Cuvier took charge of it. After Napoleon's defeat in 1813, part of the collection was returned to the Netherlands, where it formed the basis for Naturalis, the present-day museum of natural history in Leiden. Starting around 1800, museums of natural history were founded in many countries. Whether independent or affiliated with universities, they continued the work of collecting.[125]

Linneaus's taxonomic system demanded the constant addition of new species, genera, and families. Linnaeus was also interested in species that could be of practical use to his country. He was in contact with the Swedish East India Company in Gothenburg, which financed a variety of expeditions.[126] After returning to Sweden in 1738, Linnaeus never again traveled abroad, but he sent his students, whom he called his "apostles," to collect plants on other continents.

Pehr Kalm was sent to North America with instructions to find varieties of the red mulberry tree that would be suited to the Swedish climate. Linnaeus hoped that this would make it possible for Sweden to establish its own silk industry. Kalm completed his mission and brought home an additional ninety-odd American plant species.[127]

Fredrik Hasselqvist sent collections from the Middle East. He died during his travels, but Linnaeus was able to publish a flora of Palestine based on his specimens. At the invitation of the Spanish ambassador, Peter Löfling worked in Spain for two years and then went on to Spanish America, where he succumbed to fever at the age of twenty-seven. Linnaeus published Löfling's travel diary, which described many new plant species.[128]

Daniel Solander ended up in England, where he was employed by the young Sir Joseph Banks, a traveler, collector, and self-declared dilettante. In 1778, Banks became the president of the Royal Society, a position he held until his death. Solander accompanied Banks on Captain James Cook's first voyage on the Navy's HMS *Endeavour*, a former coal ship, from 1768 to 1771. This was a scientific expedition, for which King George III supplied both funding and the ship to the Royal Society. Its mission was to study the transit of Venus on the island of Tahiti in the Pacific Ocean and to carry out many other investigations in the fields of natural history, meteorology, and anthropology along the way. The voyage brought them not only to Tahiti, but also to Australia, New Zealand, and Patagonia. In scientific terms, it was a great success.[129] Unfortunately for Linnaeus, he never had a chance to see Solander's herbaria and drawings, since they belonged to Banks.

Finally, Carl Peter Thunberg had already spent time in Cape Colony (modern-day South Africa) and Ceylon when he was invited to accompany a group of Dutch merchants to Japan. Thunberg first learned Dutch at the Cape Colony and on Java; because the Japanese had granted a trade monopoly to the Dutch East India Company, he needed to pass himself off as a Dutchman. His later publications included floras of the Cape Colony and Japan, and he became Linnaeus's successor at Uppsala University.[130]

This is nowhere near a complete list of Linnaeus's far-flung multitude of apostles. He also had students who stayed in Sweden and studied the regional flora there. While he had few actual pupils in the Netherlands, there were adherents of his views there from an early stage: Adriaan van Royen, Boerhaave's successor in Leiden, and Johannes Burman in Amsterdam largely adopted Linnaeus's system in their own work. Burman's son Nicolaas studied with Linnaeus in Uppsala for a few months, and in 1768 he published a *Flora Indica* (flora of the Indies) that used Linnaeus's method. In 1780, he succeeded his father as professor and director of the Amsterdam botanic garden.[131] Jan Frederik Gronovius, who lived in Leiden, had befriended Linnaeus during his years in the Netherlands and later wrote a *Flora virginica* (1743) describing the flowers of the east coast of North America, and a *Flora Orientalis* (1755), both based on the herbaria at his disposal. In France, though, Linnaeus never acquired much of a following, owing to staunch opposition from Buffon.

From the sixteenth to the eighteenth century, collections of plants led to taxonomies and taxonomies led in turn to new collections. According to the nineteenth-century botanist Augustin Pyramus de Candolle, you had to have your own herbarium to really count as a botanist, and you did not even have to collect the samples yourself (Candolle had bought his set). In any case, field expertise (of which Candolle had plenty) was not enough.[132] Linnaeus's own passion for collecting was inspired by the idea that if he knew about all the plants in the world, he might be able to devise a truly natural system. In reality, the discovery of entirely new types of plants, such as the *Proteacea* family in southern Africa, made it impossible for him to develop such a system within his lifetime. Yet two generations later, though not all the world's plants had yet been discovered, new systems arose that claimed to be more natural. By that time, the organizing principles of taxonomy had changed.

The Post-Linnaean Taxonomy

Even before 1800, scientists abandoned Linnaeus's system, or at least his sexual system of categorizing the higher taxa on the basis of stamens and pistils. In 1789, the year the Bastille was stormed, Antoine-Laurent de Jussieu published his *Genera Plantarum*. Soon, this new system had been widely adopted. Jussieu proposed a botanical classification method that was "empirical" and "synthetic." Linnaeus would not have cared much for this approach. He had always despised empirical approaches, which he believed were diametrically opposed to his rigorous method based on the essential properties of the plant. Linnaeus had used inductive methods to group species into genera, but for the higher taxa he had worked from top to bottom according to Aristotle's logical, deductive method.

This account makes it sound as if Jussieu and Linnaeus simply selected two different options—as if the principles adopted by Jussieu were also available to Linnaeus, who simply chose to reject them. However, the new method of classification resulted from a more profound shift in ways of looking at plants. For Linnaeus, plants—or at least their fructifications—were combinations of thirty-eight building blocks, which were put together differently in each species. Jussieu saw plants in an entirely different way. It was not the general appearance of the plant that mattered to him, but the relationships between the parts that made up its fructification. Jussieu drew on research into plant anatomy by Joseph Gaertner, a doctor from Tübingen had studied for a time with Van Royen in Leiden. Gaertner had compared the flowers and fruits of thousands of genera to one another and identified various forms of organization—prototypes of a kind. The zoologist Georges Cuvier later said that Jussieu owed the success of his system to Gaertner's work.[133]

In one important respect, however, Jussieu remained a man of the eighteenth century; he drew a connection between the natural groups he had discovered and the *scala naturae*, the ladder leading up toward perfection. This was in keeping with family tradition: his uncle Bernard de Jussieu, who had translated Linnaeus's work into French, was also an outspoken advocate of this idea, which enjoyed great popularity in the eighteenth century. Linnaeus too had originally believed in the *scala naturae*, but later, as discussed previously, he replaced it with a two-dimensional map. It should be said that this is not a crucial difference. Like Linnaeus, Jussieu believed that the "gaps" visible between the groups would be filled by the discovery of new plants. Cuvier no longer made any such assumption.[134] It is tempting to look at Jussieu as an intermediate figure between the classical and modern epistemes that Foucault describes. My own conclusion is that Jussieu's organizing concept of the structure of the fructification places him on the side of the new episteme.[135]

It was Augustin Pyramus de Candolle who stripped Jussieu's system of its eighteenth-century elements. In his *Théorie élémentaire de la botanique* (1813), he placed much greater emphasis than Jussieu had on differences between groups of plants, and his taxa did not flow into one another continuously, but were discrete. Jussieu had believed that divisions between taxa were arbitrary because it did not matter where you drew the line between groups in a continuous *scala naturae*. In contrast, for Candolle, the taxa (including the higher ones) were real entities.[136]

For most of the nineteenth century, the majority of botanists credited Jussieu with the discovery of the "natural system," and quite a few were unaware of Candolle's corrections. Nevertheless, a holistic view of plant anatomy had taken root as the basis for classification, and would remain in place for a long time.[137]

In 1803, Jean-Baptiste Lamarck, who in his younger years had written a French flora, defined the new discipline of biology. At its heart was plant physiology, which he viewed as a kind of botanical physics requiring intimate knowledge of plants and their development. Descriptive natural history and classification had no place in it.[138] A few decades later, the pioneers of cell theory, Mathias Schleiden and Theodor Schwann, published their *Grundzüge der wissenschaftlichen Botanik* (*Foundations of Scientific Botany*, 1842). They had little to say about plant classification, defining botany as an experimental science and "the science of the configuration (*Gestaltung*) of matter in the form of the plant."[139] Around 1900, Hugo de Vries, a plant physiologist in his own right, denied taxonomists even the right to determine what species were, arguing that only experimental methods could answer this question.[140] His challenge must not have made much of an impact, since experimental biologists have never attempted to take over the field of taxonomy, but remarks like these did chip away at taxonomists' status.[141]

Until well into the twentieth century, national government bodies oversaw the management of natural history collections because of the economic importance that they ascribed to them, partly in connection with continuing colonial expansion. Yet scholarly interest at universities was fading.[142] In an ironic reversal of the values of Candolle's day, taxonomy and the more broadly defined field of systematics came to play a subordinate role, but was especially valuable in the field disciplines of ecology and geographical botany, which had been pioneered by Candolle himself and Alexander von Humboldt.[143] Recently, older ecologists have sounded the alarm about diminishing taxonomic expertise among their young colleagues.

Since the era of Jussieu and Candolle, plant classification has remained fairly stable. One would expect that biological taxonomy would now be founded on evolutionary theory, but this is not entirely the case. Currently, four schools of taxonomy are in existence, and only two of them are based wholly or partially on evolutionary considerations.[144]

The oldest of these schools, called evolutionary taxonomy, strikes compromises between purely evolutionary and functional considerations. In contrast, cladistics, an approach proposed by Willi Hennig within animal systematics in 1950, has led to the most profound changes in the grouping of animals and plants by applying evolutionary, or phylogenetic, principles of classification with utter consistency. Some of these changes are felt to be counterintuitive on functional grounds and hence have not been adopted by evolutionary taxonomy. The cladists no longer consider the reptiles a taxonomic group, for instance, because of the discovery that crocodiles share a common ancestor with the birds (as a group), and are more distantly related to lizards. Consequently, crocodiles and lizards cannot be lumped together without including the birds. Cladistic taxonomies have to be revised fairly frequently to accommodate new phylogenetic insights and new archaeological findings. Other present-day schools of systematics count stability among their priorities. This has led them to disregard any "theory," including evolutionary theory. The school known as pheneticism considers as many traits as possible and makes a point of assigning them all equal weight. Also called numerical taxonomy, pheneticism employs statistics to make patterns emerge from large data sets.

One proposed explanation for why systematics has been relegated to the margins of biology is that these days nobody goes to the trouble of rebuilding the system from the ground up.[145] Yet since the 1986 Biodiversity Forum in the United States, in which Edward O. Wilson played a leading role, we have seen a new wave of interest in the taxonomy of plants and animals.[146]

9

Statistical Analysis as a Style of Science

Whenever you can, count.
—Francis Galton

RATHER THAN DEFINING EXACTLY what statistics is, it is easier to say what statistics is about: namely, groups of entities ("populations" in modern statistical parlance). Those entities can be voters, or recruits, or observations. Statistics does not deal in individual cases, but in multitudes of individuals, preferably all more or less of the same kind.

Likewise, individual entities and cases are not central to the taxonomic or the evolutionary style. As we have seen, the taxonomic style places individuals in a roughly hierarchical framework or simply in a fixed order; it aims to form distinct groups and, at a minimum, to transform what seems like chaos into manageable diversity. The historical-evolutionary style places individuals and individual events in a historically meaningful context. Statistics, in contrast, approaches the chaos of individual cases with the aim of extracting a meaningful form of equivalence. Individuals who do not fit the mold are excluded, identified as problems, or made equivalent to one another, at least in certain respects. This frequently leads to a loss of qualitative information, which must be accepted as an inevitable by-product of the method. And there are compensatory gains: without statistics, entities such as unemployment, GNP, and IQ could not exist.[1]

What is a statistical question? The best answer is: a question answered by a statistical method. The proposition is circular, but there is no alternative. This identifies the statistical style as a distinct style of science with its own criterion of truth. You might say that by classifying a question as a statistical question and then applying the statistical method, you determine the precise content of that question. In other words, statistics "fixes the sense of what it investigates."[2]

Statistics was developed three times. In 1749, the term was brought into circulation by Gottfried Achenwall, a professor in Göttingen, who used it to refer to a kind of descriptive geography. Statistics were facts about a country, supported when possible by tables of numerical data; this meaning is not so far removed from the current sense of the word. But in fact, Achenwall's work included few such tables. The field he developed, also known as *Staatenkunde* (political studies), was intended as part of a program of training for high-level public officials, intended to impart skills they could use to further their ruler's political projects and keep him informed about the economic and military potential of his country.

The launch of statistics as an independent science can also be traced to a fairly precise date. In 1825, Adolphe Quételet, a mathematics professor in Brussels, announced his plan to search for laws that would turn statistics into a "social physics."[3] In 1844, Quételet formulated the main statistical law relating to the normal distribution, the well-known bell curve. Quételet's conception of statistics was vastly more numerical than Achenwall's, but its subject matter was still the state and society.

The third time that statistics was developed was around 1900, in the work of the British scholars Francis Galton and Karl Pearson. From that time onward, "statistics" meant the numerical analysis of any type of data, whether relating to society or to biological or physical phenomena.

Some have suggested that Achenwall's use of the word "statistics" has brought him undue attention. Almost a century earlier, the English merchant John Graunt and physician Sir William Petty had drawn general conclusions about the population of London, partly based on birth and death records.[4] They called this work "political arithmetic," a term that has not stood the test of time. Though they were inspired by questions of public welfare, Graunt and Petty did not draw a connection between the needs of the state and their statistical projects. Not long after, however, Leibniz argued in favor of just such a connection, and in the eighteenth century, the Prussians became the first to link statistics to the state. Prussia's example was followed in many European countries: in Scotland, for one, and soon after in France and the Netherlands.[5] The new state apparatuses formed by the upheavals of the French Revolution and the Napoleonic era were, to an unprecedented degree, concerned with disciplining their subjects, and they established the practice of "biopolitics," as Michel Foucault has called it.[6] The elderly, the insane, the poor, and members of religious sects were subject to regular censuses, which were motivated by fears of social unrest.[7]

The Prussians remained the pioneers in statistics for many years. When reforming their state institutions after Napoleon's defeat, they founded a statistics office that worked uninterrupted from 1805 to 1934 and served as a model for other countries. German statistics was light on mathematics and entirely free of mathematical "laws." However, the successive directors of the office were high-ranking political officials who were actively involved in the reform of the economy. From 1858 onward, they laid the foundation for a state-run social insurance system.[8]

In France, the Netherlands, and Quételet's home country of Belgium, a different style of statistics emerged. In 1825, Quételet denounced identifying statistics with mere data gathering and insisted on the need for "laws." To arrive at such laws, he forged a new form of statistics out of three components. The first was data collection, indispensable even for Quételet. The second was the concept of probability, which in its modern form dates from the seventeenth century. The third was the technique of error analysis, which Quételet borrowed from astronomy.

These three components, whose origins and development will be discussed in the following sections, led directly to a specific interpretation of what statistical phenomena actually were. This interpretation corresponded neatly to the conventional wisdom about probability around 1800. It was thought that the regularities discovered through the application of statistics concealed underlying sets of small-scale, unknown causes. In other words, statistics was a way of compensating for a lack of information about individual entities by focusing on a higher level of aggregation, a more convenient "whole." Quételet had a profound influence even in the German lands, partly because, starting in 1853, he organized a series of international statistics conferences.

In the nineteenth century, this subjectivist view of statistics and probability slowly gave way to an objectivist perspective on chance processes. The objectivist does not assume any lack of information, but regards chance processes as an expression of inherent variability in nature. For example, the phenomenon of radioactivity, discovered in 1896, was interpreted in an objectivist fashion almost immediately.[9] The objectivist view gradually gained ground, in a process that the philosopher Ian Hacking has termed "the erosion of determinism,"[10] reaching its pinnacle in quantum mechanics and the present-day field of chaos theory.

The work of Galton, Pearson, Fisher, and many of their contemporaries was based entirely on the objectivist concept of inherent variability in nature. In fact, even before their time, Darwin had assumed the validity of this concept. This shift to a new perspective on the nature of reality was linked to another major reorientation. Unlike Quételet's approach, the new statistics was radically "individualistic."[11] You might call the unemployment rate a Quételetian concept and IQ a Galtonian one. IQ does not exist as an independent figure; a specific IQ value establishes a relationship between an individual and a population of comparable individuals; without this relationship, the term is meaningless.

The subjectivist view still has its supporters, especially among admirers of the deductive style of science. The saying "If I need statistics, I have done a bad experiment" is well known among those who believe that nature should always produce simple numerical relationships of the kind observed by Galileo. But the statistical style opened new domains of inquiry not dreamt of in Galileo's philosophy.

The Concept of Probability in the Seventeenth and Eighteenth Centuries

Up until the early seventeenth century, the term "probability" was associated with what were seen as the "lower" sciences: alchemy, medicine, and so forth. In the "higher" or classical sciences, especially physics, most scientists were still serenely confident that the way to arrive at true knowledge (which they believed was attainable in their fields) was by demonstrating the necessity of certain causal relationships. This was the domain of the deductive style of science. Only in Epicurian physics—with its notion of the *clinamen* (swerve), an indeterminate variation in the motions of atoms—was there an exception to the rule of necessity, but this tradition led a marginal existence.

Aristotle used the term "probability" in reference to observed regularities in nature, things that "usually" took place in a particular way. In a discussion of biology, he said that "necessity is not present in equal measure in all the works of nature."[12] Cicero and Quintillian discussed the probability of arguments put forward in legal disputes. There are signs that may raise suspicions that a person is a murderer, such as blood-stained clothing, but they do not constitute certain proof.[13]

The term associated with fields of knowledge in which causal relationships could not be demonstrated was *opinio* (opinion). In the Middle Ages, the term "probable" was used to refer to the extent to which intelligent and well-informed people considered a certain *opinio* acceptable. Over the seventeenth century, the domain of truly demonstrable knowledge (*scientia*) shrank, and the concept of probability played an increasing role in what previously had been the higher sciences. At the same time, probability was gradually shedding its associations with opinion and authority. One noteworthy step in this transition was Newton's gravitational theory of 1687. The theologian Samuel Clarke, a friend of Newton's and his spokesperson in his controversy with Leibniz, acknowledged that the phenomenon of gravitation must be regarded as a regular and constant action, rather than a natural necessity in the classical sense.[14]

There was another reason that the distinction between *scientia* and *opinio* was becoming blurrier and less applicable: some matters previously classified as *opinio* were acquiring the status of certainties. John Wilkins, a leading member of the Royal Society, identified three types of certainty: mathematical, physical, and moral. Moral certainty was the best that could be attained in ethical and political matters, as well as in religion, history, and certain fields of natural science. The standard for moral certainty was evidence "beyond reasonable doubt."[15] The moral character of this standard is immediately apparent in the case of eyewitness testimony in court proceedings. If there is only one eyewitness, one might wonder whether the testimony is colored by self-interest or prejudice, but the larger the number of independent witnesses, the more reasonable it is to dismiss such doubts. The same argument applies, *mutatis mutandis*, to eyewitness statements about natural events, and to experiments of the kind developed by Robert Boyle.

Moral considerations—specifically, standards of fair dealing—also played a role in the drafting of commercial contracts when the parties ran significant risks. Strictly speaking, what mattered was not the probability of a certain outcome, but the expectations that the parties could reasonably have about that outcome. In the late sixteenth century, a number of moral philosophers turned their attention to the fairness of contracts and discovered that the same issues arose in dice games and other games of chance. They realized that these games could be seen as contracts, in which the risks are shared equally and the expected gains are the same for each participant.[16] It was a legal scholar, Hugo Grotius, who figured out how to calculate a reasonable rate of interest on commercial loans taken out by merchants by assessing the severity of the risks involved.[17]

In short, during the first half of the seventeenth century, a wide range of scientific and social issues were linked to the concept of probability. All these issues lent themselves to a particular rational approach, if only because they all involved multiple degrees of certainty and probability. It was difficult to find a common thread in all these strands of thought, and aside from a few attempts that failed or languished in obscurity, no quantitative approaches to probability were available.[18] That changed in 1657, however, with the publication of a book by Christiaan Huygens.

In 1655, ten years before the foundation of the Académie Royale des Sciences, Huygens spent a short time in Paris, where he heard talk of a correspondence between Blaise Pascal and Pierre de Fermat on calculating probabilities in dice games. But he was unable to get in touch with either Pascal or Fermat, or even to track down copies of the letters. So Huygens went home and solved the problem on his own.[19] He wrote his book in Dutch, though the Latin translation was published first, in 1657, under the title *De Ratiociniis in Aleæ Ludo* (*On Calculations in Games of Chance*).[20] It was to remain the standard work on the subject for almost sixty years.

Coming up with mathematical methods for calculating one's chances in a dice game may seem like a game in itself, a whimsical approach to a fundamental problem. But according to the historian of science Lorraine Daston, nothing could be further from the truth. Huygens's assumptions in his analysis of dice games, including his concept of fair play, were derived from the commercial contracts used in his day.[21] In his book, dice games served as a kind of model for these contracts. In connection with this issue, the philosopher Ian Hacking has pointed out that Huygens, who was an expert Latinist, nevertheless had good reasons for writing the book in Dutch first. Many specialized terms and other words relating to chance had a more familiar, comfortable feel in the vernacular, because of their associations with everyday concerns. Some preliminary notes by Huygens have been preserved, in which he searches for the right Latin equivalents for a number of Dutch words. In contrast, if Huygens had been solving a geometric problem, Latin would have offered him a more familiar and more appropriate vocabulary.[22]

With all the many connotations of the terms "chance" and "probability," why was it their connection to dice games that Huygens seized upon as the key to a general theory about them? The answer is probably that this subject lent itself to quantification, because expectations about risk, in both commerce and dice games, had already been quantified to some degree. In the case of dice games, this was because each player had to decide how much to wage in light of the chance of winning. But Huygens, like Pascal and his circle, placed the calculation of probability in a larger context—namely, the development of rules for good and bad reasoning in situations where not all knowledge was certain. In the broadest sense, this project encompassed any type of non-deductive reasoning—including analogical reasoning, for example.

The new method for calculating probabilities was soon applied in the insurance business. For some years, town authorities in England and the Netherlands had been selling life annuities as a source of revenue. These were a form of insurance that provided a fixed lifelong income: the buyer made one large payment at the outset and then received a certain amount annually until his death. These life annuities could take many forms. For instance, a husband and wife could buy an annuity together, and the annual payments would continue until the death of the last surviving partner. In the seventeenth century, the public authorities who sold these annuities gave no thought to the buyers' ages. Nor did it seem obvious that they should. Even parents who bought an annuity for their newborn child could not know with any certainty that the child would have many years to benefit, since child mortality rates were high in those days.

The public authorities who sold the annuities did not consult any statistical data. Instead, they based their prices on their own previous experience and on moral considerations, such as the need to combat usury. The results were not always satisfactory. Many towns lost a great deal of money on annuities, especially in England.[23]

This made life annuities a matter of national interest, which drew the attention of such luminaries as Johan de Witt, grand pensionary of Holland (and one of Europe's leading statesmen), and Johan Hudde, burgomaster (that is, mayor) of Amsterdam, then Europe's most prosperous city. De Witt and Hudde worked together to develop a new system for calculating the prices of life annuities.[24] As eminent mathematicians in their own right, the two men were familiar with Huygens's work, and they sought his advice. In 1671, De Witt reported their findings to the States of Holland (the governing council of Holland, the most powerful Dutch province). Their main conclusion was that the buyers and sellers of such annuities were playing a game of chance and were entitled to equal chances. Around 1670, it was standard practice in Amsterdam for the buyer of a life annuity to earn back his invested capital plus interest after fourteen years. If he lived longer, then he won; if he died earlier, the town authorities won. What made the collaboration between De Witt and Hudde truly exceptional was that Hudde supplied mortality data for the life annuities purchased between 1586 and 1590. Enough time had elapsed that these figures formed a sound basis for calculating probabilities of death, though it was not easy for Hudde to convert them into a useful form. In those days, no procedures had yet been developed for filtering away strange aberrations in numerical data. Hudde and De Witt had to rely on intuition and common sense.[25]

The States of Holland did not adopt De Witt's proposal, and it is not entirely clear what they did instead. They may have experimented with a fashionable new system called the tontine, which came to the Dutch Republic from abroad and had already been used in the town of Kampen. In this system, a single contract covers a group of investors, and the dividends on their investment (often a fixed annual amount) are distributed among all the living participants. In some ways, the tontine resembles a lottery. In any case, it did not bring in a great deal of revenue for the States.[26]

Elsewhere in Europe, De Witt's quantitative approach met with the same lack of interest in the thriving insurance sector. Until well into the eighteenth century, for instance, Lloyd's of London calculated insurance premiums for ships by consulting an archive of information on thirty thousand individual captains. The idea was that every captain presented unique risks. And even though more and more demographic data were becoming available, life insurance companies disregarded it for most of the eighteenth century. It was not until 1762 that a company was founded where mathematicians proficient in probability and statistics determined the business policies and the insurance premiums. Its name was the Society for Equitable Assurance on Lives and Survivorships.[27]

But at the same time that De Witt and Hudde's reasonable proposal was being passed over by policymakers, it was attracting the attention of natural philosophers. In 1686, Leibniz visited Hudde in Amsterdam to take a look at his insurance statistics.

A New Model for Probability

Jakob Bernoulli was the elder brother of Johann, the developer of post-Newtonian vortex theory. In 1713 their nephew Nicolaus Bernoulli, the son of a third brother who was a painter, published Jakob's unfinished work *Ars Conjectandi* (*The Art of Conjecturing*). This book presented a rational and wholly original model of probability, with potential applications in all sorts of new domains. Bernoulli presented a now-classic series of thought-experiments involving an urn that contains balls (or pebbles) of two colors. By removing one ball at a time from the urn, you can determine the proportion of balls of each color.[28]

Bernoulli begins by assuming that we know how many balls of each color are present: for example, 500 white and 500 black ones, a 1:1 ratio. Suppose you remove a single ball, make a note of its color, and then put it back in the urn. If you repeat this process twenty times, you will arrive at a ratio not too far from 1:1. If you repeat the process forty times, the ratio will be closer to 1:1. Bernoulli proved that the more often you draw a ball from the urn, the more closely the ratio will approach 1:1. An infinite number of draws will yield the exact ratio.

So far, so good. This model is conceptually sound, in that the number of balls in the urn could be described figuratively as the cause, and the outcome of the draws as the effect. The theorem thus proceeds from cause to effect, a process which is metaphysically and methodologically sound because it is deductive. Bernoulli explicitly points out that many situations in nature and society can be compared to balls in an urn.

But the true appeal of the urn model was that it seemed so easy to reverse the chain of reasoning. For practical purposes, the truly interesting case is the one in which we do not know the "cause"—in other words, the ratio between the two colors is not given *a priori*—and we try to determine it by drawing one ball at a time from the urn. It seemed (and still seems) intuitively clear that as the number of draws increases, so does the probability that the outcome will accurately reflect the "cause." Nevertheless, this "inverse probability theorem" does not follow from Bernoulli's actual theorem. In fact, there is no way it possibly could, because a deductive inference (such as Bernoulli's theorem) can never be inverted to yield an inductive one (such as the inverse probability theorem). Bernoulli himself probably thought this *was* possible. Latter-day thinkers—Abraham de Moivre and Thomas Bayes in the eighteenth century, and Pierre-Simon Laplace and Siméon-Denis Poisson in the nineteenth—found other ways of deriving inverted Bernoulli theorems. Apparently, each one of them believed that he could improve upon the efforts of his predecessors. Even today, statisticians are not in complete agreement about this issue.

The inverted Bernoulli theorem not only implies the formal possibility of induction but also suggests a number of real-world applications. Shortly before his death, Bernoulli discussed these applications in a correspondence with Leibniz. Leibniz suggested that between two draws the number or color of the balls in the urn could change; in other words, nature is inconstant. Bernoulli rejected this notion out of hand. Leibniz asked Bernoulli what he should assume about the true proportion if after fifty draws he had seen 23 black and 27 white balls. Was it 23:27? Bernoulli responded that 1:1 was the best estimate because simplicity is a property of nature.[29]

In the eighteenth century, almost everyone shared Bernoulli's views that nature was inalterable and that the relationship between cause and effect was fixed and deterministic. The only difficulty arose when there was no prior knowledge of the cause or causes at work. It was not until the late nineteenth century that scholars suggested that the relationship between cause and effect might be subject to chance processes. This idea attracted particular interest among physicists such as Ludwig Boltzmann and Rudolf Clausius.

But around 1800, in his more sophisticated version of the inverted Bernoulli theorem, Pierre-Simon de Laplace remained committed to a deterministic view of nature. Laplace summed up his views about knowledge and probability as follows:

> An intelligence that, at a given instant, could comprehend all the forces by which nature is animated and the respective situation of the beings that make it up, if moreover it were vast enough to submit these data to analysis, would encompass in the same formula the movements of the greatest bodies of the universe and those of the lightest atoms. For such an intelligence nothing would be uncertain, and the future, like the past, would be open to its eyes. The human mind affords, in the perfection that it has been able to give to astronomy, a feeble likeness of this intelligence.[30]

Laplace derived his optimism in part from his own successes in astronomy, a field in which he had solved a number of important problems relating to an apparent instability in the solar system. Accordingly, his term "intelligence" should not be taken to refer to God; rather, it represents an ideal for humanity itself. Laplace went on to remark that "without any doubt" the regularity of nature that astronomy had demonstrated could be found "in all phenomena."[31] *All* phenomena—those were his literal words. In short, Laplace believed that if he were omniscient, he would not need probability theory or error analysis.

Eighteenth-century probability theory, including Laplace's version, remained oriented toward individual cases. Though such cases were merely "chance" (that is, contingent) facts, they had the advantage that their properties could be traced back to specific causes. In the words of the Scottish philosopher David Hume, "what the vulgar call chance is nothing but a secret and conceal'd cause."[32] An emphasis on individual cases is also visible in Laplace's version of the inverted Bernoulli theorem; in his formulation, what was at issue was the probability that, after a given number of draws, the next individual ball drawn from an urn would be of a certain color. As the number of preceding draws increased, so did the accuracy of the prediction. Laplace's student Quételet developed a very different view. He took a collection of individuals to be a greater whole that truly existed in its own right. This invested both probability theory and collections of data with a new kind of significance.

Nevertheless, Bayes, Laplace, and their contemporaries also made great strides. By helping to build confidence in the possibility of developing a rational procedure for induction, they managed to put up a defense against the revival of Skepticism occasioned by Hume's work. Paradoxically, the loss of faith in the certainty of *scientia* made it possible to place more confidence in a working method such as induction, which was uncertain by its very nature. In limiting cases (that is, with very large samples—in theory, infinitely large), induction was even seen as leading to "moral" certainty.

Astronomy and Error Analysis

In 1823, Adolphe Quételet received a grant from the government of the Netherlands (which then included present-day Belgium) so that he could go to Paris and acquire the knowledge he needed to set up an astronomical observatory in Brussels. In Paris, Quételet worked with Laplace, who was then writing his great work *Traité de mécanique céleste* (*Celestial Mechanics*), and he learned the technique of error analysis that Laplace had developed.

The earliest version of what later became known as the normal distribution was described by Abraham de Moivre in a 1733 publication. De Moivre asked what the probability was that if you tossed a coin 1,000 times, it would come up tails 480 times and heads 520. The usual calculation method in those days, which went back to Bernoulli, was incredibly time-consuming, but de Moivre played a role in developing an exponential function that provided a close approximation. Forty years later, Laplace saw that he could use this function for a very different purpose: not to calculate *a priori* probabilities, but to infer a cause *a posteriori* from a series of data or observations. He tested this technique on birth records and noted that significantly more boys than girls were born in Paris, but also that the surplus of boys was smaller among illegitimate children. Laplace concluded that married couples abandoned more girls than boys.[33] His next application of de Moivre's function was to observations of atmospheric pressure. Individual observations showed great variability, apparently resulting from the unreliability of the barometers used. But if the number of observations available was large enough, then the error could be reduced to a manageable level. This method made it possible to observe a difference in pressure between the morning and the afternoon, and Laplace concluded that there must be a real physical reason for that difference.

The year 1807 saw the discovery of the "method of least squares" for drawing the best-fitting curve for a set of astronomical measurements, a technique that quickly spread to almost universal use. The German mathematician and astronomer Carl Friedrich Gauss showed that the function discovered by de Moivre and Laplace provided a probabilistic basis for this method.[34] The general idea of applying probability theory to observations of a star is that the star can be assumed to occupy a stable position. Nevertheless, each measurement yields a slightly different value. The most reasonable conclusion is that this is a result of imperfect measuring devices, the human senses used to read those devices, atmospheric disturbances, and so on. As long as the number of observations is large enough, all these minor factors cancel each other out. The theory of coin-tossing relies on essentially the same reasoning: a multitude of very minor factors produce variations from a precise 1:1 ratio. In any event, this is how Gauss's contemporaries viewed the application of probability theory to error analysis in astronomy.[35]

We have seen that Quételet first encountered the theory of error analysis in the astronomical context. But in 1844, as a "social physicist," he transformed de Moivre and Laplace's original function into something very different. He was able to do so because he had stumbled across a highly exceptional set of figures in a peculiar article in a medical journal, which consisted entirely of tables stating the height and chest size of more than five thousand Scottish recruits.[36] He disregarded the data on height and went to work on the chest measurements, which he argued were analogous to five thousand measurements of a single soldier's chest. From these measurements, he could infer not only an average, but also the spread of the "measurement error" around that average, which took the form of a bell curve, or normal distribution.

Through this process, Quételet discovered *l'homme moyen*, the "average man." This was not just an arithmetical average, a mathematical abstraction. Quételet assumed, in accordance with the standard interpretation of error analysis, that hundreds of factors, such as illness, malnourishment, and overeating, caused deviations from the naturally ordained standard. *L'homme moyen* was thus a *normal* man—and even, in Quételet's view, the *ideal* man.[37] Quételet also came to believe that the normal distribution applied to many phenomena in human society, such as crime and suicide. What seemed accidental or even irrational at the individual level showed lawlike regularities in the aggregate. Inspired by Victor Cousin's social philosophy, Quételet viewed the normal person as a *moral* standard comparable to Cousin's *juste milieu*, the golden mean between extremes, an idea that could be applied in many domains. He saw deviations from the standard as undesirable and believed that average people contributed to the stability of society. The very existence of "society" was another conclusion that Quételet drew from statistics, as we shall see below.[38]

In the decades that followed, more and more phenomena were found to conform to the bell curve. In 1859, the physicist James Clerk Maxwell drew a comparison between the demographic makeup of a country and the gas in a container. In both cases, it was not feasible to describe the whole in terms of individual people or atoms. Statistics had proved its effectiveness in the demographic case, and Maxwell believed it could do the same in the field of thermodynamics. Thus was born statistical mechanics, inspired by Quételet's social physics.[39]

Quételet and other statisticians believed that their method's greatest achievement was to transform the seeming chaos at the level of individual observations—what is now sometimes called the micro level—into an orderly higher-level system governed by fairly straightforward laws.

The Avalanche of Data

The unusual provenance of the figures that Quételet used to arrive at the concepts of the normal distribution and the average person suggest that such material was hard to come by. But it was not long before he could find numerical data much closer to home. Hacking describes the period from 1800 to 1850 as one of great enthusiasm for counting and measurement for its own sake, without a theoretical purpose. In fact, from 1820 to 1840 there was a veritable "avalanche of numbers." Many scientists were content to deduce "laws" from their collections of figures that had no meaning except as descriptions of an apparent regularity. Quételet, for example, made the "discovery" that lilacs flower when the sum of the squares of the mean daily temperatures since the last frost reaches $4264°\,C^2$. This is the type of observation at which Hacking sneers.[40]

Nevertheless, quite a few statisticians—Quételet among them—had something grander in mind than inductive generalizations. They used the inductive method not as an end in itself, but as a means to develop a new variety of abstraction. Statistical data were *indices* of a greater whole.[41] Nowadays, we are used to thinking in terms of greater wholes of this kind. Using statistics, we transform a certain number of people who have been observed to have no jobs into a single quantity: "the" unemployment rate.[42] This type of holistic thinking can be traced back to Quételet, for whom the "average man" that he discovered in his data was nothing less than a higher-level individual. Similarly, in his statistical studies of suicides and crimes, Quételet discovered the existence of a higher-level entity called "society." The sociologist Emile Durkheim later drew on Quételet's work in arguing for the existence of what he called "social facts." The sociology of the irrational masses replaced Huygens's psychology of the rational, calculating individual.[43]

Alongside Quételet's style of searching for regularities, there is also a statistical tradition centered on collecting numerical data. Simon Vissering, who became a professor of political economy at Leiden University in 1850, was a typical gatherer of figures, whose aim was to learn more about "political forces." Accordingly, Vissering saw Achenwall as the founder of statistics. Vissering was also active in a "statistical movement," the Society for Statistics, whose members included not only public officials and industrialists but also physicians.[44] Earlier in the nineteenth century, the medical profession had seen the origin of the hygienic movement, which propagated a new ideal: the improvement of what its members called public health. From the 1850s onward, that concept was placed on a firmer statistical basis.[45]

In the course of Vissering's career, he was increasingly influenced by some of Quételet's ideas, and he raised his sights from compiling collections of data to discovering "laws of cause and effect" and practicing the "inductive method." What he meant by this last term was, for instance, calculating meat consumption on the basis of the revenue from excise taxes on slaughtered animals.[46] Like unemployment, meat consumption is what would today be called a macroeconomic variable.

The Society for Statistics called upon the Dutch government to set up a central statistics office, a proposal staunchly supported by the socialist member of Parliament Ferdinand Domela Nieuwenhuis, a well-known radical, who believed it would improve understanding of labor relations. The country's Central Statistics Office (known today as Statistics Netherlands) was founded in 1899.[47]

Quételet and his followers were not the only ones who saw the compilation of large collections of data as a step toward identifying and understanding larger wholes. The explorer and geographer Alexander von Humboldt trekked out into the wilderness and measured almost everything he could measure by the means at his disposal. In 1799, when he left Paris for South America, his luggage included two hygrometers (for measuring humidity), four kinds of eudiometers (for gauging the level of oxygen in the atmosphere), six thermometers, a cyanometer (for determining the blueness of the sky), two chronometers, two barometers, a theodolite (a telescope-like instrument for measuring angles), four sextants of different sizes, quadrants, achromatic and reflecting telescopes, an ebullioscope, an electrometer, and an inclinometer (for measuring the strength of geomagnetic forces).[48]

Humboldt never once used statistical methods.[49] Instead, he strove for the clearest possible visual presentation of his data, in the hope that this would provide insight into the whole through a process of abstraction. Along the way, he became the first to treat the "landscape" as a higher-order physical reality. Previously, the term had been used only for natural features united in a composition by a painter.[50]

What Humboldt, and other practitioners of what historians have called "Humboldtian sciences," had in common with statisticians such as Quételet was their familiarity with *Staatenkunde*, Achenwall's variety of statistics.[51] Humboldt's teachers in the field of natural history, the father and son Johann and Georg Forster, came from the Göttingen milieu and had learned about Achenwall's work there. In the final years of the eighteenth century, Humboldt, Laplace, and Humboldt's student Augustin-Pyramus de Candolle were all members of the same exclusive learned society, the Société d'Arcueil. This organization entrusted Candolle with the task of exploring the regions of France that were as yet unknown to botany. He was expected not only to document the plant life of those regions but also to present the implications for agriculture, in a spirit of Enlightenment utilitarianism.[52]

Humboldt had developed a method known as botanical arithmetic, which involved calculating the ratio, in a given region, of all the plants in a single family to the total number of plant species. In typical grasslands, for example, grasses represented a high proportion of the total species. This enabled Humboldt to express his conjectures about ecological and geographical correlations in the most precise possible terms. Candolle—and, to an even greater degree, his son Alphonse, the author of *Géographie botanique raisonnée* (*Methodical Botanical Geography*, 1855)—took this numerical form of botanical research to great heights. Critics of their research style called it "tabular statistics," mere data gathering, which could be thought-provoking but had no clear theoretical goal. The "botanical statisticians" (who did not perform statistical analysis of their data in the modern sense) hoped for the emergence of a "Newton of the plant kingdom" who would reveal the grand underlying system. Alphonse de Candolle himself, however, was anything but a theorist. He merely suggested a variety of factors that might explain the patterns he had identified in the geographic distribution of plant species. Four years later, when the botanical Newton actually appeared in the form of Charles Darwin, his work proved to lean heavily on Candolle's style of reasoning.[53]

Francis Galton, Ronald Fisher, and the Variability of Nature

In the hands of Francis Galton, the normal distribution underwent another transformation. Like Quételet, Galton was fascinated by this curve, and like Quételet, he found new applications for it. But Galton also differed from his predecessor in a fundamental way. He had given up interpreting the bell curve as the product of numerous small-scale, independent "causes," or factors. Instead, he saw it as a reflection of the natural variability in a single cause. To illustrate this point, Galton used an experiment, based on a physical analogy, which was easy to perform just about anywhere: he dropped small metal balls from the top of a vertical board with a regular pattern of pegs, known as a Galton box, bean machine, or quincunx. The balls naturally fell into the columns at the bottom in a normal distribution.[54]

The significance of this demonstration for natural philosophy was that Galton gave statistical laws, and statistics in general, a new, autonomous status. He no longer regarded the variability of statistical data as a superficial phenomenon overlying a pattern of fixed, deterministic causality, nor did he believe that statistical laws could be reduced to causal explanations if only one had sufficient knowledge. Instead, he saw nature itself as variable. It should be emphasized that he did not prove this point, and it is not even clear that he sought to prove it. In 1936, however, John von Neumann presented a proof in the domain of quantum mechanics that there are no "hidden variables." If there were, the possibility of an underlying deterministic reality would remain open.[55] More recently, since the late 1970s, chaos theory and the thermodynamics of systems far from equilibrium have demonstrated something similar with respect to macroscopic phenomena. As noted above, Galton probably did not have any evidence for irreducibility; it suffices to observe that he saw nature as irreducible *de facto*, a new phase in "the erosion of determinism." Thinkers up to and including Quételet saw probability as the chance of being correct; in other words, "probability" was an epistemological term, relating to the status of our knowledge. From Galton's day onward, probability described the frequencies with which variations occurred in nature.

Galton also departed from Quételet's views in other respects. For Galton, the average or normal value had no metaphysical, higher-order quality. He was more interested in the tails of the normal distribution, especially the upper values, which he saw as representing excellence. He associated the normal value with the middling and mediocre, and noted with regret that there was a tendency to slide back from excellence into mediocrity, to regress to the mean. This observation of Galton's was a statistical formulation of a genetic process. Galton was Darwin's first cousin and kept up with his relative's publications. Though he was not a naturalist or biologist himself, evolutionary theory and genetics were central to his thinking. Galton laid the conceptual groundwork for the infamous nature-nurture debate that was to rage throughout the twentieth century, and was an unequivocal supporter of the nature camp.[56]

Galton was also a tireless promoter of eugenics, an ideology he had developed that taught that the human condition could be improved by genetic means, using a breeding program under scientists' supervision. In earlier decades, Georges-Louis Leclerc, Comte de Buffon, had expressed similar ideas. Drawing on Darwin's ideas about heredity and evolution, Galton founded a movement called the Eugenics Education Society, which campaigned for the social exclusion of mentally disabled people. In political terms, Galton was a conservative. Though racial politics was not one of his main interests, he had firm ideas about racial superiority and inferiority. Galton's American supporters went so far as to lobby for a ban on immigration by nonwhites, with some success.[57]

In his book *Hereditary Genius* (1869), Galton tried to demonstrate statistically that exceptional talents in areas such as music, law, wrestling, and rowing were hereditary. In a later work, *English Men of Science* (1874), Galton tried to identify the traits that contributed to scientific genius. Never before had the bell curve been applied to anything as difficult to define as intelligence. Galton showed that people who excelled in their professions were more likely than others to have close relatives with similar talents. He also observed, however, that the children of great thinkers displayed regression to the mean: they were almost never as exceptional as their fathers. He had no choice but to acknowledge the difficulties that this presented for his eugenics program.[58] In any event, Galton greatly expanded the arsenal of statistical methods of analysis. From that point on, statistics was a standard technique in genetics.

In fact, genetic research gave rise to one of the most powerful methods in statistics: the search for correlations. Galton believed he had discovered that the smaller scientists' heads were, the more energy they had. His explanation was that the genetic carriers of these traits had a mutual affinity and so were often jointly transmitted.[59]

The Galtonian program of research was carried forward by the mathematician Karl Pearson and the physicist Ronald A. Fisher, who placed Galtonian statistics on firmer mathematical foundations than Galton himself had. While Galton had studied mathematics at university level for several years (alongside medicine), he had never completed this course of study. After inheriting his father's fortune at the age of twenty, he set off to travel in Africa (writing a book about the trip that won him a fellowship in the Royal Society) and devoted some time to meteorology, which taught him to see patterns in large collections of data. In the 1890s, Karl Pearson—who was already a mathematician of some reputation—became interested in Galton's research. In 1911, Pearson became the first to hold the chair in eugenics that Galton had founded at the University of London. Around this time, researchers developed many of the statistical tools used in numerous fields of science today, such as the chi-square test, Student's *t* test,[60] factor analysis, and so forth.

Pearson did much of his statistical work in the context of genetics and evolutionary theory, and the same was true of his successor, Fisher, whose accomplishments included the standard methods for analyzing variance and covariance (discussed below). Fisher succeeded Pearson in the Galton Chair in 1933; at the same time, Pearson's son Egon took charge of another section of the same department. Egon Pearson also achieved great fame as a statistical theorist and developer of new statistical techniques.

It should not, however, be supposed that these three men produced a homogeneous, consistent body of work. The opposite was true: there were few points on which Egon Pearson and Fisher agreed. Furthermore, around the same time, a third school of statistics emerged with its roots in the "subjectivist" approach of Thomas Bayes. Yet statistics textbooks, the first of which appeared in 1937, presented the field as a unified whole. The contrasting schools were merged into a hybrid most reminiscent of Fisher's work.[61]

Nevertheless, Galton, Fisher, and the Pearsons showed many similarities in their philosophical outlook. For all of them, statistical variation was much more than just deviation from an average. Recall that Quételet had interpreted variation in this way, as departure from the norm; this is shown, for instance, by his use of Laplace's technique of error analysis to describe such variation. In contrast, the new biological interpretation viewed individual variation as the basis for evolutionary change and development.[62] This crucially affected the interpretation of concepts such as correlation. Scientists often tried to account for correlations by finding their hidden causes, but in Pearson's eyes, correlations pointed to processes without causes.[63]

It has been suggested that Karl Pearson's turn away from causal explanation sheds light on a well-known controversy between Pearson and the biologist William Bateson. Shortly after the so-called rediscovery of Mendel's laws in the field of genetics, Bateson had emerged as a champion of "Mendelism." Mendelians had developed the distinction between the genotype and phenotype to clarify the relationship between an organism's genetic structure, which was hidden from view, and the expression of that structure in its outward characteristics, which were plainly visible. In contrast, Pearson and his supporters restricted themselves to visible characteristics, but their descriptions were much more complex and subtle than Bateson's.[64]

Ronald Fisher and Statistics as a Scientific Methodology

Before going to Cambridge, Fisher had spent fifteen years as a statistician at an agricultural research station in Rothamsted, England. There he found logbooks describing agricultural experiments carried out over some sixty years. In one exceptional series, thirteen wheat fields had been observed from 1852 to 1918 without interruption, and the observations had hardly been subjected to any form of analysis or interpretation.[65] These observations formed the basis for Fisher's new experimental design methodology.[66]

One of Fisher's main contributions to statistics was his test of significance, which was a direct response to the demands of experimental practice. Fisher often criticized statisticians of a mathematical bent for not being firmly grounded enough in fieldwork. His significance test offered the solution to a two-pronged problem. The point of such a test is not only to make sure that we do enough experiments, but also to keep us from wasting time and money on pointless ones. Let us take a simple agricultural example: can a certain manuring technique increase crop yield? Suppose we have two fields that are comparable in many respects (for instance, with regard to irrigation). We can compare them by using the manuring technique on one field and not on the other. We may find that the manured field yields one hundred bushels of wheat per acre and the non-manured field only ninety. But the comparison teaches us nothing, because past experience has shown that a difference of this size could be the result of natural variation. A farmer who adopts the new manuring method because of this experiment may be wasting his money.

It is worth noting that this problem did not arise in classical physics. No one ever asked Galileo for statistical validation of the experiment in which he rolled a ball down a ramp, the basis for his law of falling bodies. Physicists generally assume that one falling body is an acceptable model for all of them. Of course, physicists do consider it important for experiments to be replicable; normally, they address this issue by carefully controlling the conditions under which the experiment takes place.[67]

But in domains where entities show a high degree of variation, larger numbers of comparative experiments have to be done. Fisher developed a whole set of criteria for experimental design, which involve various restrictions on the conditions under which comparison is possible. Experimenters who follow his protocol can say with reasonable certainty that their findings apply to the greater whole. This is the *inductive* step that statistics allows us to make. When this is the case, the results are "significant"; that is to say, they have some meaning. Their significance is only provisional, however. Fisher cautioned experimenters always to take account of new experimental results when evaluating the results of earlier experiments. In principle, conclusions reached at an earlier stage are always open to reconsideration, since an inductive argument never establishes its conclusions with absolute certainty.[68]

Fisher's methods have other noteworthy limitations. One frequent error is to suppose that statistical significance implies practical relevance. But in the above example, even a statistically significant result may not be of any interest to farmers—for example, because manure is too expensive for the new technique to be worthwhile. One last limitation of statistics is that it tells us nothing about the nature of correlations between variables, such as manuring and crop yield. To explain such correlations, the experimenter is free to develop a simple theoretical model or a complex one. Fisher's fundamental innovation, which he presented in his very successful textbooks *Statistical Methods for Research Workers* (1925) and *The Design of Experiments* (1935), was to design experiments from the outset so that statistical testing was possible. However, Fisher's statistical methods are easier to apply to simple observed correlations than to complex theoretical models, to such a degree that one might well ask: when we apply these methods, are we still investigating anything of interest?

Of course, we cannot hold this against Fisher; his contribution to experimental method was far too great. In fact, he has been called the greatest pioneer in the methodology of experimental science since Galileo and Boyle. Furthermore, his methods were more than adequate for use in his own fields of study, agricultural research and genetics. But no methodology can be imported into a new field without bringing along assumptions and constraints from its original field of application.

That sometimes leads to problems, as shown by an example from the medical sciences. Fisher was never willing to accept the statistical correlation between smoking and lung cancer, because the research that established this correlation had not followed his protocol. He had conducted his own studies for the tobacco industry, and he argued that a single genetic factor could conceivably cause both a predisposition to lung cancer and a desire to smoke. However, clinical experiments to rule out this factor would be both impractical and unethical.[69]

One renowned statistical technique of Fisher's is the attempt to "reject the null hypothesis." In our agricultural example, the null hypothesis would be that manuring has no effect. The rival statisticians Egon Pearson and Jerzy Neyman were critical of this technique and developed an alternative with two competing hypotheses. Suppose, for example, that a manufacturer is testing the quality of a medication. The first hypothesis (H^1) is that it is good enough, and the second hypothesis (H^2) is that its quality is inadequate. The manufacturer faces a *moral* choice: whether to be more concerned about incorrectly accepting H^1 (with the risk that an unsafe medication will come onto the market) or about incorrectly accepting H^2 (with the risk of unnecessarily halting production and losing money). The trade-off between these two risks determines what confidence intervals the manufacturer chooses, and this choice determines how likely each hypothesis is to be rejected. Note a passing resemblance to the approach to probability taken by Christiaan Huygens, who sought a fair system for apportioning the risks of a poor decision between the parties involved.

Neyman and Pearson contended that their method was also epistemologically superior to Fisher's, in that accepting a hypothesis did not imply that it was true, but simply led you to behave *as if* it were true. Instead of inductive reasoning, Neyman spoke of "inductive behavior." In the 1960s and 1970s, the null hypothesis came under heavy criticism from other quarters. A number of statisticians showed that if the sample size is large enough the null hypothesis can *always* be rejected. The paradoxical implication of this finding is that if your aim is to reject the null hypothesis, you can spare yourself the trouble of collecting empirical data.[70]

Random Sampling

In 1825, Quételet set out to estimate the total population of the Netherlands (including present-day Belgium) by counting the number of births in just a handful of parishes and Protestant congregations. (During the *ancien régime*, birth records had been kept by churches rather than government authorities.)[71] Before that time, Graunt had tried out a similar method as part of his political arithmetic, and in 1785 Laplace had refined the technique. But Quételet met with stiff resistance from Baron Charles-Louis de Keverberg de Kessel, a member of the Council of State and former governor of the province of Antwerp. Keverberg questioned whether it was plausible that the birth rate was equal throughout the country, citing differences between thinly and densely populated areas as well as the unequal distribution of poverty. His criticism appears to have been influential; for the rest of the nineteenth century, no one tried to calculate the total population by means of sampling. Censuses of the entire population were accepted as the standard, even by Quételet.[72]

Nevertheless, a number of statisticians kept working on techniques for drawing conclusions about a whole (such as the entire population of a country) from information about only part of it. Around 1840, for example, the French engineer Frédéric Le Play developed a method known as the monograph: an exhaustive description of a "case" (such as the household budgets of workers in a particular street) that could be described as typical. It went without saying that, in case studies of this kind, deviations from the norm and chance occurrences were disregarded as far as possible.[73]

Monographs of this type were produced until the turn of the century, when they were superseded by an entirely new working method. At a series of international statistics conferences between 1895 and 1904, the Norwegian official Anders Nicolai Kiaer presented his sampling method, along with the concept of representativeness. But he was not terribly strict about selecting his samples randomly; instead, he sought a technique for checking their representativeness after the fact. Soon after, the Englishman Arthur Bowley championed rigorous random sampling, developing criteria for this method and introducing the key concept of the confidence interval. Yet typological thinking remained appealing to statisticians for some time. The Italian Corrado Gini used seven major demographic variables to identify Italian regions with scores comparable to those of Italy as a whole. His hope was that these regions would be representative of the entire country in other ways as well, but he eventually had to give up on that idea. It was Jerzy Neyman, in 1934, who ultimately developed the concept of the statistical representativeness of a sample that is still in use today.[74] Since that time, statisticians have taken it for granted that random sampling offers the best chances of representativeness.

This is not to say that typological thinking has disappeared entirely. The *pars pro toto* variant has been discredited; serious practitioners of the social sciences no longer believe that "America is Jonesville writ large." But using Jonesville as a "model" for a series of other small American towns is much more acceptable. "Types" can be treated as individuals and ordered systematically.[75] Galton himself saw composite portraits as a valid way of constructing types through generalization. The example he used was a photo of a "typical" military officer. Since the mid-1990s, neurological researchers have been producing images of "normal" brains and, for instance, the brain of a "typical" schizophrenic patient.[76]

Statistics in the Social Sciences

The statistical analysis of populations changed the intellectual landscape, and not just in the realms of biology and agricultural science. Statistics opened up new domains of empirical research in an array of disciplines, including many social sciences. Its impact on psychology was especially great. While statistics has also become fundamental to disciplines such as sociology and political science (consider voting research, for example), in psychology it is equated with the scientific method. And while biologists normally hire mathematicians to do their statistical studies, psychologists and political scientists crunch the numbers themselves. The computer program SPSS (Statistical Package for the Social Sciences) has also led to an unprecedented degree of uniformity.

In short, the embrace of statistics has had a transformative effect on many fields, notably psychology. One might even say that statistics has brought about the redefinition of psychology's subject matter. Traditional psychology focused on individuals and their functioning, just as psychiatry still does today. But between 1914 and 1950, the statistical approach taken by Galton and Fisher profoundly altered the field, causing individuals to be defined solely in relationship to a population.[77]

This transformation took place in several stages. From 1914 to 1936, the Galtonian paradigm was dominant in applied psychology but poorly represented in more fundamental psychological research. Applied psychology was mainly concerned with testing employees and recruits and, in later years, with assessing whether children and adolescents were better suited to theoretical education or to practical, occupational training. These tests were intended as tools for officials such as school principals to use in selection processes, rather than to help individuals gain greater insight into themselves or, more fundamentally, to shed light on thought processes or the psychological basis of action. What Kurt Danziger has called methodological behaviorism ("intelligence is what intelligence tests test") was central to this form of applied psychology.[78]

After 1936, starting in the United States, the Galtonian style began to dominate general psychology as well.[79] Today it is the universal standard, having almost entirely supplanted more qualitative approaches such as Gestalt psychology and the Piaget school. This has had tremendous consequences for psychology's subject matter. For one thing, when general psychology adopted the techniques of statistical analysis used in applied psychology, it also adopted the objective of applied psychology, namely to provide information relevant to political or management decisions. Moreover, general psychology's wholehearted embrace of the statistical method implies that the only questions that can be asked and the only theories that can be tested are those relating to a large set of data.[80]

Parallel developments have taken place in sociology and political science, though not to the same degree as in psychology. Voting research developed out of market research, and sociological surveys were originally developed by progressive civic groups as a way of coming to grips with the issue of poverty.[81]

Since 1936, the use of statistics in psychology has displayed a number of paradoxical features. Even before that date, psychology aspired to resemble the "hard" sciences as closely as possible, at least in its methodology. At first, this aspiration took the form of an emphasis on the experimental method, but after 1936 the emphasis gradually shifted toward statistical analysis. The many disagreements between the founders of modern statistics had an unfortunate impact. It became standard practice for psychologists *not* to name the originators of the statistical techniques that they used.

Furthermore, the statistical method became more rigid, narrowing its scope to a fairly small number of techniques, the most important of which was the significance test involving the rejection of the null hypothesis. By 1955, significance tests were being used in 80 percent of the scientific articles in four leading psychological journals. In 1962, the *Journal of Experimental Psychology* announced that articles in which the null hypothesis was not rejected would not be considered for publication. Fisher's brand of statistics thus gained a virtual monopoly in the psychological field. Experimental approaches to psychology that did not lend themselves to this type of statistical testing were forced into the margins.[82]

The peculiar thing is that because statistics, as an experimental model, came to psychology by way of agricultural science, its probabilistic origins were almost entirely forgotten. The goal of statistical research in agriculture is to arrive at a yes-or-no decision (whether or not to spread manure, for example). This suits the needs of managers and policymakers—who might, for instance, have to decide whether or not to adopt a new curriculum—but it does not do justice to the provisional character of scientific conclusions, especially those with a probabilistic basis, as the historian Gerd Gigerenzer has pointed out.[83]

Similar issues are raised by another technique in widespread use, analysis of variation. In psychology and other social sciences, the primary application of this technique is to distinguish between what are known as independent and dependent variables. This terminology avoids direct reference to cause and effect, thus creating the superficial appearance of skepticism about the causal implications of statistical analysis. But in fact, Gigerenzer argues, this technique is used to identify causal connections. Again, what lurks behind the methodology is the desire to emulate the natural sciences and, more specifically, the Laplacian ideal of determinism.

This is a paradoxical way to end the history of nineteenth- and twentieth-century statistics.[84] The value of statistics for psychology lay in its ability to accomplish through methodology what had never been accomplished through theory: the unification of the field.[85] As a result, psychology no doubt enjoys greater authority. Yet the field has paid a price: some of its traditional subject matter falls outside the scope of its new methodological criteria, and has thus been left out in the cold.

Fisher's statistical methodology is one of the clearest twentieth-century cases in which the ideal is to mechanize the production of knowledge and to eliminate personal (and hence subjective) opinion from the process of drawing "scientifically justified" conclusions. This ideal implies that every problem has a unique solution, which can be discovered by correctly applying statistical methods. Fisher reinforced that belief, despite the reservations about the statistical method that even he sometimes expresses in his work.[86] The methodology known as inferential statistics (a collection of procedures for inductive reasoning) has taken up the burden formerly shouldered by the deductive style in the natural sciences. Fisher claimed the justification for this methodology was "as satisfying and complete, at least, as that given traditionally [for] the deductive processes."[87] But that suggests that he misunderstood the nature of induction.

Fisher's belief—that if all the relevant figures were available, it would be possible to arrive at a unique, correct conclusion in a mechanical way—also underlies other decision-making techniques that rely heavily on quantification. Cost-benefit analysis is a case in point. This technique was developed in the United States in the 1920s and 1930s for use in political decision making about major infrastructure, partly in the hope of moderating the clash of opposing interests. Within a few decades, it became a standardized method and the subject of thousands of pages of regulations. One fairly late modification of cost-benefit analysis, which took place in the 1980s, was to assign economic value to distinctive landscapes that might be under threat from such projects. This move took the sting out of the fundamental objections raised by environmental activists by objectifying their concerns.[88]

Superficially, both cost-benefit analysis and standardized aptitude tests represent the triumph of the "experts." Yet if some groups of experts have benefited, others have been ousted from their advisory or decision-making roles. The losers have been the experts who traditionally drew on their professional experience to make context-dependent decisions about individual cases.[89]

This tendency has its roots in broader social developments. The quantitative basis of statistics and its standardized methods inspire confidence in outsiders and offer an alternative to depending on scientists' personal opinions. The statistical style of science has thus acquired an even greater aura of objectivity than the experimental method.[90]

10

The Evolutionary Style

At the time when Nature with a lusty spirit
Was conceiving monstrous children each day
—Baudelaire

CHARLES DARWIN PUBLISHED *On the Origin of Species* in 1859—a very late date, in at least two respects. He had arrived back in England in 1836 from a five-year voyage around the world on the HMS *Beagle*. Upon his return, he immediately began to flesh out the ideas in his notebooks, and he wrote up his theory of evolution the following year. But it took Darwin more than twenty years to go through his notes and write his book. In fact, he might have taken much longer if an acquaintance, Alfred Russell Wallace, had not decided to publish a book of his own on the same topic. Darwin then had no choice but to hurriedly publish his own work. But the field of biology as a whole was also very late to develop a mature theory of evolution, considering that the nineteenth century is often seen as the age of history, of historical thinking. Around 1800, a historical style emerged in a number of sciences: in astronomy, in geology, and, above all, in narratives of the political and cultural history of humanity. This makes it tempting to look for nineteenth-century forerunners of Darwin. Were there any?

In an ideological sense, the eighteenth-century world had been prepared for the theory of evolution by ideas about "transformism." But these ideas were embedded in scientific frameworks that no longer seemed relevant to later generations. The nineteenth century called for facts of a new kind, and for a long time such facts were not available, or not in sufficient numbers. One of the great ironies of nineteenth-century science is that the man whose intellectual style makes him Darwin's most plausible precursor, the anatomist Georges Cuvier, explicitly rejected the theory of evolution. The standard historical account of evolutionary theory explains the late publication date of the *Origin* mainly by reference to Darwin's concerns that he might offend religious sensitivities. To be sure, there was hostility to Darwin's views, but not primarily among theologians. And Darwin himself had no desire whatsoever to challenge Christianity. Furthermore, we should not forget that many of the concepts used in evolutionary theory predated the *Origin*. Darwin had many sources on which he could draw.

For instance, the idea that Earth had a history was accepted long before Darwin's day. In the seventeenth century, Robert Hooke in England and Niels Stensen (Steno), a Danish scientist working in Italy, had discovered fossil shells and sharks' teeth at high elevations in the mountains. Hooke and Steno noticed that the shells resembled known living species, but with certain differences, and this observation raised all sorts of questions. Similarly, in the seventeenth century, the philosopher Leibniz had pondered the origin of human languages. Leibniz, like most people at the time, had been reluctant to believe that Adam had spoken Hebrew in Eden. His hypothesis—namely, that Hebrew was at most somewhat closer to the common ancestor of all languages—was in fact a kind of evolutionary theory, though not for the plant and animal kingdoms.[1]

In searching for Darwin's predecessors, we must proceed very cautiously. None of Darwin's forerunners formulated their ideas in the same terms as the later theory of evolution, and most or all of them were influenced by the theories and preoccupations of their own day. The scientific standards of the seventeenth and eighteenth centuries (leaving aside Leibniz) did not allow for evolutionary explanations in a modern sense. Both ancient and biblical sources, as well as new geological and paleontological findings, were interpreted in accordance with the dominant, mechanistic mode of thought.

Some eighteenth-century theories implied that one species could somehow transform into another, but the way they described this process did not meet the standards of early nineteenth-century biologists.

The history of Earth continued to fascinate both scholars and the general public. But researchers could not reconcile their new findings about the structure and functioning of organisms with the concept of evolution from one species to another. Jean-Baptiste Lamarck was the only person to attempt this feat, and his theory (published in 1807) was too redolent of eighteenth-century ideas, and its factual basis too weak, especially for English tastes. The geologist Charles Lyell made short work of it. It was not the theologians whose opposition Darwin feared, but the geologists and paleontologists. Darwin had the greatest respect for Lyell and had pored over his book during the *Beagle* voyage. After his return, Darwin became Lyell's friend and protégé. A year later, Lyell published the fifth edition of *Principles of Geology*, again expressly denying that varieties of the same species could develop into different species.[2]

In July 1844, Darwin's ideas about the evolution of species had crystallized, and he penned a manuscript on the subject. But no sooner had he begun to try out his ideas on a few friends and colleagues than the anonymous work *Vestiges of the Natural History of Creation* hit the shelves, soon becoming a best-seller. Robert Chambers, the author, was a journalist. He relied heavily on Lamarck's theory but incorporated ideas similar to Darwin's, as Darwin himself admitted after reading the book. For example, Chambers drew connections between findings in a wide range of fields: anatomy, embryology, paleontology, and geology. Nevertheless, Chambers's "proofs" were generally too outlandish to be taken seriously. The criticism from the scientific community was devastating, and for years afterward evolutionary theory was a lost cause, at least among scholars. Darwin was glad that the criticism had not been leveled at him, and had every incentive to build a much stronger case for his own theory.[3]

When Darwin opened his last bottle of specimens from the *Beagle* voyage, in 1846, he found a very unusual barnacle. Anticipating the criticism he might unleash if he speculated about the origin of species without first proving his skill as a zoologist, he decided to describe and catalogue all the species of barnacles in the world, living or fossil—their anatomy, their sex lives, their taxonomy, their ecology, and everything else about them. He spent eight whole years on nothing but barnacles.[4]

Yet Darwin saw no choice but to publish when Alfred Russell Wallace—an adventurer whom he had met once, briefly—sent him a manuscript from the jungles of the Moluccas in 1858. Wallace's essay presented a theory of the transformation of species that was very similar to Darwin's. Darwin considered allowing Wallace to establish precedence, but his friends, especially Lyell, convinced him not to. Instead, two separate papers, one by Darwin and the other by Lyell, were formally presented at the Linnean Society in London and the work of Wallace (who was still roaming southeast Asia) was duly acknowledged. A year later, Darwin laid a still stronger claim on the theory of evolution when he published the *Origin*. When Wallace returned to England a few years later, he seemed content with how things had worked out for him, to Darwin's relief.[5]

Myths of Origin

The debate about whether Adam spoke Hebrew shows that when we turn to the early history of the theory of evolution, there is no getting around the Bible. The creation stories in Genesis and the story of Noah and the flood had a profound influence on the natural sciences throughout the seventeenth century and for much of the eighteenth. In the early sixteenth century, natural philosophers had begun taking Bible stories literally rather than interpreting them as allegories; this reflects a scientific urge to establish the facts. The first botanical gardens, in Padua (1545), Leiden (1587), and Amsterdam (1638), were products of the far-flung commercial empires of Venice and Holland, as well as of the desire to recreate the earthly paradise. The intention was to gather all the plant varieties that had been present in the Garden of Eden and were later spread around the world, and to plant them all in one garden again.[6] In the sixteenth and seventeenth century, many attempts were made to reconstruct how Noah's ark must have looked and how the many different animals on board could have fit into its stalls and decks. In the seventeenth century, as more and more species of animals were discovered, the models of the ark grew ever larger, until the reconstruction project was given up in despair.[7] Around the same time, the reconstruction of the Garden of Eden was abandoned for similar reasons. Nevertheless, the Bible remained significant in many fields of study.

It was not uncommon to return to the allegorical mode of interpretation; for instance, each day of creation could be interpreted as a different era. One of the last great scientists to draw inspiration from the Bible was Thomas Burnet, the author of *The Sacred Theory of the Earth* (1681), which portrays the flood as the epochal event in the earth's history. Burnet assumed that not only Bible stories but also all sorts of mythological tales contained a kernel of historical truth. When the Icelandic Edda sagas were discovered during his lifetime, he hoped to find in them new clues to primeval history. Burnet was sometimes mocked for his outrageous assumptions. Yet some of his hypotheses—about the formation of mountain ranges, for instance—were not so far from our current understanding. And he did influence other thinkers. The theologian and mathematician William Whiston used Newtonian physics to argue that Halley's comet had caused the flood; this theory was endorsed by Newton himself.[8]

Almost a century later, the work of the great French naturalist Georges-Louis Leclerc, Comte de Buffon, included a thorough study of Burnet. Yet Buffon utterly rejected Burnet's use of the Bible. Likewise, James Hutton and Abraham Gottlob, two geologists of Buffon's day, no longer referred to the flood in their work. These Enlightenment scholars were deists for whom Bible stories had no place in science.

It should be added that Burnet made an important contribution to natural theology, a philosophy with many adherents throughout the eighteenth century. Deist scientists and philosophers could endorse the central thesis of natural theology, which was that the perfection of the design of natural phenomena, such as an insect's eye, form proof of a divine designer. While deists concluded from reason that God had not intervened in the world after its creation, theists did not fully embrace the Enlightenment mentality and went on believing in a God who was active in the world. The only theist we will mention here is Linnaeus, who hypothesized that paradise had been located on Mount Ararat.[9]

Another text had at least as much influence on the history of Western science as the book of Genesis, and that was the Platonic dialogue *Timaeus*. The *Timaeus* presents a story of the creation of the universe by a god. This god (which Plato calls a Demiurge, a term literally meaning craftsman) is said to have created the stars, Earth, the Olympian gods, the human race, animals, and plants—in other words, a world complete in every respect. He made this world the way a potter makes pots: out of earth and fire, using air and water, and according to a pattern. Plato refers to the model for this world as perfect, eternal, and necessary. In other words, Plato's god imposed a geometric order of triangles, circles, and squares on formless matter.[10] The *Timaeus* is an astonishing account of creation that goes into incredible and sometimes grotesque detail, especially in the section about the human body. But amid the bizarre particulars, one finds all sorts of hypotheses about the body's functioning. In the Middle Ages and Renaissance, this dialogue was admired not only for its claims to objective truth, but also for the rationalistic and mathematical/geometrical method that it embodies.

Plato's direct influence extends all the way into the modern era of biology, though mainly in the margins. Goethe's morphology (theory of forms) is based to a large extent on Plato. Jussieu's typological definition of species has Platonic elements, as does the search for "archetypical forms"—organisms that may have produced new species but do not have ancestors of their own—a search that went on almost throughout the nineteenth century. The mathematical and geometric investigations of D'Arcy Wentworth Thompson are also Platonic in character.[11] According to Thompson, organisms are formed by physical forces, and their forms can suddenly change if the balance between those forces shifts. Thompson applied the mathematical theory of transformation to a number of species with astonishing results, tracing their differences to simple changes in the shape of a coordinate system. Like the seekers of ancestral forms, Thompson—whom Stephen Jay Gould has called "a glorious anachronism"—is an illustration that even after Darwin, non-Darwinian theories could still play a role in biology.[12]

As science developed, the account in Plato's *Timaeus* remained influential, because it could be merged with the biblical narrative into a kind of rationalized creation myth. The deuterocanonical book known as the Wisdom of Solomon sings the praises of a God identified with the Platonic Demiurge: "But thou hast ordered all things in measure, and number, and weight."[13] The *Timaeus* was the only Platonic dialogue that remained in circulation throughout the Middle Ages, rather than being rediscovered in the fifteenth or sixteenth century.

In *Le Monde* and *Principia Philosophiae*, Descartes presented a *Timaeus*-like reconstruction of creation. He began with a primordial chaos, which the laws of nature then molded into the differentiated cosmos we know today in an evolutionary process. Steno and Hooke relied on Descartes's reasoning to explain their fossil findings. Rejecting the idea that the flood had been a historical event, Hooke speculated about slow changes in Earth's rotational axis as a possible cause of volcanic eruptions and of rising and declining sea levels, processes that could in turn lead to changes in plants and animals.[14]

Hooke saw his own observations foreshadowed in the *Metamorphoses* of the Roman poet Ovid, in part because the transformations described by Ovid had taken place after the creation of the world and humankind. Ovid's stories were generally about individuals who turned into other living things, such as birds and bushes. Hooke took these stories more literally than medieval and Renaissance readers, who had interpreted this very popular book allegorically: through their transformation into beings that could not speak (except for the birds, who could bemoan their fate in song), these individuals became subject to the laws of nature.[15]

Mechanism and Preformationism

Descartes's and Hooke's ideas were a far cry from modern evolutionary theory. Descartes did acknowledge that Earth and its living creatures had a history, that they had looked different in the past. But the idea that an initial state and a handful of unchanging laws of nature can lead to an end state has a great deal to do with classical physics and very little to do with evolutionary theory. The present-day view is that biological evolution is not headed toward a predetermined goal, but is, by its very nature, an open-ended process. The same is held to be true of human political and cultural history. But Descartes and the Cartesians applied the doctrine of mechanism not only to physics but to all fields of study.

For instance, around 1685, the Dutch scientist Jan Swammerdam offered mechanistic rebuttals of a number of claims made by William Harvey, who was renowned for discovering the circulation of the blood and the function of the heart. Harvey defended the possibility of spontaneous generation, which was incompatible with the mechanistic worldview. According to Swammerdam, Harvey's view that new individuals can develop out of fertilized eggs (in a process called epigenesis) was utterly without merit. What sorts of mysterious powers, Swammerdam asked, could induce undifferentiated tissue to form structures? Die-hard mechanists could not tolerate the idea of obscure forces.[16] True, even Descartes himself supported the concept of epigenesis, but his attempts to explain the phenomenon failed to convince his contemporaries.

Swammerdam was a major advocate of the theory of preformationism, according to which a fully formed individual is already present in the egg. (In a variation on this theory proposed by Antoni van Leeuwenhoek, an "animalcule" was contained in the sperm.) During the development of the embryo, this preexisting form gradually unfolds. To demonstrate the plausibility of preformationism, Swammerdam conducted careful dissections, revealing a complete moth beneath the skin of a silkworm. As a microscopist, Swammerdam was unequaled, and he was greatly admired for his observations (though they were not entirely accurate by today's standards). His works were published in translation until well into the eighteenth century, and preformationism remained influential, supported by individuals such as Charles Bonnet and Lazarro Spallanzani (until Spallanzani's death in 1799). Bonnet contributed by discovering parthenogenesis (asexual reproduction, without fertilization of the egg cell) in the louse.[17] Preformationism had the advantage of being compatible with mechanism and Neoplatonism. The only disadvantage was that it placed women, not men, at the center of creation.

Preformation was also embraced in the plant kingdom by botanists such as Marcello Malpighi and Nehemiah Grew, both of whom were involved in the discovery of plant sexuality. Malpighi and Grew were ovists, believing that the fully formed plant was present in the female ovule, but other botanists were animalculists, supporting the seed theory.[18]

Eve's Ovaries and the Cyclical Conception of Time

One peculiar consequence of preformationism (which its advocates were quite willing to accept) was that all humanity had been present in Eve's ovaries. The logic is simple: if the individual is fully formed in the egg, then that individual has a fully formed ovary with a smaller egg in it, and so forth . . . This makes history no more than the unfolding of a predetermined developmental process, the turning of the pages of a book that has already been written. With a few exceptions (in particular, the philosopher Immanuel Kant), eighteenth-century thinkers saw time as something that living beings and Earth experience passively.

This preformationist mentality was also found outside the field of natural history. For instance, the well-known art historian Johann Wickelmann asserted that there were well-defined, progressive stages in the history of art, proceeding from the archaic to the grand and beautiful, reaching the mature stage, and finally descending into decadence. Each stage was nested inside the last like one Russian doll inside another.[19]

This shows that, in a sense, preformationism could accommodate a cyclical conception of time modeled after the youth, adulthood, and old age of an individual. It was generally supposed that civilizations went through a similar cycle. One well-known example was the Roman Empire, as described by Edward Gibbon in *The Decline and Fall of the Roman Empire* (1776). There was a widespread fear that eighteenth-century civilization would likewise go into decline, though there was also the hope of "progress" through technological advances. Many people were quick to draw parallels between the worlds of nature and culture. Buffon (who was, incidentally, an opponent of preformationism) saw the nonwhite races of man as degenerate versions of an original, perfect type, and tried to "regenerate" the peasants on his own estate through arranged marriages, much to the dismay of the people involved. Buffon's well-known article "L'asne" in his *Histoire naturelle générale et particulière* (*Natural History: General and Particular*, 1753), which modern readers may be inclined to read as a fascinating speculation on the possibility of evolution, should in fact be seen in this context. Buffon conjectures that a donkey might be a degenerate horse or a horse an improved donkey. But in that case, one could just as well claim that an ape is a degenerate human! Buffon could not resolve this issue, but finally concluded that there was no evolution from the fact that no new species had appeared since Aristotle's day.[20]

Transformism and Genealogy

In 1774, the young botanist Antoine-Laurent de Jussieu reorganized the plant collection at the Jardin du Roi (the current Jardin des Plantes) in Paris, with the consent of Buffon, the director of the garden. Jussieu based the new classification system on the degree of perfection and complexity shown by each plant. It was a linear system, a ladder, with fungus at the bottom and trees at the top.

This was not the first time in the eighteenth century that the *scala naturae*—the ladder of nature, or great chain of being—had been invoked.[21] Charles Bonnet had classified the animal kingdom the same way, from polyp to human. His book *La palingénésie philosophique* (*Philosophical Palingenesis*, 1769) presents a dynamic chain allowing for the possibility of all sorts of transformations. And Bonnet was not the only eighteenth-century scholar with this view; he had derived his ideas fairly directly from Leibniz. The philosopher Diderot had also come out in favor of transformism (and hence for a kind of evolutionary theory). (As mentioned previously, Buffon was opposed to this idea.) There are dynamic versions of the great chain of being in which all living creatures can reach the next rung, as it were, by striving toward God. Apes can become humans, and humans angels. Bonnet's dynamic version is somewhat more complex, because it is not the organisms as they appear to us that ascend the ladder but the germs from which they develop. And because the environment plays a role in determining what will emerge from those germs (how they will be expressed, to borrow the language of modern biology), Bonnet predicted that in the future, Earth would be populated by creatures totally unlike the ones known to him and his contemporaries. This claim of Bonnet's is an especially clear illustration that transformism does not imply a *genealogy*. A germ lies dormant and comes to maturity within an organism, but does not descend from that organism. Bonnet believed that germs could move from organism to organism, as long as they found a suitable host.[22]

Jean-Baptiste Lamarck's transformism was also closely connected to the *scala naturae*. Nonetheless, he anticipated the theoretical possibility of small sideways evolutionary steps, for the purpose of adjusting to the local environment, rather than continual progress up the ladder. Lamarck, who came up with the term "biology" in 1800, and whose magnum opus was not published until 1815, still thought in Enlightenment terms.

Despite the appearance of dynamism in Bonnet's and Buffon's theories, the changes in form about which they speculated were the result of living beings passively reacting to changes in the environment. A parallel can be drawn to views about the climate in human history. Montesquieu was the last, but also the most influential, in a line of thinkers who argued that civilizations were shaped by the climate. According to this view, it is no coincidence that the most advanced civilizations, such as that of Western Europe, lie in the areas with moderate conditions. Warmer regions lead to laziness, while in the north, progress is obstructed by the cold.[23] Along these lines,

Buffon, who liked to compare himself to Montesquieu, saw variation in climate as an explanation for the differences between the races of man (but with the underlying similarities outweighing those differences). It was not until the end of his life that Buffon began ascribing greater influence to the ways in which people could change their natural environment.

Bonnet and Lamarck ascribed a predominant role to environmental factors. And they and Buffon all believed that a given set of environmental conditions would always have the same effect on living beings, regardless of time and place. This betrays the Cartesian mode of thought, according to which general laws of nature always operate in the same way.

Baroque Mechanism

In the seventeenth century, Gassendi had developed an alternative mechanistic system that was linked to a vitalistic conception of atoms. Though it did not attract as much attention as Descartes's system, which became the new scientific orthodoxy, it had a lasting impact and inspired Newton's tentative conjectures about "living" matter. The real issue, of course, was whether a force of attraction, a variety of action at a distance, could truly be attributed to matter. While Newton preferred not to reveal that his thoughts were turning in that direction, a half-century later the astronomer Pierre-Louis Moreau de Maupertuis had no qualms about action at a distance. Maupertuis, the first French Newtonian, was famous for his heroic expedition to Lapland, where, as he waded through the snow, he made measurements proving that Earth was flatter at the pole. But he failed to pull off a heroic act of a different sort; he had hoped to convince King Louis XV to purchase one of the large Egyptian pyramids and blow it up so that its internal structure could be investigated.[24]

Maupertuis was one of the very few eighteenth-century opponents of preformationism. His refutation of the theory began with the observation that for God, there was no difference between creating all living beings at the beginning of time and creating them one by one shortly before they were born. In the latter case, however, he would have to perform a miracle every time. Wouldn't it be better, he said, to look for a material explanation? Maupertuis also made empirical arguments against preformationism, unearthing forgotten observations of Harvey's that were in conflict with the theory. He thus managed to breathe new life into the hypothesis of epigenesis. Although the empirical support for his new version of epigenesis was flimsy, the idea remained an interesting one. Maupertuis's explanation of how material particles in male and female sex cells could form organs did not involve mechanical laws, but chemical processes of attraction and affinity between the particles. Maupertuis believed that individual particles possessed the qualities of desire, aversion, and memory.[25] Memory is particularly relevant here, because Maupertuis claimed that the matter in the seed carried the memory of the father or mother. He believed that during reproduction, particles from every type of bodily organ gathered in the sperm, a theory called pangenesis that had previously been advanced by Gassendi. In the nineteenth century, Darwin's book *The Variation of Animals and Plants under Domestication* (1868) would propose a similar theory. Maupertuis's version of pangenesis allowed for the possibility that one species could transform into another; the particles could regroup in slightly different ways from generation to generation, eventually producing new species.[26]

Yet even these bold speculations remained within the intellectual horizons of the Baroque.[27] The transformations envisioned by Maupertuis were shifting permutations of what was "already there."[28] The elements that made up the body could recombine, but they themselves did not change. If change in the basic elements had been possible, the result would have been a dynamic ecology of bodily elements. Leibniz may have considered something of this kind, but Maupertuis most definitely did not.

There was another eighteenth-century thinker for whom the balance tipped away from preformationism and toward epigenesis: the philosopher Immanuel Kant. Later, in his *Kritik der reinen Vernunft* (*Critique of Pure Reason*, 1787), he would pave the way for an entirely new mode of thought. But there was no sign of that in his "pre-critical" period. "Von den verschiedenen Racen der Menschen" (*On the Different Races of Man*, 1775) is a somewhat odd work in which Kant speculated about what he called evolutionary theory,[29] advancing a number of claims and conjectures about the origins of human "races," which he suggested had all descended from a common stock. He explained the differences between races and varieties of humans as the result of modifications due to environmental conditions—specifically, climate and soil. By means of "hidden inner predispositions," humans could develop in different ways, but Kant believed that this did not (and could not) lead to the formation of new species. People living in the "ice zone" are smaller, he wrote, because otherwise they could not keep themselves warm. He also argued that people living in hot and humid climates had swollen noses and lips in response to environmental conditions. Kant acknowledged that the details of his examples (and there were many others) were purely speculative, and that it was not for him to judge their scientific merits. But he insisted on the more general point that once racial differences had emerged as adaptations to the climate, they would persist even after secondary migrations to different climates. And he offered another conjecture: in ancient times, Africa and Asia had been separated by a sea, and this was why the Negro and Hindu races had developed so differently. At this stage in his argument, Kant expressed the hope that the *Naturgeschichte*, the "history of nature," would someday become an independent science based on understanding rather than personal opinions (apparently including his own). He did not conceive of *Naturgeschichte* as the history of humankind alone; at several points, Kant applied his reasoning to plants and animals, too.

As intriguing as Kant's early speculations may be, his "hidden predispositions," like Maupertuis's theory, do not move beyond the notion that all possible variation is "already there." This is a far cry from the view explicated in his first *Critique*, which focuses on the innovative responses of individuals to their environment.[30]

Life versus Living Beings

Michel Foucault remarked that the main difference between the natural history of the seventeenth through the eighteenth century and the biology of the nineteenth is that the former deals with living beings, while the latter deals with life.[31] The new biology sought to understand the essential nature of life, whereas in natural history, the question had never been posed in those terms. Biology contended that life is capable of independent activity and enters into interaction with its environment; in contrast, the living beings of natural history were at the mercy of environmental factors. Their nature followed directly from the way in which their parts were put together, as Julien Offray de la Mettrie theorized in his celebrated treatise *L'Homme-machine* (*Man a Machine*).

The historic mode of thought that began its ascendancy in the early nineteenth century places much greater emphasis on the innovative potential of an individual or group. It assumes that life is an active principle; it is through interaction between many active lives that the particulars of one unique situation give rise to another unique situation. This makes it impossible to predict the future, but post hoc reconstruction is always possible in principle. This historical, reconstructive perspective is what distinguishes the theory of evolution from an "ordinary" theory, making it a style of science in its own right.

The roots of this new historical thinking lay in epistemology. John Locke had described the human mind as a passive recipient of sense impressions. In Locke's work, the term "impression" retains its vivid figurative meaning: the mind receives an impression just as a piece of wax is stamped by a signet ring. He and other Enlightenment philosophers saw the mind as a loosely connected set of mental functions (such as reason, imagination, memory, and will), which ideally worked together but could also be at war with one another. The most serious danger was a runaway imagination; only by keeping the imagination firmly in check could a person obtain knowledge. But more than associational knowledge was not possible in any case. The self was a nebulous thing, perhaps even an illusion. In the eighteenth century, a relatively large number of people kept a diary to ensure the continuity of the self and improve their chances of waking up as the same person the next morning.[32]

But Kant changed the terms of the debate irrevocably. According to his *Critique of Pure Reason*, all the mental functions were under the control of the will, which he also referred to as the "I." This gave the imagination a free hand to transmute knowledge from a set of loose associations into a coherent whole. In this system, observations became part of mental processes that presupposed an active, organizing "I." Within his epistemological framework, Kant thus assumed active observation of and response to the environment. But Kant did not extend this epistemological argument into a naturalistic argument. His later *Kritik der Urteilskraft* (*Critique of Judgment*, 1790) contains no more than a few vague allusions to a "positive history of nature."[33]

Instead, it was Georges Cuvier—appointed in 1795 to the chair of comparative anatomy at the *Muséum d'histoire naturelle* founded two years earlier—who transformed the epistemological line of reasoning in the *Critique* into a naturalistic one. Cuvier replaced the Buffonian term "structure" with "organization" to indicate that he saw living beings as organic wholes that relate to their environment functionally. Animals that responded similarly to their environmental conditions could therefore be expected to have similar forms of organization, and he argued that this was in fact the case.[34]

Cuvier unseated humankind from the throne of creation by doing away with the great chain of being. His alternative classification system to the linear classification of the animal kingdom involved four major branches corresponding to four basic organizational patterns, none of which could be reduced to any other: vertebrates, mollusks, articulates, and radiates. Each of these branches was also associated with a different way of life, none of which was superior to the others. This established the principle of the family tree, which only later, in Darwin's work, took on an evolutionary meaning.[35]

According to Cuvier, similarities between organizational patterns could best be determined by a comparative anatomist at the dissection table. This implied that a scholar such as Cuvier was superior to an explorer like Alexander von Humboldt, who was praised as a hero during his lifetime. A traveler gathered no more than fleeting impressions and was limited by the route he happened to take on his travels, while a scholar in his study had the entire cosmos within arm's reach.

But what this criticism overlooked was that Humboldt himself was also a kind of comparative anatomist, not of individual plants or animals, but of entire landscapes and vegetation types. As if in rebuttal, Humboldt wrote, "This is why I left Europe with the firm intention of not writing what it is usual to call the historical account of a voyage, but to publish the fruits of my researches, in works which would be purely descriptive. I have arranged the facts, not successively in the order of which they have presented themselves, but according to the relations which they have between themselves."[36] Darwin later wrote in his *Autobiography* (1887) that Humboldt's *Personal Narrative of Travels to the Equinoctial Regions of the New Continent, during the Years 1799–1804* had profoundly influenced his own decision to sail on the *Beagle*. One thing that Darwin absorbed from Humboldt was the ecological concept of organisms in harmony with their environment (though he had a different opinion about the extent of that harmony). The evolutionary play is performed in the ecological theater, as the ecologist G. Evelyn Hutchinson later said.[37]

Darwin was already personally convinced that one species could transform into another when, in 1838, a treatise by Thomas Malthus suggested to him what the mechanism of change must be; namely, the process he later called "natural selection." Malthus was a professor of history and political economy at Cambridge. His *Essay on the Principle of Population*, which appeared in 1798, argued that poverty, war, and epidemics were part of the ordinary course of life, the simple consequences of the natural law that "population, when unchequed, increases in a geometrical ratio. Subsistence [that is, food and other necessities of life] increases only in an arithmetical ratio."[38] The series 1, 2, 3, 4, 5 . . . fell further and further behind the series 1, 2, 4, 8, 16 . . . , until a shortage of food for all those hungry mouths led inevitably to competition and death. This applied to both humans and animals, according to Malthus, who saw his theory as a corrective to naive Enlightenment faith in progress. Malthus was widely read, and his insights had already been absorbed by a variety of philosophers and scientists whose work was familiar to Darwin. Charles Lyell had pointed out the ubiquity of violence in nature. William Paley, the main exponent of natural theology around 1800, had also incorporated Malthus's idea into his own thinking. One would expect natural theology to be in conflict with Malthus, because it emphasized the harmonious interconnectedness of God's creation. But Paley saw Malthus's law as a mechanism that corrected occasional imbalances.[39]

As a young man, before his voyage on the *Beagle*, Darwin was enthusiastic about Paley's work. During the voyage, however, he was struck by the ferocity of nature. Even Lyell had exaggerated the extent to which organisms were adapted to their environment, Darwin noted.[40] In this respect, Darwin drew quite different conclusions from his travels than had Humboldt, who never changed his mind about the harmony of nature. In the 1980s, the adventurer and biologist Redmond O'Hanlon retraced Humboldt's journey along the Orinoco and Amazon Rivers. O'Hanlon's tale of his travels makes it difficult to imagine how Humboldt could have maintained such a sunny view.[41] But Humboldt was looking at the relationship between climate and vegetation zones, what we would now call the landscape or ecosystem. In contrast, Darwin focused on reconstructing historical processes in nature; at first, most of these processes were geological in nature.

In short, when Darwin finally turned to the work of Malthus himself, little of what he read was new to him. Nevertheless, it was only then (as he later recollected) that the building blocks of evolutionary theory fell into place. At this juncture, it is important to realize that Darwin had been exposed, not long before, to the work of Adolphe Quételet, the Belgian statistician.[42] Darwin's scientific work could be summarized as the application of an economic and statistical model to biological populations.[43] Natural selection of individuals with even a very small advantage over other individuals in the struggle for survival (a "marginal" advantage, in economic terms) causes such large-scale redistribution of matter and energy that it brings about an ongoing historical process, which continually creates new forms. No other factors are necessary to bring about this process, which has no direction, objective, or design, and in which unpredictable coincidences play a central role, a point Darwin emphasized to others.[44] Twenty years later, in 1858, Wallace arrived at the same conclusions independently of Darwin, also after reading Malthus.

For Darwin, nature was not a chaotic war of all against all, but a complex society in which diverse organisms negotiated a division of labor. In this sense, Darwin's vision of nature was ecological through and through, in much the same way as Humboldt's had been. (The word "ecology" was not yet in use; it was coined in 1866 and did not become common currency in the scientific world until 1890.) Yet Darwin's vision was somewhat more individualistic than Humboldt's. Humboldt saw communities of species that obviously belonged together, whereas Darwin saw each species as holding a particular office in nature, an idea that corresponds fairly precisely to the later ecological concept of the niche. But in an evolutionary sense, no species could ever count on remaining in office for the long term. It should be added that in Darwin's view, the number of offices in nature was not fixed, but expanded in the course of evolution.[45] Both Darwin and Wallace saw cases in which species avoided competition through specialization and geographical divergence. Darwin was also very interested in the cooperative evolution of flowers and insects, and studied some striking examples of this phenomenon in detail.

The Crisis of Darwinism and the Origin of Genetics

Variation (among individual organisms) and *natural selection* (by the environment) were key terms in Darwinian thought. That remains true in contemporary neo-Darwinism, even though ideas about the mechanism of evolutionary change changed considerably in the twentieth century with the rise of genetics and population biology. Darwin himself saw natural selection as his greatest discovery and the cornerstone of his theory. He knew very well that this term was metaphorical and he went into great detail about the similarities and differences between the way people breed animals and cultivate crops (artificial selection) and nature's working methods.[46] Incidentally, this illustrates another major departure from the eighteenth-century belief in the overriding influence of the environment or climate; in Darwin's work, the set of species present at a given time is the product of those species' long *history* of interaction with the environment and each other. Buffon, in contrast, argued that each type of climate was associated with a specific set of organisms, as if this relationship were set in stone.

Darwin and his more orthodox followers believed that the creative aspect of evolution was mainly due to the mechanism of selection. The term "variation" was used to refer to small differences of the kind always found within a population since individuals are never completely identical. In other words, Darwin defined a population in statistical terms. The mixture of traits through sexual reproduction was continually producing small differences like these. But under normal circumstances—that is, in a stable environment—mixing traits in this way leveled off large differences between individuals, because the offspring of divergent individuals tended to be closer to the norm. Darwin pondered whether changes in the environment (caused primarily by geological processes) were the only type of condition that would give divergent individuals a reproductive advantage.[47] At this point, the comparison with artificial selection came into play. Animal breeders simulate an altered environment through their selection criteria.[48] This type of selection is the only way that evolution could take place in small steps. "Selection" thus became the central concept in Darwin's view of evolution. But did this leave enough room for creativity?

Darwin escaped this problem by postulating that organisms had the ability to migrate and seek new environments on their own initiative. This proposal went well beyond the assumptions made by earlier naturalists, who had frequently invoked hypothetical land bridges to explain the occurrence of organisms in regions that were far from one another. For some years, Darwin performed experiments to learn more about this subject; for instance, he looked into the survival rates of plant seeds after long immersion in sea water. What he hoped to discover was that natural variation resulting from sexual reproduction could, on its own, lead to new species.[49]

But the statistician Francis Galton, Darwin's cousin and an enthusiastic supporter of the theory of evolution, believed that selection for small variations could not lead to lasting new forms, let alone to new species. In his opinion, the statistical phenomenon of regression to the mean made this impossible. The only alternative, as far as Galton could see, was that evolution had proceeded in leaps and bounds, but he had no idea how that might have worked. The leaps-and-bounds theory had other points in its favor. The physicist Lord Kelvin had been trying to estimate the age of Earth based on the time it must have taken to cool to its current temperature. His calculations suggested that Earth was not very old, and even Darwin admitted that they did not made gradual evolution seem any more probable.[50]

By the late nineteenth century, the Darwinian theory of evolution was plagued by so many uncertainties that non-Darwinian evolutionary theories gained a foothold in the scientific community. Paleontologists were struck by two things: first, a kind of logic that seemed inherent to certain developmental sequences, such as the transition from the primitive to the modern horse; and second, the dramatic gaps in the fossil record. In the 1930s, this led to Otto Schindewolf's theory of orthogenesis, which stated that variation tends to move in a predetermined direction. He attributed only a minuscule role to selection by the natural environment.[51] Another view that attracted great interest around 1900 was neo-Lamarckism, which, unlike orthogenesis, assigned a dominant role to the environment, and which held that traits acquired during an organism's lifetime can be passed on to its offspring. Darwin had played into neo-Lamarckism's hand with his theory of pangenesis, published in 1876. Numerous reputable ecologists declared their support for neo-Lamarckism. All in all, the turn of the century was a period of genuine crisis for Darwinism.

The Investigation of Hybrids

Darwin did not attribute any special role in evolution to crosses between individuals of different physical types. In his view, the hybrid offspring displayed a mixture of the two parents' traits. When the differences were not terribly great—for instance, when the parents belonged to two different varieties or races of the same species—a mixture of this kind was a perfectly ordinary phenomenon and could not be expected to work miracles for his theory. Mixing the traits of two distinctly different individuals—in the extreme case, individuals of different species, such as a horse and a donkey—was unproductive in a different sense; the offspring were often much less fertile or completely infertile.

One hundred years earlier, Linnaeus had seen things differently. Later in his life, he had embraced the view that God did not create species but genera—that is, he created one species within each genus, and the others came into being later through hybridization. This idea met with incredulity from many of his contemporaries. Joseph Kölreuter, the director of the Margrave of Baden's botanical gardens at Karlsruhe, attempted to disprove Linnaeus's hypothesis. His most noteworthy result involved crossing different species of carnations. The plants in the first generation resulting from these crossings (the F_1 generation, in today's terminology) resembled one another and, in physical appearance, were somewhere between their two parents in many respects. But the plants in the second (F_2) generation were diverse in appearance and resembled the original, unmixed species more than they did the F_1 hybrids. From these experiments and others, Kölreuter concluded that new species could not come into existence. That was not the only battle he fought; alongside his hybridization experiments he also challenged preformationism.[52]

Kölreuter's work drew opposition and was partly responsible for establishing plant hybridization as a line of research. Carl Friedrich von Gaertner, the son of Joseph Gaertner, performed tens of thousands of experiments with plant hybrids, including many replications of Kölreuter's experiments. For his report on this research, he won a prize from the Royal Netherlands Academy of Sciences in 1837. Darwin later said that a number of Gaertner's findings had been useful to him, but Gaertner himself was unable to reach any general conclusions.[53]

Gregor Mendel came from a family of small farmers. He was offered the chance to join the Augustinian abbey in Brnó, a Moravian city which is now in the Czech Republic but was then part of the Austro-Hungarian Empire and known as Brünn. At the abbey, Mendel trained as a secondary school teacher. Part of his studies took place at the University of Vienna, where he spent numerous periods between 1853 and 1856. One of Mendel's professors in Vienna introduced him to the work of Kölreuter and Gaertner, but in a new and very different intellectual context. This professor, Franz Unger, was a supporter of the concept of evolution (even before Darwin published his magnum opus) and did not believe in the immutability of species. It seems certain that he passed on this belief to Mendel. Unger was also a great admirer of Matthias Schleiden's recent cell theory, which had changed the field of biology radically. In 1856, while Mendel was in Vienna, it was shown that male and female germ cells merge during fertilization, a milestone for cell theory.[54]

Mendel must have believed, despite the objections, that it was possible for hybridization of plants to produce new species. But his own experiments with pea plant varieties, which he conducted in the monastery garden in Brünn starting in 1858, yielded little evidence for this belief. On the contrary, it was the pea plant varieties that differed in only one respect, such as their color (green or yellow) or their skin (smooth or wrinkled), that later provided Mendel with the foundation of his theory of heredity: a 3:1 ratio of traits in the F_2 generation. While all the hybrids in the F_1 generation were green (Mendel called the color green "dominant" and yellow "recessive"), the yellow returned in one-quarter of the F_2 cases. The descendants of those yellow F_2 plants remained yellow; no trace of the "mixture" with green remained. Furthermore, some of the green F_2 plants had only green descendants, without a trace of yellow. Mendel concluded that the visible F_2 ratio of 3:1 concealed an underlying ratio of 1:2:1. To arrive at these numerical ratios, Mendel had to count. That was an important step, something Kölreuter and Gaertner had never done. Another major innovation of Mendel's was his concise notation for the results of his experiments: "A" for a dominant trait and "a" for a recessive one. He described the F_1 hybrid as Aa, meaning that the yellow color was concealed by the green. In this notation, the 1:2:1 ratio in the F_2 generation was described as A:2Aa:a.[55]

Mendel published his results in 1865. The main conclusion was that the characteristics of organisms did not really mix, a view that would have surprised Darwin and most cell biologists of Mendel's day. Another implication of Mendel's research was that organisms were not organic "wholes," but collections of distinct traits. Darwin might have agreed with this, but many other leading biologists maintained a "holistic" view of heredity until after 1900.

With one of them, Carl Nägeli, Mendel carried on a correspondence that lasted several years. Nägeli did not entirely trust the results of Mendel's pea plant experiments. He wrote that in later generations of pure A, something of a must still be present, and even persuaded Mendel to conduct the relevant experiments on various species of *Hieracium*, herbaceous plants in the sunflower family, and *Salix*, the family of shrubs and trees that includes willows. These experiments actually did produce results that seemed to suggest the formation of new species through hybridization.[56]

Historians now believe that Mendel did not see his 3:1 ratio as a step forward. The theory of evolution he had in mind was not Darwinian, but

Linnaean, even though he had read *The Origin of Species*. He never contemplated anything like the later theory of genetics, though ironically enough, Nägeli's thoughts did run along those lines. Nägeli, too, had read the *Origin*, and criticized it for what he saw as an exaggerated emphasis on natural selection. Nägeli thought there might be an innate developmental tendency in a certain direction, a view comparable to the theory of orthogenesis mentioned above.[57]

Darwin never learned of Mendel's 3:1 ratio. He owned a reference work published in 1880 that mentioned Mendel's findings, but the relevant pages of Darwin's copy are uncut.[58] It was not until 1900 that Mendel's 1866 article was stumbled upon by three evolution researchers, who were quick to recognize its importance. Unlike Mendel himself, they invested the 3:1 ratio with the significance it still possesses in modern genetics. The main rediscoverer of Mendel was the Dutch plant physiologist Hugo de Vries. A few years later, the British biologist William Bateson, who had been alerted to the 1866 article by De Vries, turned Mendel's hybridization results into the foundation for much of modern genetics. Bateson coined the word "genetics," came up with the F_1 and F_2 designations, revised Mendel's description of the F_2 hybrids to read AA:2Aa:aa (generously attributing this version to Mendel), and invented the term "allelomorph" (later abbreviated to "allele") for one of the possible forms of a trait or its carrier. Another few years later, in 1909, Wilhelm Johannsen introduced the term "gene" (which replaced the earlier "factor"), along with the paired terms "genotype" and "phenotype," the first of which came to stand for the genetic makeup of an individual and the second for that individual's observable characteristics.[59]

Bateson (like almost all later historians of science) credited Mendel with the insights that an individual inherits two material factors, one from his father's side and one from his mother's; that these two factors combine; and that the dominant one causes the individual to have a particular trait. If the individual produces germ cells, these factors are then separated again. The British historian Robert Olby, in collaboration with the Dutch biologist J. Heimans, a student of De Vries's and his successor (at one remove) as professor in Amsterdam, has presented strong evidence that Mendel must not have made a distinction between characteristics and their material (that is, genetic) carriers. Their primary evidence is Mendel's notation, A:2Aa:a. If these symbols referred to underlying genetic carriers, then A and a would have to be represented as pairs: AA and aa. It is tempting to read Mendel's use of the terms "dominant" and "recessive" as allusions to a material carrier, and it is true that one member of what we now call a pair of genes is dominant over the other. But Mendel may have meant that the color green was dominant over yellow in the external appearance of the plant.

Olby also points to developments in biology between 1866 and 1900. Around 1885, the cell biologist Eduard Strasburger in Bonn had described not only cell division but also the simultaneous splitting of the cell's nucleus, a process called mitosis. Another cell biologist, the Austrian Walther Flemming, had noticed the role played in this process by small threadlike structures in the nucleus, which readily absorbed dye. These were later called chromosomes. Many scientists agreed that the carriers of hereditary traits must be located in the nucleus. A year earlier, fertilization had been observed to take place through the fusion of the spermatozoon with the egg cell. Then, in 1887, the Belgian biologist Edouard van Beneden observed that reproductive cells only have half the usual number of chromosomes, while fertilized cells have a full complement.[60]

Mutations and Variations

Hugo de Vries had studied plant systematics and biogeography at the University of Leiden with W. Suringar, an avowed opponent of the theory of evolution, and experimental plant physiology with Julius Sachs in Würzburg.[61] In those days, any ambitious botanist eventually found his way to Sachs, whose botanical laboratory was held to be the finest in the world. There De Vries met Francis Darwin, and made contact with Francis's father, Charles. Sachs was one of the earliest Darwinists, but like his mentor Nägeli, he gradually moved toward a view more like orthogenesis.[62] De Vries, on the other hand, identified himself all his life as a supporter of Darwin's theory of evolution. Of all the Darwinist biologists, including the British ones, only Sachs agreed with the theory about the material carriers of hereditary traits proposed by Darwin in *Variations of Animals and Plants under Domestication* (1868).[63] According to this theory of pangenesis, small particles called gemmules formed throughout the body and collected in the reproductive cells. These gemmules were responsible for hereditary traits, each one corresponding to the body part, cell, or tissue that originally created it. Together, they could form a unique new individual. But De Vries was not convinced by Darwin's speculations about gemmules traveling through the body. In *Intracellulare Pangenesis* (*Intracellular Pangenesis*, 1889), he renamed them "pangenes" and located them in the nucleus, arguing that they were passed on unchanged from generation to generation.

Starting in 1882, De Vries had set up a series of crossbreeding experiments, including some that involved hybridization. By 1894, these had led him to the conclusion that a variety of traits could be passed on independently of each other. A year earlier, he had discovered the statistical work of Quételet and Galton, and he theorized that what he called the "1.2.1 law" was a consequence of the binomial distribution. In his studies, De Vries had observed a 3:1 ratio of inherited traits on several occasions, and he drew a link between this distribution and the statistical one. In 1900, he read Mendel's 1865 article. Later that year, De Vries published an article in which he (rightly) claimed that he had independently arrived at Mendel's laws before seeing them confirmed by Mendel. What he derived from Mendel's article was the idea of *paired* characteristics, an idea that Mendel himself probably never had.[64] Nevertheless, De Vries soon lost interest in Mendel's laws and went in search of something greater. Not long after, he actually found what he was seeking.

De Vries was also able to interpret the results of his other crossbreeding experiments with the help of one of Galton's ideas, namely, regression to the mean, which implied that evolution could not take place simply through repeated selection for particular characteristics within a population. Like Darwin, De Vries was very interested in the work of professional breeders and plant cultivators. He noticed that they valued aberrant traits, or "monstrosities." This was where the potential for new species began.[65] So De Vries was fully prepared when, in a meadow near the Dutch town of Hilversum, he made his well-known discovery of an aberrant group of evening primroses. He used these flowers in many experiments at the Hortus Botanicus in Amsterdam, and they were central to his two-volume work *Die Mutationstheorie* (*The Mutation Theory*, 1901 and 1903).

De Vries's new mutation theory was a stunning success, especially in the United States, which he visited many times. He flattered his American hosts by pointing out that the evening primrose was originally from the Americas, and had come to Europe as a garden plant before escaping into the wild. De Vries's American supporter Daniel Trembley MacDougal published an edition of his work that made much easier reading than the voluminous *Mutationstheorie*.[66] Many biologists believed that De Vries's theory solved major problems with Darwinism. Among them was Thomas H. Morgan, who carried out a famous series of experiments on fruit flies (*Drosophila*) not long after.[67] Fortunately for De Vries's reputation, Morgan used the term "mutation" to include much smaller differences than those De Vries had observed in evening primroses. In his work with fruit flies, Morgan had observed small divergences of the kind he described.[68] Later, when De Vries's evening primroses turned out not to be mutants but hybrids (a cross between two American species, one from the East and one from the West Coast, which originated in the English port of Liverpool), the term "mutant" nevertheless remained in use for more or less the phenomenon that De Vries had initially had in mind.

MacDougal believed that mutation theory presented the possibility of steering evolution in favorable directions by artificially inducing mutations.

One of MacDougal's colleagues experimented with exposing plants to radioactivity and studying the changes in their chromosomes.[69] MacDougal seized upon these results to support his vision of creating new life forms.

Field biologists opposed De Vries's mutation theory. Over the next few decades, they lost the battle for the right to participate in the debate on heredity and evolution. De Vries, MacDougal, and Wilhelm Johannsen denied that studies of the geographic distribution of organisms could contribute to the understanding of evolution, and they also questioned the value of paleontological studies. Ultimately, though, De Vries found himself left behind by history as well. Though the Danish biologist Wilhelm Johanssen made grateful use of De Vries's mutation theory, his work, along with Morgan's, opened the way to the integration of mutation theory with Mendelian laws of inheritance. This was not a step that De Vries had wished to take.

The Neo-Darwinian Synthesis

Darwin had leaned slightly too far toward selection, and mutation theory restored the balance between the phenomena of variation and selection. Nevertheless, many biologists had great difficulty with the idea of mutation, and so evolutionary theory remained in crisis. It took another thirty years before "Mendelian" genetics and mutation theory were united with Darwinian natural selection in an evolutionary model known as the neo-Darwinian synthesis. Field biologists such as Theodosius Dobzhansky and Ernst Mayr played a major role in this process, accepting the models developed by Fisher and others as a framework for interpreting their own findings.[70]

The neo-Darwinian synthesis integrates a wide variety of mechanisms and phenomena that bear on the evolutionary process. Even today, almost all biologists regard it as largely successful. Nevertheless, it is still very much in flux. Its relevance was narrowed somewhat in the mid-twentieth century (mainly the 1950s and 1960s) by a nearly exclusive emphasis on problems of natural selection and the adaptation of organisms to a specific environment (a concept known as fitness), problems that were formulated at a fairly high level of abstraction. This has been referred to as the "hardening" of evolutionary theory.[71]

Factors causing this process may include the nature of the studies that led to the evolutionary synthesis, which were more often laboratory experiments than field work. On the other hand, the "physics envy" that was at its height in the period in question led to a quest for "laws" of evolutionary theory, and more generally for a model of explanation that was hierarchically organized, deductive, and comprehensive. This quest was provoked in part by the philosopher Karl Popper, who described the theory of evolution as a metaphysical construct that could not be falsified and was therefore unscientific. Even so, Popper saw it as a viable research program, which could form a framework for theories that were falsifiable (for example, the theory of a specific biological mechanism responsible for a particular adaption).[72]

Other deductivist philosophers developed axiomatic systems based on the neo-Darwinian synthesis and contended that this theory was better than its Lamarckian, saltationist, and orthogenetic competitors. In 1970, the philosopher Mary Williams presented a complete axiomatization of evolutionary theory based on three abstract concepts: a replicator (such as a gene), a tester (such as an organism), and an evolver (such as a species).[73] One of her axioms corresponds to what is known more loosely as the survival of the fittest:

> If (a) any Darwinian subclan, D, has a subclan D_1, and (b) D_1 is superior in fitness to the rest of D for sufficiently many generations, than the proportion of D_1 in D will increase.

This axiom presupposes another axiom about *fitness*:

> For each organism, there is a positive real number that describes its fitness in a particular environment.[74]

On the basis of this reconstruction (only a few elements of which have been presented here), the philosopher Alexander Rosenberg has classified the theory of evolution as a deterministic theory with a few probabilistic elements, not fundamentally different in status from Newtonian mechanics. In response to Ernst Mayr, who maintains that the theory of evolution consists primarily of historical explanations of unique events, Rosenberg claims that historical explanations conceal natural laws—possibly many at once. In the Anglo-American literature, this view is known as the covering law model.[75] If Rosenberg is right, then biology is a special case of physics, and the evolutionary style of science is not an independent mode of explanation but can in principle be reduced to the deductive style.

The problem with this view is that it has no practical relevance for either historians or evolutionary biologists, as can be shown by both systematic philosophical arguments and more pragmatic lines of reasoning. On the systematic side, consider that if we wanted to reduce unique events to one or more laws, we would inevitably have to develop new laws for each event. Furthermore, for each event, a theory or model would have to be developed stating how the laws in question interact.[76] Williams's axiomatization applies only to an idealized mathematical construct, not to the differentiated ecological reality that forms the stage for the evolutionary history of life.

On the pragmatic side, the criteria used by historians and evolutionary biologists in their research have nothing to do with a covering law model, and a great deal to do with the plausibility of a historical narrative, often in comparison to rival historical accounts. Rosenberg's work comes out of a tradition in which philosophers of science were supposed to develop criteria that good scientists were then expected to follow.

It would, by the way, be incorrect to think that the philosophical views described above have no relevance to the history of evolutionary theory. The point, in this context, is not whether they actually influenced past scientific practice or whether they formed an expression of it. The greatest weakness of the neo-Darwinian synthesis, at least during part of the 1960s and 1970s, was its assumption that organisms were passive, when in fact they are "ecologically flexible."[77] In this respect, Darwin's own biological work was much richer than "hard-line" neo-Darwinian evolutionary theory. But one can interpret ecological flexibility even more radically than Darwin himself did.

Evolutionary biologists have recently proposed that organisms actively construct their own ecological niches and that this fact is just as central to a proper understanding of evolution as is natural selection.[78] Niche construction is a process in which an organism alters the environment for its own benefit. Conventional evolutionary theory viewed this simply as one way for an organism to adapt to its environment (birds and ants build nests, for example). But niche construction also alters the conditions under which natural selection takes place, especially when it leads to large and potentially dramatic changes in the environment (for instance, when beavers build dams). Scientists who have developed this concept point out that species can make choices when selecting a new habitat, and that individuals can learn and pass on what they have learned to other members of their species (and, to some extent, even to other species). Birds, for instance, have learned to open milk bottles by pecking off the tops.

Evolutionary Processes in the Physical World

The nineteenth century, the century of history, brought a historical dimension even to the development of physical systems. This was not an obvious step to take. One feature of mechanical processes, even in the Newtonian universe of action at a distance, is that in theory it does not matter whether they move "forward" or "backward." Time is a neutral dimension. Every event is part of a cycle repeated for all eternity in a cosmos that runs like clockwork.[79] Laplace believed that, with full knowledge of the present, it was possible to calculate both the past and the future with certainty. In a mathematical sense, the difference between past and future was a minus sign; the universe was in eternal balance. Laplace's contemporary Etienne

Bonot de Condillac suggested that in descriptions of relationships between objects, the temporal factor could be eliminated. It was his belief that even though, when making an analysis, the human mind perceived the qualities of an object as following one another in succession, in fact they existed simultaneously. Even events that seemed to take place over time had not been so intended by God, Condillac asserted; the connections between things were timeless.[80]

By definition, the super-equation that Laplace believed would describe the universe had to include a law of conservation, because apparently the universe remained eternally identical to itself in some fundamental way.[81] But what was it, precisely, that was conserved?

Later, nineteenth-century historians identified Simon Stevin's prohibition on the existence of a perpetual motion machine, in 1586, as the earliest version of a law of conservation. According to the historian Philip Mirowski, the perceived necessity of a law of conservation stems from a moral judgment with its roots in economic life: "Nothing can be created out of nothing." In the economic context, this is a statement about value. Conservation principles lend themselves to quantification; Mirowski suggests that they are an ethically responsible bookkeeping system for nature from the days of merchant capitalism.[82]

Descartes formulated the first true law of conservation—namely, the conservation of mass multiplied by velocity, mv in modern notation. This led the dispute between the Cartesians and Leibniz in 1686 referenced previously. Leibniz claimed that *vis viva* (living force, or mv^2) is conserved in mechanical interactions such as collisions. It was not until the mid-nineteenth century that the word "energy" was proposed as a replacement for *vis viva*; after that, the new term gradually gained ground.

The years around 1800 saw a series of discoveries that involved transformations between what we now recognize as different forms of energy. In 1800, Volta designed the first battery and generated a powerful electric current from chemical energy. Volta thought that his battery was actually a perpetual motion machine, which stopped working only because of some kind of blockage. Earlier, in 1712, Thomas Newcomen had constructed a steam engine for the mining industry, which had been greatly improved around 1765 by James Watt. This engine converted heat into mechanical energy.[83]

In 1834, William Hamilton formulated a very general conservation law. In this case, what was conserved was the sum of *vis viva* and an undefined variable (now called potential energy). In the 1840s, the law of conservation of energy was purportedly discovered simultaneously by at least four people: Julius Mayer, James Joule, Hermann Helmholtz, and the lesser-known Ludwig A. Colding. Mayer and Helmholtz derived this law mainly from biological phenomena; Mayer discovered that venous blood is much redder in the tropics, and Helmholtz proceeded from his view that an organism was a kind of combustion engine, a theory designed to explain the source of body heat in animals.

The law of conservation of energy is known as the First Law of Thermodynamics. Around 1850, two scientists were working toward what is now called the Second Law. In his theoretical analysis of earlier work by Sadi Carnot on the steam engine, Rudolf Clausius not only rediscovered the First Law of Thermodynamics, but also noticed an entity that he called entropy, which unlike energy did not remain constant but tended toward a maximum. Clausius also drew on research by William Thomson, the later Lord Kelvin, who had observed around 1844 that during the various types of transitions from one form of energy to another—in a combustion engine, for instance—not all the energy remains available for work.

This is now a well-known phenomenon, described in terms of thermal efficiency, but in Thomson's mind it had moral implications. He described whatever was not available to perform work as "waste." From the economic perspective, machines should be productive, and waste means idleness or careless use of energy.[84] Thomson's thinking was inspired by the use of the term "work" for the operation of machines. The word had first been used in this way not long before. James Watt, the inventor of the modern steam engine, had spoken of its "mechanical effect," and terms such as "mechanical power" and "laboring force" were also proposed.[85]

Thomson quickly drew a general cosmological conclusion from the Second Law: the level of entropy in the universe can only increase over time. Given that entropy is a measure of chaos—in other words, of unusable heat—the universe as whole seemed to be headed toward what Clausius called "heat death." In both isolated physical systems and the universe as a whole, this gave processes an irreversible time dimension, a kind of evolution.

Of course, this evolution was not Darwinian. Darwin inspired many people to see progress in nature, the emergence of ever higher forms of life. Compared to that vision, the Second Law of Thermodynamics is very pessimistic indeed. But Thomson (who watched Clausius take credit for discovering the Second Law) put a much less pessimistic spin on the phenomenon, claiming that Clausius had gone much too far in extrapolating from the dissipation of mechanical energy to a cosmic principle. Clausius might be right, Thomson said, if the universe and matter were finite. But that was not what Thomson believed.[86]

In general, the English and Scottish had little tolerance for pessimism in those days. The combustion engine, along with the irreversible changes that it wrought, was seen as a symbol of progress and the Industrial Revolution. If things fell apart, this would merely serve to make room for a new world. Thomson had confidence in the creative potential of energy.[87]

But he too acknowledged the certainty that our sun would someday burn itself out. That meant an end to life on Earth as well. In 1862, Thomson wrote in a popular magazine: "As for the future, we may say . . . that inhabitants of the earth cannot continue to enjoy the light and heat essential to their life, for many million years longer, unless sources now unknown to us are prepared in the great storehouse of creation."[88] There may be a large theoretical difference between the heat death of the universe and the demise of the sun, but for ordinary people it amounts to the same thing, and in this respect, Darwin was an ordinary person.[89] Moreover, Thomson had addressed his article directly to Darwin, whose evolutionary theory Thomson believed was not up to scientific standards. In the same article, Thomson also confronted Darwin with the supposed fact that the sun and Earth were too young to allow for the immensely long time that evolution would have needed to generate life as we know it. After Darwin's death, this argument seemed to support Hugo de Vries's theory of large evolutionary leaps due to mutation.[90]

Yet throughout his life, Darwin never abandoned the idea of slow, gradual evolution. He was concerned about Thomson's counterargument, but it was the looming death of the sun that really upset him. Darwin wrote in his *Autobiography* (1876), "Believing as I do that man in the distant future will be a far more perfect creature than he now is, it is an intolerable thought that he and all other sentient beings are doomed to complete annihilation after such long-continued slow progress."[91] The debate flared up again in 1896, after radioactivity was discovered by Henri Becquerel, and raged even more fiercely after 1903, when Pierre Curie informed the world of the slow but continuous emission of energy by radium. Charles Darwin did not live to hear this news, but his son George responded that same year, pointing out the significance of the discovery for the theory of evolution. Soon afterward, Ernest Rutherford threw his weight behind this argument, asserting at a well-attended lecture that, in view of radioactivity, the sun's lifespan must extend much further into both the past and the future than Thomson had deemed possible. Thomson was still alive and even present at Rutherford's lecture, but unlike his younger colleagues, he was not convinced.[92]

Nevertheless, the pessimistic interpretation of the Second Law gained the upper hand in the early twentieth century, and from that time onward, it was Clausius's version that set the tone. Many physicists and biologists continued to question the compatibility of the Second Law with Darwinian evolution. Before the Second World War, there were three camps: vitalists who believed that the Second Law did not apply to living creatures, hard-line physicists who thought it should be possible to use the Second Law to explain life itself, and an intermediate group that believed in an as yet unknown law or explanatory principle that would reconcile these two points of view.[93]

Incidentally, the term "vitalism" should be used with some caution and not interpreted solely in the sense given to it by the embryologist Hans Driesch and the philosopher Henri Bergson. But Bergson's views did inspire the Dutch physicist Jan Burgers (who later joined the MIT faculty and became an American citizen). In a debate with the biologist Lourens Baas Becking during the Second World War, Burgers defended the view that life was not subject to the Second Law.[94] According to Burgers, "life" was "the name for coordination of spontaneity on a large scale." He believed that an organism's reactions could "not be fully explained by any tradition of pure inheritance" and that "a causal explanation of life was impossible."[95] The epigraph to Burgers's book was "Life creates matter."

Burgers ignored a proposal made by the physicist Erwin Schrödinger in 1944, which later became much more influential. Schrödinger saw living beings as islands of low entropy in a vast ocean of ever-increasing entropy. "It is by avoiding the rapid decay into the inert state of 'equilibrium' that an organism appears so enigmatic . . . How does an organism avoid decay? . . . What an organism feeds upon is negative entropy."[96] But Schrödinger had not figured out how living organisms accomplish this miraculous feat.[97] In 1955, the chemist Ilya Prigogine in Brussels developed a model of life that reflected Schrödinger's view, based on an open system of chemical reactions of the kind found in a living cell.[98] Such a system is in a steady state, but "far outside of [chemical] equilibrium." Matter enters and leaves the system, and the entropy of the incoming matter is lower than that of the outgoing matter. This is consistent with the requirements of physical law, in that the system and its environment, considered as a whole, show an increase in entropy. Yet at the same time, the entropy of the system considered separately can decrease. Prigogine thought that a system of this kind would necessarily develop in a certain direction, namely, toward minimal entropy production.[99]

In 1968, however, Prigogine's ideas about cellular systems began to change significantly. He had learned of the work of two Russian scientists who had studied a strange and visually spectacular system of chemical reactions. On the basis of the Belousov-Zhabotinskii reaction named after them, Prigogine designed a model for a system whose development was open and determined by chance. Sometimes the continual minor fluctuations around the average value for a variable in the system would not die out or fade away as usual, but grow stronger. There was no telling when this would happen, but the process would bring the system into a new steady state. Strengthened fluctuations of this kind were not necessarily destructive; they could lead the system further from thermodynamic equilibrium and, in that sense, be creative. When a fluctuation reaches the threshold at which it may lead to a new state, this moment can be interpreted as an "event." The physical systems in question have histories marked by a series of such events. It is impossible to tell in advance, through deductive reasoning, which events will be important to the future of a system. "By definition, an event cannot be deduced from a deterministic law," Prigogine and the philosopher Isabelle Stengers write.[100] This implies that the development of a cellular response mechanism is a historical process, like Darwinian evolution. If this is a potential model for the development of all physical systems "outside of equilibrium," then the theoretical distinction between physical and biological systems is superseded, in favor of the latter. Prigogine was, as it were, developing a metaphysics of fluctuations to counterbalance the iron logic of the Second Law.

Present-day students of biology who expect to learn all about evolutionary theory are usually disappointed. Those who persist in this desire, and make their way to a research department or museum working on the evolutionary reconstruction of life, will soon find that the organization's research managers have other priorities. It is not that evolutionary theory has disappeared from biology, but it is no longer central to the field.

In 1938, the term "molecular biology" was invented by Warren Weaver at the Rockefeller Foundation in the United States. It stood for a new vision that the American historian Lily E. Kay has called the "molecular vision of life."[101] In the latter half of the twentieth century, this vision has gone from triumph to triumph. In molecular biology, Darwin's metaphor of the family tree is of marginal relevance. A different cluster of metaphors has supplanted it: the book of nature, the code, information, and the computer program.

In the 1970s and 1980s, molecular biology developed into biotechnology. With the recombinant DNA technique, pieces of one individual's DNA can be inserted into another individual's genome. Because DNA is "universal," it does not matter whether the donor and recipient belong to the same species; mice can be equipped with bacterial genes or human ones. The polymerase chain reaction (PCR) technique, presented in 1985, makes it possible to identify strands of DNA.[102] The use of these two techniques, recombinant DNA and PCR, has swelled to industrial proportions, and they are routinely applied in countless laboratories all over the world. The sequencing of the human genome and those of other organisms could never have been achieved so quickly without PCR. In cancer research and other branches of biomedicine, nearly all experiments are now carried out on transgenic mice or other transgenic laboratory animals.

The enormous detail in which biomedical research seeks to describe molecular phenomena—not only genes, but also proteins and all the machinery of the cell—stands in stark contrast to the concept of the gene in population genetics, the study of evolutionary changes in populations. The historian and molecular biologist Michel Morange has remarked that to population geneticists, the gene is a black box. Conversely, molecular biologists have shown little or no interest in the evolutionary aspects of the functioning of genes in the cell. The theory of evolution and molecular biology (including biotechnology) have turned their backs on one another.[103]

Yet an evolutionary perspective can help to critically assess the most popular metaphor in molecular biology, that of the code. The proposition that DNA contains the code for our biological processes, as if it were software for our "wetware," suggests an ontological difference between the genome and the rest of the cell. But this idea is utterly misleading, Morange tells us. DNA, RNA, and proteins are macromolecules that work together. In the history of life, the cell predated DNA. Furthermore, embryologists helpfully remind us that the formation of the first cell with the potential to become a new individual, the cell created by the union of two germ cells, is not "controlled" by DNA. At the earliest embryonic stage, the DNA is not yet active. Rather than claiming that the cell is controlled by DNA, it would be just as reasonable, if not more so, to say that DNA is controlled by the cell. As it has been said, "The failure to embrace evolution is the Achilles' heel of molecular biology."[104]

11

Science in the Twentieth Century

> *I have to keep going, as there are always people on my track. I have to publish my present work as rapidly as possible in order to keep in the race.*
>
> —Ernest Rutherford

ONE MIGHT NOT EXPECT a hierarchy to exist for the six styles of knowing, since it is not as if any one style forms the foundation for any of the others. Yet through much of the twentieth century, there was, in fact, a hierarchy of styles, with the deductive style at the apex. This bolstered the status of physics as standing at the apex of the disciplines.[1]

Nevertheless, much of what was presented as theoretical and deductive science consisted of relatively modest middle-level theories. In other cases, supposed premises or first principles involved a concealed analogy, a hypothetical assumption underlying the deductive framework. It was in the decades that followed the Second World War that the deductive style held the greatest prestige, even in sciences that seemed far removed from physics. Textbooks in chemistry, ecology, and many other fields bore titles beginning with "Fundamentals of" and "Principles of." In the United States, the historical approach to geography lost ground to the expanding discipline of "new geography," which developed a deductive framework for "geographical space" so that the spatial order could be interpreted as subject to general laws.[2] Even the social sciences aspired to resemble physics, a phenomenon known as "physics envy."[3]

There are counterexamples, such as schools of economics that focused on compiling large data sets and analyzing them statistically, and anthropologists who engaged in interpretative work. Taxonomy survived in private organizations, such as the American Museum of Natural History and the Missouri Botanical Garden, and government institutions such as Kew, the British Royal Botanical Gardens, and the Dutch National Herbarium.

Within the field of physics itself, the volume of experimental research outweighed that of theoretical physics. In retrospect, we cannot conclude that experiments served only to test theoretical ideas. Current historical reconstruction stresses the relative independence of experimental practice.[4] At the time, however, a different view prevailed. It is no exaggeration to say that every scientist and social scientist was aware of the cultural prominence of high theory. Even engineers accepted the view that technology was derived from theoretical science by way of "applied science." Karl Popper, arguably the most influential philosopher of science of the twentieth century, said: "The theoretician puts certain definite questions to the experimenter, and the latter, by his experiments, tries to elicit a decisive answer to these questions, and to no others. All other questions he tries hard to exclude.... Theory dominates the experimental work from its initial planning up to the finishing touches in the laboratory."[5]

Starting around 1980, however, the deductivist ideal came under a great deal of criticism. This shift took place even within physics, and in closely related disciplines such as chemistry, whose practitioners went on paying lip service to physical chemistry and quantum mechanical explanations for some time, but worked with a chemical theory of the atom that had many pragmatic features and few strictly deductive ones.[6] Moreover, several philosophers of science identified strict limits to the applicability of deductive thinking.[7] At the universities, and certainly in society at large, technology now enjoys a status equal to that of science.[8] Several sciences seem to have moved to a seemingly one-sided focus on the acquisition of data, to such an extent that some have expressed a fear of "drowning in data." Although it is tempting to be nostalgic for a healthy dose of theory, the current data revolution in the sciences poses interesting challenges of a new kind.

Shifts in the relationships between the styles of science during the nineteenth and twentieth century show some correspondence to changing conceptions of the "utility" of scientific research. The Humboldtian view was that pursuing research for its own sake, "in solitude and freedom" (*Einsamkeit und Freiheit*), fostered a higher form of utility and contributed to the formation of individuals in the service of the state. In the period just prior to the First World War, utility was identified with technological progress and other societal payoffs, and although clearly separated from science, this changed the very meaning of "research." The post-1945 period saw these new concepts gain strength, even as their scope expanded to an unprecedented degree. Around 1980, the ties between technological development and the emancipatory process of "progress" were loosened, and the definition of "utility" was left to a multitude of different societal actors.

The Formation of Disciplines in an Educational Context (1800–1914)

In 1939, the historian of science J. D. Bernal asked why the idea of "pure science" arose in the early nineteenth century, "at the time when science should have been most obviously connected with the development of the machine age."[9] One noteworthy illustration of the new ideal of purity was the mathematics of Joseph-Louis Lagrange and Augustin-Louis Cauchy. Lagrange did not use diagrams and geometrical constructions in his algebraic work, and Cauchy set new standards of "rigor" in mathematical proof. The existence of pure mathematics implied its separation from physics and the other sciences, which from then on were seen as domains of applied mathematics.[10]

Physics and chemistry parted ways in a similar manner around the same time. During the eighteenth century, chemistry had been considered a specialized branch of physics (or natural philosophy, as it was usually called). One major achievement of the "chemical revolution" of Antoine-Laurent Lavoisier was to demarcate the two fields, making it possible for them to emerge as separate disciplines. Like chemistry, physics became a more specialized subject than *physique générale* had been in the eighteenth century.[11]

The new specialization infuriated the English poet Samuel Coleridge, who in 1833 forbade the scientific practitioners of his day to bear the distinguished title of "philosopher." William Whewell took up the suggestion and, as a substitute, proposed the word "scientist," in an ironic reversal of the original meaning of the term *scientia*.[12] Whewell continued to refer to himself as a philosopher, however, as did many others, including the experimentalist Michael Faraday. But in all major European countries, the final rupture between philosophy and science occurred around this time. In Germany, the death in 1831 of Georg Friedrich Wilhelm Hegel, the philosopher of absolute idealism, signaled a parting of ways between philosophy and *Wissenschaft*. Historians took the lead in the revolt against philosophy; two of the defining characteristics of what was later called historicism were its positivist reliance on facts and its insistence that all cultural phenomena should be explained and understood as purely historical.[13] In France, the anti-metaphysical positivism of Auguste Comte, the inventor of sociology, was a major ideological influence on the *Société de biologie*, founded in 1848.[14] The sciences of the nineteenth century were, above all, empiricist.

Despite the early disciplinary separation of physics and chemistry, it took many decades before the two fields hardened into their twentieth-century forms. The Royal Society did not establish specialized branches until 1838, and even then, electricity was grouped under chemistry, and the chemistry commission was chaired by Faraday. University professors had considerable freedom of choice and exploration in their research, and the schools of thought that they established only gradually negotiated shared disciplinary identities through journals and learned societies. By 1873, firm disciplinary boundaries had been established through most of Europe.[15]

From the early nineteenth century on, the universities enjoyed the patronage of the new nation-states. University alumni went on to work in state bureaucracies and the educational sector. Industry was relatively unimportant as a source of employment for university graduates. Between 1800 and 1899, only 7 percent of the alumni of Cambridge University went into industry or commerce.[16] Chemists educated in Giessen or Göttingen employed their newly acquired laboratory techniques in education or (in a few cases) as government agricultural experts.[17]

Logical Positivism, Deductivism, and Relativity Theory

In the early 1800s, laboratory science was synonymous with chemistry, but later in the century, physicists began working in laboratories too. Physics professors in the German states, France, England, and the Netherlands requested laboratories from their ministers of education, and received them for status reasons and to help them attract bright students. In a paradoxical way, it can be said that experimentalism underscored the status of physics as a pure science, in that all its concepts and principles were derived entirely from physics itself. The rapid development of precision measurement in the second half of the nineteenth century boosted physicists' confidence in observation. As several historians have noted, physics became "modern" by rejecting mechanicism (which was based on mechanical analogies and therefore on metaphysics), along with the classical concepts of matter and force. The experiential notion of energy took their place.[18]

Around 1900, several physicists and some chemists rejected the very notion of the atom, claiming that it was not grounded in experience. Moreover, the physical chemist Wilhelm Ostwald wanted to confine science to defining relationships between measurable quantities: energy, space, and time.[19] This view was quite similar to the philosophy of logical positivism. The philosopher Rudolf Carnap and the mathematician Hermann Weyl considered observed phenomena (which retained the individuality of the observer) as stepping stones toward invariant numerical relationships between phenomena. Logical positivists postulated that these numerical relationships would be true for all individuals and hence "objective." Maxwell's equations seemed to fit this view perfectly. Originally derived in conjunction with a mechanical model of electromagnetic ether, they had survived a number of theory changes in physics. According to Henri Poincaré, these equations had acquired the status of abstract mathematical structures and were no longer dependent on theories.[20]

As Albert Einstein developed his ideas about relativity theory, he came close to this position. His special relativity theory of 1905 is based on two "principles," one of which is the constant speed of light. (The other is the validity of the Maxwell equations.) Earlier physicists had taken the constant speed of light to be an empirical outcome, but Einstein elevated it into a principle, an axiom. This calls to mind the deductive ideal in its pure form, as presented in Aristotle's *Posterior Analytics*, according to which first principles are the result of repeated and generalized observation, guided by intuition. There is a sense in which the speed of light *must* be constant; otherwise, we would have a universe in which time could flow backward, in which we could receive a letter not yet put in the mail.[21] In 1933, by which point Einstein's thought showed hardly a trace of positivism, he wrote, "I hold it true that pure thought can grasp reality, as the ancients dreamed."[22]

Until 1921, Einstein stressed the overriding importance of observed facts when writing to colleagues and friends, and his tone was sometimes almost positivistic. He seemed willing to submit relativity theory to a kind of experimental test, through observation of a solar eclipse. Yet in 1907 he had refused to "to let the 'facts' decide the matter" in the case of an apparent refutation of special relativity theory. In 1919, upon receiving a cable from Arthur Eddington on the island of Principe that the measurement of the curvature of light by the sun's gravitational field during a solar eclipse had confirmed relativity theory, Einstein responded with indifference. He was certain that his theory was too beautiful ever to be refuted, even if contrary evidence were found.[23] In a way, Einstein was right. His theory of relativity was a masterly reformulation of nineteenth-century physics from a new deductive standpoint. But Daston and Galison point out that the theory of relativity should be taken as a whole, complete with principles, observations, and conventions.[24] In form, relativity theory relies on axioms much like Aristotelian first principles. In content, it is a hypothetical theory in a modern sense.

The "Utility" of Science Redefined (1914–1945)

Just prior to the First World War, university education started to acquire a new function: training employees not only for government bureaucracies and the educational sector, but also for research institutes. These institutes, which in Europe were usually publicly funded, contributed to the technological advancement of industry.

In 1870, the total number of students at all German universities was 13,000, and this figure had not changed significantly since 1830. But a trend of steady, robust growth led to a total enrollment of 61,000 in 1914. This expansion was facilitated by the German authorities in a number of ways. The *Gymnasium* was no longer the only path to the university; other secondary degrees gained recognition as valid qualifications.

The obvious place to seek an explanation of this wave of expansion is in the upward mobility and new social privileges of the lower middle class. This was one factor, but its impact was limited because the cost of education rose in this period. Another, more significant factor was the rising number of students who entered universities "through the side door," from the *Besitzbürgertum* (commercial and industrial middle class) rather than the traditional *Bildungsbürgertum* ("cultured" middle class). The *Besitzbürgertum* was starting to demand its share of the cultural capital that until then had been enjoyed only by the *Bildungsbürgertum*, and this change reduced the gap between the political and industrial elites. Furthermore, large companies developed bureaucracies that were not unlike government bureaucracies; a university diploma gave a competitive edge in the former, just as it had in the latter for some time already.[25]

Around 1870, an emerging German industrial lobby urged the governments of the German countries to promote technological and "applied" research at the universities. This lobby was successful to a limited degree. Many German professors saw applied research inside university walls as a threat to their academic freedom. But they could not prevent the *Technische Hochschüle* (institutes of higher education in natural science and engineering) from obtaining the right to grant doctoral degrees in 1900. Government and industry began to work together directly at the Physikalisch-Technische Reichsanstalt (PTR, established 1887) and the Kaiser-Wilhelm-Gesellschaft (KWG, established 1911). The PTR was for the most part publicly financed, but also received a large donation from the German industrialist Ernst Werner von Siemens. It soon became a model for the U.S. National Bureau of Standards and the National Physical Laboratory in the United Kingdom. In the KWG's budget, by contrast, business contributions outweighed those of the German authorities.

The PTR and KWG, along with the laboratories of IG Farben and other chemical industry leaders, were the first major research institutes to be entirely independent of the universities. They did influence university research, however, simply by providing a job market for graduates. By doing so, they at least partially transformed the character of the universities, shifting the emphasis from molding the character and morals of the students to producing knowledge. The meaning of "research" changed into the definition we hold today.

After the First World War, the United States overtook Germany in industrial research. In the period prior to the war, the United States had produced the enduring archetype of the inventor as lone genius, personified by Thomas Edison. This was a myth that Edison carefully nurtured. Yet research laboratories like his, which worked for large companies such as Western Union, began conducting large-scale experimental research during his day or soon after.[26]

During the first half of the twentieth century, industrial working methods were adopted by universities in the United States. Some university researchers were actually working for industry, initially just conducting routine tasks.[27] But in the period that followed (1930–1970), considerable intellectual latitude was granted to researchers in the employ of leading industrial research labs (such as Bell Laboratories, established by the American telephone company AT&T, as well IBM, and the "NatLab" at Philips).

In both America and Europe, university professors were loyal to the institutions for which they worked, but *employees*, all the same. Some American professors were much quicker than their European counterparts to develop into entrepreneurs. The most successful of these academic entrepreneurs transformed the Massachusetts Institute of Technology (MIT), Stanford University, and the California Institute of Technology (Caltech) from relatively little-known institutions into huge research enterprises. European professors could, as Dominique Pestre writes, permit themselves to be "intellectuals" or "scholars" not only in the humanities but also in the natural sciences. They usually saw themselves, and were seen, as torchbearers of culture. Henri Poincaré and Marie Curie in France, Julian Huxley in England, Einstein and Max Planck in Germany, and Hugo de Vries in the Netherlands had a direct rapport with an interested public that valued their exploration of the secrets of matter and life as a contribution to culture and progress.[28] In short, they were heroes. When Hendrik Lorentz died in 1928, the entire Dutch nation mourned.[29]

But these scientists were also torchbearers of culture in another sense. As scholars, they represented the values of the Enlightenment and modernity. That made it their duty to serve the public interest in times of war and peace. During the First World War, Marie Curie offered medical assistance to French troops and the International Red Cross on the battlefield; the German chemist Fritz Haber developed the deadly weapon mustard gas, and Edison was a senior advisor to the U.S. Navy.[30] After the war, Lorentz and Einstein were active within the League of Nations. Many prominent scientists served as advisors to national governments; one example is Max Planck, who found himself in a highly precarious position after Hitler came to power.[31] The tremendous expansion of the scientific community after the Second World War was accompanied by a similar expansion of scientists' advisory role. The sociologist Zygmunt Bauman referred to scientists, in this capacity, as "intellectual legislators," heirs to the Enlightenment *philosophes*, whose task was to speak truth to power.[32] The alliance of truth and power (science and the state), which was possible thanks to their formal separation, created the conditions for the never ending march of progress.[33]

Yet, in another sense, science had become a profession like any other. The notion that there was a group of people engaged in the impartial pursuit of truth cried out for a sociological explanation. In 1919, Max Weber discussed the nature of science as a vocation in the matter-of-fact context of an ordinary academic career. In 1942, Robert Merton identified values, or norms, of science, two of which were communalism and disinterestedness. Merton contended that his theory was not psychological but sociological in nature; these norms were systemic requirements imposed by the social subsystem known as science rather than personal traits of scientists. Individually, scientists were just like anyone else.[34]

Scientific Philanthropy in the United States

Also around the start of World War I, some American universities developed major research programs in which funding was not linked to education. Europeans often associate the United States with close ties between academia and the private sector. The importance of such ties should not be overstated, however, as we can see from the case of the university that was most successful, throughout the twentieth century, at raising funds from the business sector: MIT. Admittedly, in 1913, MIT was granted a substantial endowment by AT&T to be spent on research without restriction; but it received still larger donations from the Carnegie Institution of Washington, founded by the wealthy philanthropist Andrew Carnegie.[35]

Opposing camps within MIT disagreed about the extent to which researchers should focus on the needs of the business sector. On one side of the debate was Arthur Noyes, who in 1903 had used funding from Carnegie and other benefactors to establish the Research Laboratory of Physical Chemistry. On the other was William Walker, who set up the Research Laboratory of Applied Chemistry five years later. Each of these institutes had its own staff and budget, but Walker's laboratory earned its revenue from contracts with companies. Walker was highly successful and managed to put Noyes on the defensive, so much so that in 1919, Noyes resigned. It was partly due to the presence of Walker and likeminded individuals that

MIT received an endowment of more than 10 million dollars from the rich industrialist George Eastman. But after the economic crash of 1929, private funding dried up, while money from philanthropic institutions such as the Rockefeller Foundation bypassed MIT because research there tended to be "applied" rather than theoretical. By this time, MIT had already begun to lose the brightest students to other institutions. Alarmed, MIT's governing corporation appointed Karl Compton, a physicist at Princeton, as president of MIT, and he shifted the emphasis back to basic research.[36]

At that point, relatively few American universities harbored major research institutes within their walls. But in 1932, financing for scientific research within universities became available due to a change in policy at the Rockefeller Foundation. A newly appointed officer, the engineer Warren Weaver, managed to convince the foundation's managers that a system of relatively small grants, awarded to a much larger number of scientists than in the past, would be a far more effective way of achieving the modernization that he believed necessary in the field of biology. His main intent in establishing this grant system was to convince physicists and chemists to apply their advanced techniques to biological problems. What issued from these efforts was the subdiscipline dubbed "molecular biology," a term first used in the Rockefeller Foundation's 1938 annual report. Weaver can also be credited with inventing modern "science policy," though that term later came to be associated primarily with the objectives of national governments.[37]

Weaver further anticipated the science policy of the postwar years in his expectation that an investment in what he regarded as basic science would yield social returns in the long run. Weaver and the Rockefeller Foundation were not dealing in contract research in which scientists would perform specified tasks as they did for the private sector. Charitable foundations have generally had well-defined cultural and scientific objectives and have not required a quid pro quo from scientists and their universities. The objective of the Rockefeller Foundation, as redefined in 1927, was to promote the "advancement of knowledge," especially the "basic knowledge" (in the natural sciences) underlying the medical sciences, agriculture, forestry, and engineering.[38] In other words, the rhetorical distinction between fundamental and applied science, typical of science policy from 1945 to 1975, was a central feature of the Rockefeller strategy even before that time.

And the Rockefeller Foundation was permeated with the culture of the private sector. The lucky scientists who received funding were well aware that their grant proposals were actually business proposals and that they were competing for financial assistance. The historian Robert Kohler has concluded that this mode of support imbued science with business methods and managerial values.[39]

After the Second World War, officials at the new National Science Foundation adopted the Rockefeller method for awarding grants.[40] Similar organizations elsewhere, such as the Netherlands Organization for Pure Scientific Research and the Belgian National Fund for Scientific Research, also studied and followed the example of the Rockefeller Foundation. The key elements adopted in the Dutch and Belgian context were the emphasis on "pure," rather than applied, research; support for university researchers, rather than the establishment of independent research institutes; and "great flexibility" in the expenditure of research funds.[41]

Governments and the Benefits of Basic Science (1945–1980)

If 1945 marks a watershed for science, it is that in the decades that followed, government support for "basic science" reached an unprecedented level. This change was most apparent in the United States, where, in the prewar period, government support had been more or less confined to the agricultural research stations at the land-grant universities, and legal barriers prevented federal support to scientific research at the universities. But 1945 was not so sharp a breaking point in respect to the longer-term societal "utility" of what was thought of as basic, disciplinary-based research.

The Debate on the National Science Foundation

During the Second World War, Harley Kilgore was a young senator for the Democratic Party and an outspoken supporter of President Roosevelt's New Deal. When he learned about the enormous war-related research effort taking place at MIT and elsewhere, he had many concerns about the conditions under which this research was taking place. Eventually, he developed a plan for the manner in which the federal government should support fundamental research after the war. In late 1945, he presented a complete bill to establish a National Science Foundation (NSF).[42]

When the NSF was founded in 1950, it differed in many ways from the organization that Kilgore envisioned. Kilgore wrote his bill to make the NSF responsible for planning scientific research. He wanted to ensure that research would have social relevance, be evenly distributed across the country, and remain under the direct control of the U.S. president. Kilgore's main opponent was Vannevar Bush, who had become president of the Carnegie Institution of Washington in 1939, had received Roosevelt's support for the establishment of the National Defense Research Council (NDRC) in 1940, and was thus in control of enormous budgets. The largest NDRC contract was with the MIT Radiation Laboratory, where radar was being developed. In short, Bush was already a well-established figure in science policy by 1944, when he heard about Kilgore's activities. At Bush's urging, Roosevelt assigned Bush the task of developing a plan for the organization of postwar science. By July 1945, he had completed his report, titled *Science: The Endless Frontier*. Bush's sympathies lay with the participants in contract research: the private sector and the entrepreneurial universities. Giving them authority over government funding was out of the question, of course, but he wanted to make sure that scientific research would not be controlled by politicians. Bush's proposed solution was to put elite-level scientists in charge of distributing financial support to their peers. Bush's report met with a warm reception in the United States, which of course had a great deal to do with the prestige garnered by physicists during the war.

The Second World War and the Cold War

The Second World War was won by radar and ended by the atom bomb, to paraphrase the historian Daniel Kevles. That fact brought science, especially physics, unprecedented standing. The ties between science and the state became stronger during World War II (in the Manhattan Project), and stronger still during the Cold War. In particular, the launch of Sputnik by the Soviet Union in 1957 led the United States to spend more than ever on fundamental scientific research and to greatly improve mathematics education in secondary schools.

Those who recall the 1945–1970 period as the golden age of unfettered, pure research, or as the age of scholarship in the ivory tower, are overlooking the stunning extent to which academic science relied on defense contracts and fed into military programs.[43] Consider just one figure: in 1960, the U.S. federal government financed 70 percent of the country's industrial research in electronics.

In the early 1950s, two hundred American universities received funding from the Pentagon. Most of it went to just a handful of institutions, with MIT leading the pack. Universities also educated the many physicists and engineers needed in the defense industry. MIT, in particular, exercised a special kind of influence; many of its graduates became professors and instructors at other universities and reshaped the curriculums there to reflect the MIT perspective.[44]

Under these conditions, how could the postwar physics community still trumpet its commitment to "pure" science? The development of the atom bomb and radar had convinced the U.S. military that funding "free" scientific research would make the greatest long-term contribution to America's defense. That same belief was held in industrial circles: as long as scientists were given the means to proceed with their research, the results would naturally lead to technological advancement.⁴⁵

But the NSF had a great deal less money to spend than the military. The scientifically inclined U.S. Navy, which believed that the Army had done little to earn the prestige it gained from the atom bomb, showered money on scientists through its Office of Naval Research (ONR). Most of them were only too happy to accept this funding, even if their aims were antimilitaristic. The ecologist Arthur Hasler, who had found an explanation for salmon's homing instinct, received funding from the ONR to research the foraging behavior of bottom-feeding fish. It was thought that Hasler's research might be relevant to submarine navigation close to the sea floor. But Hasler was not asked to address military applications in any concrete way, and he was very pleased that the navy was spending its money on ecology rather than submarines.⁴⁶

In the 1960s, a new term was coined to describe the emerging context for experimental work: "Big Science."⁴⁷ Although the term can refer to a variety of phenomena, it was most closely associated with enormous research laboratories financed directly or indirectly by the public sector. The American space organization NASA is one striking example, other examples include magnetic fusion research at Lawrence Livermore; Brookhaven National Laboratory with its large particle accelerators; and CERN, the European Organization for Nuclear Research. Many of the Big Science institutes (often affiliated with universities) had the task of contributing to national defense. Individual researchers, however, often experienced the research climate there as typical of fundamental science.⁴⁸

Many physicists shared Hasler's perspective, but in retrospect, a number of them felt cheated. The historian Paul Forman has expressed sympathy with their point of view: "We, like them [the physicists], saw the enormous expansion of basic physical research as a good in itself, a praiseworthy diversion of temporal resources to transcendental goals." "Transcendental" science is science whose truth goes beyond the limits of human existence, an ideal form of knowledge that was the dream of many eager physicists. Forman's final conclusion is sharp: the outpouring of military funds had a substantial influence on "unfettered, pure scientific research," pulling it in a military direction, and a small majority of physicists were well aware of this influence at the time.⁴⁹

Did the Cold War have an equally great impact on fields of science other than physics? Outside of a few exceptional cases, the answer appears to be no; or at least the emphasis did not shift toward national defense objectives. Instead, military funding seems to have had the greatest impact on "ideological" factors. In DNA research, for instance, concepts such as "code" and "information" derived their appeal in part from the Cold War setting. America's "war on cancer," declared in 1971 during Richard Nixon's presidency, is a very different reminder of the wartime rhetoric associated with science in those days.

The relationship between the Cold War and fundamental research has not been as thoroughly studied in Western Europe as it has in the United States. Although Western Europe was literally at the foot of the Berlin Wall, it did not join the United States on the ideological front lines. This made it possible for European physicists to employ the rhetoric of pure science without dirtying their hands. In Great Britain, for example, universities were not involved in defense research. But if we look at total spending on research and development by the British ministries responsible for defense, both before and after the war, we see that it was several times the amount that went to universities. In the 1960s, scientists and historians of science pointed out that this spending related to "technology" rather than "science." The historian David Edgerton has argued that drawing a line between science and technology, helped scientists to maintain the image of clean hands.[50] In British universities at that time, research did not have the status and urgency that it was to acquire in the 1980s. Greater priority was given to passing on an intellectual tradition within a relatively small social elite. At Scottish universities, which are among Europe's oldest, physics was known as "natural philosophy" until the 1980s.[51] The social superiority of science to technology was at the core of this tradition.

The pride and joy of European science was CERN, the institute for high-energy physics in Geneva. At the time it was established in 1952, it was a fantastically expensive initiative that seemed to have no other purpose than to facilitate European cooperation in the field of research. CERN's innocent nature seemed to be underlined by the decision to focus on high-energy physics—the exploration of the fundamental particles of matter—rather than the nuclear research that had led to the atom bomb. Economic factors were also at play; many of Europe's finest scientific minds were relocating to the United States, and CERN was intended to curb this brain drain.

But John Krige and Dominique Pestre, historians of CERN, have found that this picture may be too idyllic.[52] Fully aware of the sensitivity of the issue, they exercise the greatest discretion in discussing CERN's military significance. But they do cite Werner Heisenberg—the representative of the German government involved in the process of establishing CERN and a confidant of Konrad Adenauer, then-chancellor of the Federal Republic of Germany—who remarked that CERN would foster expertise that would be of great use in sectors such as the defense industry. Various other physicists involved in CERN expressed similar views. As it happens, years later, CERN became a force for *détente*, opening its doors to researchers from what was then still the Soviet Union.

In retrospect, public confidence in science reached a high water mark in the 1960s, as did the self-confidence of the scientific community. There was great optimism about what science could achieve,[53] and scientific research grew at an unparalleled rate; there seemed to be no end in sight. Two successive presidents of the U.S. National Academy of Sciences, Frederick Seitz and Philip Handler, independently predicted in 1963 that by the year 2000 half of the gross national product would be devoted to science and technology.[54] There was no more than a sporadic awareness that science and technology could not only solve problems, but also cause them. A year earlier, Rachel Carson had raised environmental issues in *Silent Spring*, but it would take many years more before her message began to hit home.[55]

Starting in the late 1960s, captains of industry—especially the defense and chemical industries—began to lose confidence in "fundamental" science and cut their scientific research budgets. The Pentagon also started asking questions about the effectiveness of its financial support for fundamental science. In 1962, the philosopher Michael Polanyi had described the practice of pure science, tongue in cheek, as wasting money with impunity in the name of truth.[56] Later, the science journalist Daniel Greenberg identified quite a few cases of wasteful spending on major scientific projects.[57]

The erosion of the image of scientists as disinterested servants of truth had another dark side. Before 1970, almost no one even contemplated the possibility that scientists might commit fraud. In the decades that followed, this changed completely. In 1989, the U.S. government set up the Office of Scientific Integrity—later part of the Office of Research Integrity (ORI)—within the Public Health Service in response to a string of alleged fraud cases that had come to light since 1974 and received considerable media attention.[58] The regulation of the search for truth could apparently no longer be fully entrusted to the scientific community itself. The first major case the OSI handled involved suspicions of "research misconduct" by Theresa Imanishi-Kari, an assistant professor at MIT and coworker of the Nobel laureate David Baltimore. The charges were partly based on reports by the Secret Service, which investigated the case at the request of a congressional subcommittee. The affair had started in 1986. A full ten years later, it was ruled in appeal that the OSI had failed to prove its allegations. Imanishi-Kari and Baltimore were exonerated.[59]

The "Baltimore case" had begun with a falling-out between Imanishi-Kari and a postdoctoral researcher working with her. The argument related to the trustworthiness of data supporting part of an argument in a paper already published in a scientific journal. Depending on which side one takes, the data were either too sloppy or open to differences in expert judgment. Baltimore's view was that further research was necessary in order to supply new data and new opportunities for interpretation. This view is congruent with the exposition of the experimental style. Baltimore's adversaries at the OSI and in the congressional subcommittee wanted to ascertain the validity of just the facts under consideration. If their "truth" could not be upheld, they were fraudulent. There was no public interest in the data other than the question of whether tax dollars had been well spent in their production. The fact that respected science journalists took sides against Baltimore gives one pause for thought. The journalists stressed the desirability of holding science "accountable"—but accountable in terms of what? Apparently, the journalists had a different view of science than Baltimore had, one that is closer to the data-oriented styles of science.

Mission-Oriented Science

The end of the period from 1945 to 1980 was marked by a belief among government leaders and officials that science could be shaped and guided by public policy. In other words, it was thought that politicians could assign scientists a "mission." This reflected a public demand for science to become socially relevant.[60] As for national policy on research funding, Kilgore seemed to be scoring a victory over Bush, some twenty years after the fact. In 1971, a new NSF operation began: Research Applied to National Needs (RANN).[61] After 1971, physics received less financial support, while significant funding was injected into health and environmental research. The National Institutes of Health became a bureaucratic colossus with large headquarters and departments at various universities.

Preceding RANN, and in some ways a forerunner to it, was a large research venture into ecology called the International Biological Program (IBP). The House Subcommittee for Science, Research and Development of the U.S. Congress approved the IBP after hearings in 1967 on the "adequacy of technology for pollution abatement," and earmarked funding for its establishment through the NSF. The basis for the subcommittee's commitment to the IBP was a strong belief in systems ecology's technological potential. It recognized systems ecology's perceived ability to manage large-scale, total ecosystems (for instance, the deciduous forests of the eastern part of the United States, taken as a single whole). "Total ecosystems" referred to a comprehensive view that encompassed literally all biotic and non-biotic relationships in an ecosystem. The managerial ethos of systems ecology was broadly conceived and of a general and abstract nature. It attempted to elucidate the entire inner structure of ecosystems by developing comprehensive computer models. These models were believed to be physical analogies of ecosystems. On that basis, ecologists believed that the models could support policy by producing results to questions of pollution abatement and enhancing "productivity" of ecosystems.[62]

The IBP was a monodisciplinary Big Science ecology program. In the United States, it divided into five projects, housed at universities and national laboratories with teams of up to 130 scientists under each principal investigator. At the end of the IBP, only two biome projects were able to complete a comprehensive model: the grassland biome model, called ELM (developed at Colorado State University) and the eastern deciduous forest biome model, overseen by an ecology group at Oak Ridge National Laboratory. By the time of the IBP's termination, the Oak Ridge modeling team had ceased to believe in the IBP's original mission. They saw their model as a theoretical plaything and a heuristic tool. But the grassland biome modelers took the mimetic qualities of their model very seriously, notably the conditions it specified for the ecosystem to maintain its overall stability. Unfortunately, the model received rather bad press in the scientific journals after IBP's termination.[63]

Although IBP ecologists had promised useful applications during the congressional hearings, the fulfillment of these promises was postponed. The ecology of the IBP was pure science. At Oak Ridge, a separate department was set up to serve the then-growing market of environmental impact statements. Stanley Auerbach, the principal investigator of the eastern deciduous forest biome project and director of the division of environmental sciences at Oak Ridge, maintained the distinction between pure and applied science within his own division.

The IBP was of a short duration (1968–1974). In 1990, ecology received a new opportunity to participate in a Big Science operation, namely, the International Geosphere-Biosphere Programme (IGBP, operational since 1990, currently planned until 2013). The IGBP is an interdisciplinary program that investigates global climatic change in which ecology is one of several contributing disciplines. There are many marked differences between the IBP and the IGBP; several features of the latter are typical for post–Cold War science. Despite its large size, the IGBP is not really Big Science, as its teams are much smaller than those in the IBP. In this way, the IGBP more closely resembles traditional "Little Science" that takes place at universities. Yet its planning activities take place continually; the Organisation for Economic Cooperation and Development (OECD) has coined the term "Distributed Mega-Science" to describe the IGBP's operations.

NASA started the chain of events eventually leading to the establishment of the U.S. Global Climate Research Program and the IGBP. An initial attempt in 1982 involving NASA's leadership failed to win support, but several years later, middle-level officials at NASA formed an alliance with their counterparts at the National Oceanic and Atmospheric Administration (NOAA, into which the former U.S. Weather Bureau has been merged) and the NSF. Much to the surprise of everyone who knows the landscape of U.S. government agencies, these individuals worked together well. They obtained the support of their own agencies, lent support to the mobilization of elite earth scientists around a few high-profile planning committees (respecting and frequently blurring the boundaries between science and policy), and they received a favorable reception for their ideas in the White House. NASA benefitted from the strong orientation of the global change program toward the development and use of state-of-the-art technology in earth monitoring equipment in space.[64] The earth scientists in NSF programs receive free data, many of them more than they can handle.

For many practitioners of the sciences in the IGBP—ecologists, oceanographers, climatologists, geographers—the flood of geospatial data seemed a mere instance of "technology push." In defense of the agencies involved, it might be said that in climate science, data are extremely important. On one hand, there is public interest and support for climate change research, on the other hand, questions about the reliability of data supporting the existence of global warming trends and the "human fingerprint" in these trends have been contentious at times.

The data orientation of the IGBP has an impact on the field data collected by ecologists as well. They are asked: are the observations and outcomes of the field experiments generalizable beyond the local features of the ecosystem which they happen to study? The larger context of climate change research makes the basis for this question obvious, and scaling-up of the investigations is a requirement from which no ecologist in the IGBP can escape. In principle, two sorts of generalizations can be made: one from (published) case studies, the other directly from the data assembled for the case studies. The first, while indirect, would leave intact the original context of the data. The data revolution, however, invites the second type of generalization.[65]

The reuse of data by other researchers places high demands on the format of the data. Data should be calibrated and standardized. Tacit knowledge around field data needs to be made explicit. The NSF now requires the separate publication of data and of metadata (data about these) from ecologists receiving funding through NSF programs. Making data accessible to other researchers requires a degree of standardization and organization that is not usually achieved by individual researchers. For geospatial data, it has been noted that producers and users are in the process of becoming separate communities.[66] A similar fate might be in store for ecological field data as well.

The Neoliberal University in the Twenty-First Century

In 2007, the *Albany Times Union* reported that a professor at the State University of New York (SUNY) at Albany would be earning a salary of more than $700,000 a year. Admittedly, this was no ordinary professor, but the chief administrative officer of the University's College of Nanoscience and Technology, who had persuaded IBM and other companies to invest billions in the college. Nevertheless, the news raised eyebrows. The salary was three times that of New York State's governor, and SUNY is funded from the state budget.[67]

A less fortunate incident befell Kary Mullis, inventor of the polymerase chain reaction (PCR), for which he received the Nobel Prize in 1993. At the time of his discovery, Mullis was working for a private firm, the Cetus Corporation, in Emeryville, California. After paying Mullis a $10,000 bonus, Cetus sold the patent rights of the PCR technique to Hoffman-LaRoche for $300 million. For the record, it would be unfair to Mullis to portray him as a pitiful victim, not to mention inconsistent with the picture he paints of himself in his 1998 autobiography *Dancing Naked in the Mind Field*, in which his comment on the affair is, "Screw Cetus."[68]

Compare the production of monoclonal antibodies by César Milstein and Georges Köhler at Cambridge University. Milstein and Köhler published their results in the scientific journal *Nature* without applying for a patent. The research council that had supported their work did not patent the discovery either; nor did Britain's National Research Development Agency, which announced that it was not "immediately obvious" what the "patentable features" of the technique might be. Today, monoclonal antibodies are a billion-dollar industry.[69] As Sheila Jasanoff has pointed out, this story would have played out quite differently today. Even the tiniest technological advance is now armored with patents and aggressively marketed in the hope that it will make money. In economic terms, this is a justifiable strategy. When fundamental scientists present a proof of principle of a potentially useful mechanism, it is often argued that this is only the first step down a long and expensive developmental path. Private investment is essential and can only be obtained if the basic finding is protected, rather than being in the public domain where anyone can use it. But there are important arguments against this position as well. Patenting is simply not the core business of universities, and it draws them into a type of entrepreneurialism to which they are ill-suited. Moreover, the practice of patenting has escalated to such a degree that it may hamper rather than promote innovation.[70]

Kary Mullis and the chief administrative officer in Albany are examples of entrepreneurial scientists, and they, rather than Milstein and Köhler, represent the new scientific norm. One more than symbolic indicator of this change is the Bayh-Dole Act, passed in the United States in 1980 as a bipartisan initiative supported by both Democrats and Republicans. This act not only permits universities to retain the revenue from their own patents, but is also generally interpreted as requiring them to cooperate with the private sector in order to maximize that revenue.[71]

Ironically, the Bayh-Dole Act has had a tremendous influence on the development of university research, but not because of its primary objective, increasing revenue from patents, and not even because it encouraged cooperation between universities and industry. In 2005, American universities spent a total of $45.8 billion on research, to which the federal government contributed $29.2 billion. The private sector contributed another $2.3 billion. This is a tiny figure in comparison to the total amount budgeted by businesses for in-house research and development ($191 billion of their own money in 2005, augmented by $20 billion or so attracted from federal agencies). The private sector has played a diminishing role in academic research since the 1990s, when this form of cooperation reached its apex.[72]

Universities that have pinned their hopes on patent licensing are probably in for a rude awakening. Once in a great while, some lucky institution hits the jackpot. In 2005, Emory University earned $525 million from an AIDS drug developed there, and Stanford made $336 million in a licensing deal with Google for Internet search technology. But as the science journalist Daniel Greenberg has observed, these universities held the golden tickets. Many more have entered the licensing lottery with big dreams and ended up disappointed. The fact is that the large majority of universities are giving serious thought to the costs that they pay for technology transfer.

When contracts between universities and the private sector relate to anticipated rather than actual discoveries, there is a different sort of price to be paid. Berkeley made a $25-million-dollar deal with biotechnology giant Novartis, granting it the right to read doctoral dissertations prior to publication and a strong voice in the university committee that determines research priorities. The case received national attention when a critical Berkeley employee who had opposed the agreement with Novartis came under fire because his results were not to the company's liking. This same employee was later denied tenure.[73]

By now, American universities have come to the unhappy realization that working with the private sector has other substantial costs. Furthermore, nothing poses so much of a risk to a university's credibility and good name as collaboration with the pharmaceutical industry, especially on clinical research. The editors-in-chief of leading medical journals such as the *New England Journal of Medicine* have repeatedly warned that the system of impartial review of scientific articles is under severe strain. Nevertheless, in 2002, that journal was forced to rescind its ban on reviewers with ties to the drug industry, because impartial reviewers were no longer available. Another, even greater problem for journals is that of ghost authorship. A firm offers a finished article to a professor; all he has to do is sign his name to it. If the article is published in a top journal, he receives a $20,000 reward. According to a 1998 estimate, 11 percent of the articles in medical journals are ghost-written.[74]

If Bayh-Dole has accomplished anything, it has been to promote academic entrepreneurship as a norm at every level—among central university administrators, deans of colleges, and, last but not least, individual researchers. At MIT and Stanford, that business ethos long predated Bayh-Dole. Now it has become a universal standard, and not just in the United States, but even for professors in small humanities fields at public universities in Europe. In 2003, the Dutch universities required all faculty members to take part in fund-raising, and made accomplishments in this area a key criterion for promotion. What began as a means to the end of supporting academic research has now become an end in itself. The result has been a run on funding agencies and external sponsors and clients.

The Neoliberal Society and the Public Sphere

Neoliberalism, relative to the classical liberalism of the nineteenth century, inverts the relationship between the state and the economic sphere. Classical liberalism accepted the primacy of the state, but fenced off a space in which the state was to leave economic actors alone (the domain of *laissez-faire*). Neoliberals proceed from different premises. For them, a society consists of economically motivated actors; everyone is a *Homo economicus*; and the principle that drives the actions of both businesses and individuals is competition. The state serves only to facilitate the activities of these parties by making sure that their meeting ground, the market, runs in an orderly fashion. As the master of the market, the state is pro-active and cannot afford to adopt *laissez-faire* policies. In other words, for neoliberals, competition is a basic principle, an axiomatic point of departure.[75]

In both Europe and the United States, various factors prevented national governments from pursuing a purely neoliberal course after World War II.

Aside from the cultural persistence of more socially orientated views, the main factor was the Cold War. War and preparations for war require a strong state.

The British prime minister Margaret Thatcher and U.S. president Ronald Reagan began setting up neoliberal-style government programs as soon as they entered office in 1979 and 1980, respectively. The most important signal that the state was withdrawing somewhat from society were cuts in both taxation and government spending. Starting in 1980, government subsidies in various countries for noncommercial scientific activities at universities dropped consistently every year. In some countries, the process was gradual, though no less constraining. In others, like Great Britain, it took place in leaps and bounds. In 1981, British universities saw their total budgets cut by an average of 11 percent (and some by 20 to 40 percent). More than twenty-five years later, they have not yet fully recovered.[76] The fact that, in spite of these cuts, more academic research is now taking place in Britain than before 1980 is largely due to earmarked funds for special, well-defined purposes. This type of research is only weakly linked to university education.

The withdrawal of governments from public space and their neoliberal faith in market forces also found expression in reduced efforts by the state to influence investment. The philosopher Jean-François Lyotard, who noted this phenomenon in 1979, saw it as an effect of what he called the "diminishing appeal of socialism."[77] These tendencies were closely connected to, and took place in the same period as, the disintegration of the military-industrial complex (the brief revival of the Cold War that appeared to take place under the Reagan Administration during the Star Wars episode was not a significant barrier to this trend). In the 1970s, the major high-technology defense firms had diversified their product lines so that they no longer depended on an exclusive relationship with a single government (that of the United States) and could embrace the neoliberal agenda of competitiveness. In the field of health care, too, the "medical-industrial complex" (centered on the NIH) gave way to a leading role for "Big Pharma" firms, which produced drugs as if they were consumer goods.[78]

Research and development to enhance the competitiveness of businesses became a top government priority during the presidencies of George H. W. Bush and Bill Clinton in the United States and Tony Blair's tenure in Great Britain. But while in the United States many success stories were told of profitable technology transfers from academia to industry, European governments wrestled with the perceived inability of the private sector to put academic research findings to practical use.[79] One of the results of this situation was increasing pressure on universities to pay more heed to the wishes and needs of "the market" and to get involved in the market themselves, as in the field of biotechnology, where they were encouraged to set short-term goals of developing new drugs, foodstuffs made from genetically modified crops, and so forth. In 2006, the Dutch education minister, Maria van der Hoeven, expressed her astonishment that members of university faculties were still using public funds to investigate topics of their own choosing, the type of research that used to be called "curiosity-driven," noting that for policymakers, the added value of such activities is unclear.[80]

Taking stock of the changes in scientific research over the last twenty years, we see three major dichotomies: deductive and performative science (science for its own sake and result-oriented science), science as high and as low culture, and science in the private and the public domain. These issues are tightly intermeshed, but it is useful to examine each one separately.

From Deductive to Performative Science

Without drawing any conclusions about a causal relationship, Lyotard proposed the "hypothesis" that the withdrawal of governments from the public domain and the growing importance of the market in the field of knowledge production coincided with a crisis of legitimacy for deductive science (and for science as the basis for *Bildung*).[81] This observation, made in 1979, has been repeated many times since then.[82] To put it differently, the balance between the styles of science has swung from a situation in which the deductive and the hypothetical-analogical styles enjoyed cultural supremacy to one dominated by the experimental style and, in particular, by combinations of (or alliances between) various styles of science and the technological style. This is true outside academia, in both the public and private sectors, and it is also true within universities.

Among the natural sciences, physics is no longer held up as a model for other disciplines. One symbolic turning point came in 1993, when the U.S. Congress canceled plans for the Superconducting Super Collider, an extremely expensive megaproject. The experiments that were to be conducted in that project would have been relevant only to high-energy physics and not to other areas of science.[83] This type of narrow disciplinary focus has had its day. The new ideal of interdisciplinarity opens up the possibility of borrowing theoretical and practical ideas from just about any other field of study. More generally, the disciplinary structure of science is under pressure. Where this structure has remained solid, it owes this success largely to a stable influx of students, but there are no guarantees that student numbers will remain dependable in the future.

Forty years ago, science was organized into clear-cut disciplines, each overseen by its elites; today, it is run by managers. Evaluation of the quality of scientific work based on discipline-internal standards has made way for quantitative performance assessment. The weakening of the ties between science and the state has cleared the way for a breathtaking loss of professional autonomy for individual scientific researchers. Financing organizations and university bureaucracies are becoming more powerful and are using that power to create fully formed programs of activity to which researchers are forced to adapt.

These new developments will affect different fields of science in different ways. The rewards for sciences that can offer immediately applicable results will be far greater than those for less pragmatic disciplines. Biotechnology and nanotechnology have become the "stars" among the sciences. The close integration of science and technology is no longer dependent on the use of large teams, as it was in the 1950s and 1960s, when this type of integration took place mainly at very large research institutions such as national laboratories in the United States and industrial research laboratories.[84] These days, technoscience can be conducted even by fairly small academic research teams.

Bildung and the Cultural Value of Science

The mind-set according to which science had a higher cultural value than other disciplines has been discredited. This is partly because the economic, political, and cultural elites no longer coincide.[85] The crisis of science, and of high culture in general, is that it no longer has the support of the economic and political elite. It therefore has to rely on support from a much larger public, which it has so far been able to acquire only through "the market" or through new forms of private patronage.

After the Second World War, European governments imitated the American "social contract" between science and the state (encouraging autonomous, "free," deductive science, with fairly little concern for anticipated revenue from industrial applications); this seemed to them to be highly compatible with the "higher" objective of unfettered science. The Dutch government wrote that public support for science should not be guided by a "dubious utilitarianism."[86] This view reflects one of the central values of Bildung and suggests that in 1950, the Humboldtian objective of cultivating a general social elite was still alive and kicking. In a neoliberal society, however, offering students Bildung, moral education and character formation, is no longer a public priority, because Bildung is a form of cultural capital that simply leads to greater individual opportunities in some job markets. Students are expected to find out for themselves whether or not they need it.

There is therefore no real possibility of returning to the "contract" between politics and science that existed between 1945 and 1980 and offered scientists unprecedented freedom.[87] Science will never again be the special province of a governing elite that claims to wield its authority in the service of culture and progress. Too many people have gained access to scientific information for there to be any way back. Science now belongs to stakeholders, users of technology, medical patients (individually or collectively), and members of nongovernmental organizations such as environmental groups. The intellectual legislator, to return to Zygmunt Bauman's term, cannot cope with this situation. Science no longer stands above the fray, but is down in the thick of things, subject to everyone's scrutiny. National governments, too, are falling from their pedestals. For the time being, we will all have to learn to live with controversy and uncertainty in science.

The Market and the Public Sphere

One strategy for dealing with this new uncertainty is to disregard the larger narrative that provides a context for science, instead focusing solely on the "hard facts." This is a popular approach, partly owing to the unintended impact of commercial thinking, quantitative performance assessment, and computer programming. In various fields of both the social and the natural sciences, the result has been a shift in emphasis over the last fifteen years toward gathering and manipulating large sets of data. In the optimistic 1970s, "interdisciplinarity" held the promise of an enlightening confrontation between different scientific traditions. We are now in too much of a hurry for that. More than ever, interdisciplinarity is imposed from above, and financiers are interested primarily in controlling costs through shared experimental procedures and measuring devices.

There can be no objection to holding the scientific community publicly accountable for the substantial costs of its activities. The problem is neither cost control *per se* nor the requirement of social relevance. The problem for science in general, and university science in particular, is cost control informed by a notion of utility that is defined within a neoliberal framework. That neoliberal thinking has placed the university as a public institution in the role of a service provider to powerful private parties. Another, even greater, problem is that universities are actually trying to become more like those private parties. This cannot be reconciled with their role in public life and scientific research. In their rush to serve society, universities can all too easily forget that they are the only institutions in a position to perform fundamental research and to offer *Bildung*, education that builds moral character.[88]

The challenge for science will be to find a new social role in which its value is not judged solely by its immediate practical results, and theory does not have to take a backseat to applications. Technoscience at universities is not in itself a problem, but potential dangers lie in the radical transformations of nature and society that technoscience may bring.[89] Scientists have a responsibility not only to offer facts and techniques, but also to shed light on their implications (and to train their students to do the same). The old ideal of *Bildung* lends itself to adaptation for this purpose. Ideally, scientists will embrace this ideal and step forward as mediators between the many different communities affected by developments in science and technology. At any rate, this is Bauman's image of the intellectual as "interpreter," the successor to the intellectual as legislator. Experience has shown that scholars—whether in the natural sciences, social sciences, or humanities—will not be permitted to assume this role without a struggle.[90] In that struggle, we must hope that they will continue to find a capable ally in the universities.

NOTES

Chapter 1: Introduction

1. Alistair Cameron Crombie, *Styles of Scientific Thinking in the European Tradition: The History of Argument and Explanation Especially in the Mathematical and Biomedical Sciences and Arts*, 3 vols. (London: Duckworth, 1994).

2. J. Huizinga, *Herfsttij der Middeleeuwen* (1919; repr., Haarlem: Tjeenk Willink, 1952), 247. English edition: *The Waning of the Middle Ages: A Study of Forms of Life, Thought, and Art in France and the Netherlands in the Dawn of the Renaissance* (London: Edward Arnold and Co., 1924).

3. See also Chunglin Kwa, "Alliances between Styles: A New Model for the Interaction Between Science and Technology," in *Science and Its Recent History: Epochal Break or Business as Usual?* ed. Alfred Nordmann, Hans Radder, and Gregor Schiemann (Pittsburgh: University of Pittsburgh Press, 2011). On the special characteristics of the technological style, see Eugene Ferguson, *Engineering and the Mind's Eye* (Cambridge, Mass.: The MIT Press, 1992). The case studies in Walter Vincenti, *What Engineers Know and How They Know It: Analytical Studies from Aeronautical History* (Baltimore: Johns Hopkins University Press, 1990) can be construed as examples of alliances between the technological style and other styles of science.

4. Thomas Misa, *Leonardo to the Internet: Technology and Culture from the Renaissance to the Present* (Baltimore: Johns Hopkins University Press, 2004); Mikael Hård and Andrew Jamison, *Hubris and Hybrids: A Cultural History of Technology and Science* (New York: Routledge, 2005).

5. Ian Hacking, "Style for Historians and Philosophers," *Studies in the History and Philosophy of Science* 23 (1993): 1–20; Ian Hacking, *Representing and Intervening: Introductory Topics in the Philosophy of Natural Science* (Cambridge: Cambridge University Press, 1983); Ian Hacking, *The Taming of Chance* (Cambridge: Cambridge University Press 1990); Ian Hacking, "Raison et véracité, les choses, les gens, la raison, cours 2005/2006" (lecture, Collège de France, www.college-de-france.fr/default/EN/all/historique/ian_hacking.htm).

6. John Pickstone, *Ways of Knowing: A New History of Science, Technology and Medicine* (Manchester: Manchester University Press, 2000).

7. Pickstone suggests that "analysis" is the experimental method before 1780, while "synthesis" came into being only in the nineteenth century; in so doing he may have overlooked the aspect of early experimental practice described as "making" by practitioners such as Mersenne.

8. Robert M. Young, "Malthus and the Evolutionists: The Common Context of Biological and Social Theory," *Past and Present* 43 (1969): 109–45.

9. E. J. Dijksterhuis, *De mechanisering van het wereldbeeld* (1950; repr., Amsterdam: Meulenhoff, 1975). English edition: *The Mechanization of the World Picture*, trans. C. Dikshoorn (New York: Oxford University Press, 1961).

10. Lynn Hunt, "Introduction," in *The New Cultural History*, ed. Lynn Hunt (Berkeley: University of California Press, 1989).

11. Roger Chartier, quoted in Hunt, "Introduction," 7. On the *Annales* school, see Lutz Raphael, *Die Erben von Bloch und Febvre: Annales-Geschichtsschreibung und nouvelle histoire in Frankreich 1945–1980* (Stuttgart: Klett-Cotta, 1994).

12. Clifford Geertz, *The Interpretation of Cultures* (New York: Basic Books, 1973).

13. Dominick LaCapra, *Rethinking Intellectual History: Texts, Contexts, Language* (Ithaca: Cornell University Press, 1983); Hayden White, *Metahistory: The Historical Imagination in Nineteenth-Century Europe* (Baltimore: Johns Hopkins University Press, 1973); see also Chunglin Kwa, "De retorische structuur van geschiedkundige teksten," *Krisis* 37 (1989): 54–60.

14. Roger Chartier, "Texts, Printing, Readings," in Hunt, *Cultural History*, 154–75, 159.

15. Crombie, *Styles*, 1:56–62.

16. Anna Wessely, "Transposing 'Style' from the History of Art to the History of Science," *Science in Context* 4 (1991): 265–78.

17. Long before Mannheim, Giambattista Vico had similar ideas. Isaiah Berlin attributed the belief to Vico that there is a collective style at work in the arts, institutions, language, and way of life in a society. See Isaiah Berlin, *Vico and Herder: Two Studies in the History of Ideas* (London: Chatto and Windus, 1976), xvii.

18. Karl Mannheim, *Ideology and Utopia: An Introduction to the Sociology of Knowledge*, trans. Louis Wirth and Edward Shils (1936; repr., London: Routledge and Kegan Paul, 1976), 6, 57. This translation renders *Weltwollen* less than perfectly, as "set of values."

19. Erwin Panofsky, *Gothic Architecture and Scholasticism* (Latrobe, Pa.: The Archabbey Press, 1951), 21.

20. Pierre Bourdieu, "Postface," in *Architecture gothique et pensée scholastique*, by Erwin Panofsky (Paris: Minuit, 1979), 133–67, 142.

21. Ernst Gombrich, *Meditations on a Hobby Horse* (London: Phaidon, 1963), 79.

22. Carlo Ginzburg, "From Aby Warburg to Ernst Hans Gombrich: A Problem of Method," in *Clues, Myths, and the Historical Method* (Baltimore: Johns Hopkins University Press, 1989), 17–59.

23. Ernst Gombrich, *Art and Illusion* (London: Phaidon, 1960).

24. Ibid., 20. Arnold Davidson's discussion of styles in science also has points of contact with Wölfflin's view. See Davidson's "Styles of Reasoning, Conceptual History, and the Emergence of Psychiatry," in *The Disunity of Science: Boundaries, Contexts and Power*, ed. Peter Galison and David J. Stump (Palo Alto: Stanford University Press, 1996), 75–100.

25. Thomas Kuhn, *The Structure of Scientific Revolutions* (Chicago: The University of Chicago Press, 1970).

26. See also Charles Rosenberg, "Towards an Ecology of Knowledge," in *The Organization of Knowledge in Modern America, 1860–1920*, ed. Alexandra Oleson and John Voss (Baltimore: Johns Hopkins University Press, 1979), 440–55.

27. Pascal Acot, *Histoire de l'écologie* (Que sais je 2870) (Paris: PUF, 1994), 82.

28. Michael Baxandall, *Painting and Experience in Fifteenth-Century Italy: A Primer in the Social History of Pictorial Style* (Oxford: Oxford University Press, 1972), 29–30; see also Davidson, "Styles of Reasoning."

29. This is also true of other cognitive symbolic systems, outside the domain of Western science. See Mary Hesse, "Rationality and the Generalization of Scientific Style," in *The Light of Nature*, ed. J. D. North and J. D. Roche (Dordrecht: Martinus Nijhoff, 1985), 365–81.

30. Dick Pels, "Karl Mannheim and the Sociology of Scientific Knowledge: Toward a New Agenda," *Sociological Theory* 14 (1996): 30–48.

31. Ludwik Fleck, in *Entstehung und Entwicklung einer wissenschaftlicher Tatsache: Einführung in die Lehre vom Denkkollektiv und Denkstil* (1935; repr., Frankfurt am Main: Suhrkamp, 1980) used the term *Denkstil* (style of thought) in a more narrowly defined sociological way than did Mannheim. But Fleck's concept of *Denkstil*, like Mannheim's, has an ontological dimension, much more so than Kuhn's term "paradigm," to which it is often compared. It is unclear why Fleck never cited Mannheim; see Pels, "Karl Mannheim."

32. The Greeks used the terms *episteme* (knowledge), *theoria* (speculative knowledge), *philosophia* (the love of wisdom), and *peri physeoos historia* (inquiry into nature). The Latin word *scientia* originally meant something like "the state of knowing a thing."

33. Joseph Needham, *Human Law and the Laws of Nature in China and the West* (London: Oxford University Press, 1951); Joseph Needham, *Clerks and Craftsmen in China and the West* (Cambridge: Cambridge University Press, 1970). See also Colin A. Ronan, *The Shorter Science and Civilization in China*, 5 vols. (Cambridge: Cambridge University Press, 1978).

34. On this subject, see David Goodman and Colin A. Russell, *The Rise of Scientific Europe, 1500–1800* (Sevenoaks: Hodder and Stoughton, 1991), 330–66.

35. See, for example, Morris Low, ed., *Beyond Joseph Needham: Science, Technology, and Medicine in East and South East Asia*, Osiris 13, special issue (1998); George Saliba, *Islamic Science and the Making of the European Renaissance* (Cambridge, Mass.: The MIT Press, 2007); Jan P. Hogendijk and Abdelhamid I. Sabra, *The Enterprise of Science in Islam: New Perspectives* (Cambridge, Mass.: The MIT Press, 2003).

36. See John Darwin, *After Tamerlane: The Rise and Fall of Global Empires, 1400–2000* (London: Allen Lane, 2007).

Chapter 2: The Deductive Style of Science

Epigraphs: Heraclitus, *Fragments*, 22B123 (Diels-Kranz); Jean Baptiste Fourier, "Numbers govern even fire," *Théorie analytique de la chaleur*, 1822, quoted in Enrico Bellone, *A World on Paper* (Cambridge, Mass.: The MIT Press, 1980), 55.

1. See Aristotle, *Physics* 2.192b8–34, for the classical distinction between the natural and the artificial. In another passage in the same work, Aristotle appears to

express milder views on the relationship between the two categories, but only to the extent that artifice adapts to nature, accelerating a natural process that would have unfolded more slowly on its own (2.199a15–17).

2. Crombie, 1:122.

3. Hermann Glockner, *Die europäische Philosophie von der Anfängen bis zur Gegenwart* (1958; repr., Stuttgart: Reclam, 1980).

4. Crombie, *Styles*, 1:248.

5. Aristotle, *Posterior Analytics* 1.3.72b20–25.

6. Aristotle, *Posterior Analytics* 1.11.77a26–32; Aristotle, *Metaphysics* 6.1.1026a10–33.

7. Aristotle, *Posterior Analytics* 1.2.71b10–72a24; 1.3.72b19–23; Aristotle, *Metaphysics* 6.1.1026a10–33.

8. Aristotle, *Posterior Analytics* 2.13–14.97b7–98a29; Crombie, *Styles*, 1:11, 238–41, 243. Instead of "substance," the translation "essence" is sometimes used; see A. C. Lloyd, *The Anatomy of Neoplatonism* (Oxford: Oxford University Press, 1990), 87.

9. Aristotle, *On the Heavens* 2.3–4. Aristotle used the word *ananke* (ἀνάγκη) for necessity. Pre-Socratic Greek authors used *ananke* in connection with decisions of the gods. See Empedocles, *On Nature*, frag. 115 Diels-Kranz. The goddess in Parmenides's proem *On Nature* is also said to be Ananke. The term has connotations such as authority, compulsion, and destiny; see Hans Blumenberg, *Arbeit am Mythos* (Frankfurt: Suhrkamp, 1979), 140. Compare the expression "a compelling argument." The atomists' conception of necessity is discussed later in this chapter.

10. Aristotle, *On Generation and Corruption* 2.10.337a; Aristotle, *On the Generation of Animals* 4.10.777a–b, 33.

11. Aristotle, *Physics* 8.256a19-21; Crombie, *Styles*, 1:233.

12. Heraclitus, *Fragments*, 123. In fact, Heraclitus, whose work has been transmitted only in the form of aphorisms, probably meant something very different: "The same nature that produces appearances also allows them to disappear (die)." See Pierre Hadot, *Le voile d'Isis: Essai sur l'histoire de l'idée de Nature* (Paris: Gallimard, 2004), 27. English edition: *The Veil of Isis: An Essay on the History of the Idea of Nature*, trans. Michael Chase (Cambridge, Mass.: Harvard University Press, 2006).

13. Aristotle, *Posterior Analytics* 1.7.75a38; see Amos Funkenstein, *Theology and the Scientific Imagination* (Princeton: Princeton University Press, 1986), 36.

14. Archimedes, *The Method of Mechanical Theorems*, quoted in Crombie, *Styles*, 1:176 and G. E. R. Lloyd, *Greek Science after Aristotle* (New York: Norton, 1973), 45.

15. Archimedes, *On the Equilibrium of Planes*, quoted in Lloyd, *Greek Science*, 48.

16. Dijksterhuis, *De mechanisering*.

17. E. J. Dijksterhuis, *Archimedes* (Groningen: Noordhoff, 1938). The *locus classicus* for Archimedes's technological achievements and machines of war is Plutarch's *Life of Marcellus*.

18. Aristotle, *Posterior Analytics* 1.31.87b28.

19. Aristotle, *Posterior Analytics* 2.19.100b8.

20. Victor Kal, *On Intuition and Discursive Reasoning in Aristotle* (Leiden: Brill, 1988); John M. Rist, *The Mind of Aristotle: A Study in Philosophical Growth* (Toronto: University of Toronto Press, 1989).

21. Aristotle, *Posterior Analytics* 2.19.100b7–13, trans. Hugh Tredennick (Loeb Classical Library, Harvard University Press, 1960), 261.

22. Plato, *Timaeus* 41d, e.

23. Plato, *Meno* 81e–86b; Plato, *Phaedo* 73.
24. Rist, *Mind*, 269; the formulation with "as if" is borrowed from Victor Kal, *On Intuition*, 50.
25. Wolfgang Wieland, *Die aristotelische Physik* (Göttingen: Vandenhoeck and Ruprecht, 1962), 112.
26. Crombie, *Styles*, 1:274.
27. Crombie, *Styles*, 1:96. The Babylonians had some notion of geometry, but used it primarily in support of their algebraic techniques. It is noteworthy that Babylonian astronomers worked in state institutions, like their ancient Chinese counterparts, but unlike the Greeks; see G. E. R. Lloyd, *The Ambitions of Curiosity: Understanding the World in Ancient Greece and China* (Cambridge: Cambridge University Press, 2002), 43. The earliest classical authors describing the travels of Pythagoras, namely Herodotus and Isocrates, are open to multiple interpretations.
28. Walter Burkert, *Lore and Science in Ancient Pythagoreanism* (Cambridge, Mass.: Harvard University Press, 1972).
29. Crombie, *Styles*, 1:99.
30. A concept related to the circle, namely the vortex, later played a central role in Cartesian physics, and in the 1780 Kant-Laplace theory of the origin of the solar system.
31. Funkenstein, *Theology*, 33. See also Crombie, *Styles*, 110–11.
32. Plato, *Phaedrus* 270c.
33. Hadot, *Le voile d'Isis*, 50; see also Hacking, "Raison et véracité."
34. Crombie, *Styles*, 1:121.
35. Lucretius, *On the Nature of Things* I, ll. 1022 ff. and ll. 822 ff. Translation based in part on the edition translated by Cyril Bailey (Oxford: Clarendon Press, 1921), 61.
36. Michel Serres, *La naissance de la physique dans le texte de Lucrèce* (Paris: Minuit, 1977).
37. S. Sambursky, *The Physical World of the Greeks* (London: Routledge and Kegan Paul, 1956), 86.
38. Aristotle's concept of naturalness explains why he did not regard experiments as valid ways of learning about nature; they forced nature into an artificial state.
39. Hans Blumenberg, *Die Genesis der kopernikanischen Welt* (Frankfurt: Suhrkamp, 1975), 162.
40. Luciano Canfora, *The Vanished Library: A Wonder of the Ancient World* (Berkeley: University of California Press, 1990); David C. Lindberg, *The Beginnings of Western Science* (Chicago: Chicago University Press, 1992), 76.
41. Both Plato and Aristotle use the terms *epistēmē* (scientific knowledge) and *technē* interchangeably, apparently treating them as synonyms.
42. M. Nussbaum, *The Fragility of Goodness: Luck and Ethics in Greek Tragedy and Philosophy* (Cambridge: Cambridge University Press, 1986), 138, 149.
43. This is according to the classification system presented in Werner Jaeger, *Aristotle. Grundlegung einer Geschichte seiner Entwicklung* (Berlin: Weidmannsche Buchhandlung, 1923).
44. Aristotle, *On the Generation of Animals* 5.8.788b10–29.
45. Aristotle, *On the Parts of Animals* 4.685b10; trans. William Ogle, at http://classics.mit.edu/Aristotle/parts_animals.mb.txt, accessed January 21, 2010.
46. Crombie, *Styles*, 1:342; Jacques LeGoff, *Les intellectuels au Moyen Age* (1956; repr., Paris: Le Seuil, 1976), 121.

47. L. Daston and K. Park, *Wonders and the Order of Nature* (New York: Zone Books, 1998), 136.
48. Crombie, *Styles*, 1:342.
49. Nussbaum, *Fragility*, chap. 9, also p. 250.
50. Ibid., 80.
51. Ibid., 300–302, 364.
52. Ibid., 353, 357, 372.
53. Ibid., 276; Crombie, *Styles*, 1:272.
54. Nussbaum, *Fragility*, 260.
55. See also Gernot Böhme, *Alternativen der Wissenschaft*, sec. III, chap. 2: "Aristotle's Chemie: eine Stoffwechselchemie," (Frankfurt: Suhrkamp [stw334], 1980), 101–20.
56. Aristotle, *On the Parts of Animals* 3.668a25–36.
57. Nussbaum, *Fragility*, 240.
58. Ibid., 258.
59. See Albert Einstein, "What Is the Theory of Relativity?" *London Times*, November 28, 1919. Einstein based special relativity theory on two "axioms," the constant velocity of light and the validity of the Maxwell equations in systems in motion relative to each other.

Chapter 3: The Deductive Style in a Christian Context

Epigraph: M. D. Chenu, *Introduction à l'étude de Saint Thomas d'Aquin* (Paris, 1950), quoted in Jacques LeGoff, *Les intellectuels*, 97.

1. Thomas Aquinas, *Commentary on Aristotle's Posterior Analytics*, 1.38.n258, quoted in Hacking, *Emergence*, 22.
2. Crombie, *Styles*, 1:248.
3. LeGoff, *Les intellectuels*, 20.
4. David C. Lindberg, "The Transmission of Greek and Arabic Learning to the West," in *Science in the Middle Ages*, ed. David C. Lindberg (Chicago: The University of Chicago Press), 62.
5. Lindberg, *Beginnings*, 326.
6. Paul Rose, *The Italian Renaissance of Mathematics* (Geneva: Librairie Droz, 1975), 43, 81–82; Marshall Clagett, "The Influence of Archimedes on Medieval Science," *Isis* 50 (1959): 419–29.
7. Lindberg, "Transmission," 77; Fernand van Steenberghen, *Aristotle in the West* (Leuven: Nauwelaerts, 1954).
8. LeGoff, *Les intellectuels*, 79 ff.; P. Kibre and N. G. Siraisi, "The Institutional Setting: The Universities," in Lindberg, *Science in the Middle Ages*, 120–44.
9. The trivium consisted of grammar, rhetoric, and logic, and the quadrivium of arithmetic, geometry, music, astronomy. The "three philosophies" were natural, metaphysical, and ethical philosophy. See Crombie, *Styles*, 1:315.
10. Nancy Siraisi, *Medieval and Renaissance Medicine* (Chicago: The University of Chicago Press, 1990), 86.
11. Henri de Mondeville (1250/1270–1325), physician and surgeon, tried in vain to enhance the expertise of the surgical profession.
12. Descartes, quoted in Funkenstein, *Theology*, 117; Descartes, letter to Mersenne,

June 3, 1630, quoted in Peter Dear, *Mersenne and the Learning of Schools* (Ithaca: Cornell University Press, 1988), 55–56.

13. The Greek and Hebrew conceptions of God merge to some degree in the deuterocanonical Wisdom of Solomon and the Gospel of John. See Hans Blumenberg, *Die Legitimität der Neuzeit* (Frankfurt: Suhrkamp, 1988); Gilles Quispel, *Valentinus de gnosticus en zijn Evangelie der Waarheid* (Amsterdam: In de Pelikaan, 2003).

14. Funkenstein, *Theology*, 127; Eamon Duffy, *Saints and Sinners: A History of the Popes* (New Haven: Yale University Press, 1997), 91 ff.

15. Van Steenberghen, *Aristotle*, 109.

16. Crombie, *Styles*, 1:401; LeGoff, *Les intellectuels*, 123.

17. On this incident, see also J. M. Thijssen, "Bishop Tempier's Condemnation," in *Texts and Contexts in Ancient and Medieval Science*, ed. E. Sylla and M. McVaugh (Leiden: Brill, 1997), 84–105.

18. Paul Oskar Kristeller, *Medieval Aspects of Renaissance Learning* (1974; repr., New York: Columbia University Press, 1992), 40.

19. Dante, *La divina commedia, Paradiso*, Canto 10, ll. 133–38.

20. LeGoff, *Les intellectuels*, 121–24.

21. Funkenstein, *Theology*, 129, 132–45. The term *ordinatus* probably also has connotations of predestination.

22. Ibid.; see also Blumenberg, *Arbeit*, 140–41.

23. An earlier meaning of *nomos* in Greek was "custom," referring to both social convention and the way in which nature customarily works. This meaning became more fluid in late Greek; Plato, for instance, writes that the Demiurge imposes the *nomos* of destiny (*eimarmenē*) on the world; see Crombie, *Styles*, 1:119 and Nussbaum, *Fragility*, 400. There is a passage in Aristotle's *De Caelo* (*On the Heavens*) in which Aristotle asserts that nature has imposed the perfection of the number three on us "as a law." But this is not a law imposed on nature.

24. Funkenstein also suggests another source of the concept of law: the *foedera naturae* (pacts with nature) mentioned by Lucretius (*De rerum naturae* 2.300; 5.923, 6.906). But this is mistaken. Apparently, according to Lucretius, there is a limit to the possible combinations into which the atoms enter. But this is not because they obey a law imposed from above. The atoms themselves determine the combination collectively. On this point, see Blumenberg, *Legitimät*, 178.

25. Philo, *De Josepho*, 6.29, as quoted in Crombie, *Styles*, 1:295.

26. Crombie, *Styles*, 1:300–303.

27. Ibid., 402–3, 378.

28. Jean Buridan mentions Tempier in his *Commentary on Aristotle's Physics*, bk. 3, question 7, quoted in John D. Murdoch and Edith D. Sylla, "The Science of Motion," in Lindberg, *Science in the Middle Ages*, 217; see also Crombie, *Styles*, 1:403.

29. Funkenstein, *Theology*, 192; Crombie, *Styles*, 1:407; Daston and Park, *Wonders*.

30. Blumenberg, *Legitimität*, 206.

31. Giovanni Pico della Mirandola, *Oratio de homine dignitate (On the Dignity of Man)* [1487], undelivered speech.

32. Kristeller, *Medieval Aspects*, 51; Charles B. Schmitt, *Aristotle in the Renaissance* (Cambridge, Mass.: Harvard University Press), 1983; Walter Pagel, *Joan Baptista Van Helmont* (Cambridge: Cambridge University Press, 1982), 4. Strictly speaking, Capuchin friars are also Franciscans, but they form a later offshoot of the order.

33. Sambursky, *Physical World*, 136. Sambursky sees a similarity to the modern concept of the field, but gives no evidence of historical continuity.
34. Heterodox Christian traditions, such as Gnosticism, included Stoic elements.
35. Funkenstein, *Theology*, 37–41, 66–68, 78–79.
36. Lindberg, *Beginnings*, 139.
37. Lloyd, *Anatomy*, 28.
38. These realms are called hypostases. Lloyd, *Anatomy*, 124.
39. Lloyd, *Anatomy*, 100.
40. David C. Lindberg, "Introduction," in *Roger Bacon's Philosophy of Nature*, ed. David C. Lindberg (Oxford: Oxford University Press, 1983), xli.
41. Crombie, *Styles*, 1:319.
42. Lindberg, "Introduction," xlv. On this topic, see also Joseph Wachelder, "Mirrors: Truth and Error," in *Theories, Technologies, Instrumentalities of Color*, ed. Barbara Saunders and Jaap van Brakel (Lanham: University Press of America, 2002), 215–31.
43. Lindberg, "Introduction," xlii, liv.
44. Crombie, *Styles*, 1:318, 348; Lindberg, "Introduction," iv.
45. It is useful to note that *experimentum* does not mean "experiment." One enlightening illustration of this is Grosseteste's remark that angels have *scientia experimentalis* of the divine essence, which we ordinary mortals must do without for the time being. On this subject, see also James McEvoy, *The Philosophy of Robert Grosseteste* (Oxford: Oxford University Press, 1982), 208. On the term *experimentum*, see also Crombie, *Styles*, 1:331, 337, 339.
46. According to Keith Hutchison, Aquinas's scholastic philosophy distinguished between perception and experience. Experience could be related to the effects of indirectly observable phenomena relating to manifestly occult qualities. Because the phenomena of light that are the subject matter of optics do not relate in any way to occult properties of matter, the distinction between observation and experience is unimportant in this context. Hutchison, "What Happened to Occult Qualities in the Scientific Revolution?," *Isis* 73 (1982): 233–53.
47. Crombie, *Styles*, 1:321–23.
48. McEvoy, *Grosseteste*, 216; Crombie, *Styles*, 1:324.
49. Stephen S. Mason, *Main Currents of Scientific Thought* (New York: Henry Schuman, 1953).
50. Crombie, *Styles*, 1:353, 361, 375.
51. Lindberg, "Introduction," liii; Lindberg, "Science of Optics," in Lindberg, *Science in the Middle Ages*.
52. Crombie, *Styles*, 1:375.
53. Ibid., 1:350.
54. Nicholas H. Clulee, "At the Crossroads of Magic and Science: John Dee's Archemastrie," in *Occult and Scientific Mentalities in the Renaissance*, ed. Brian Vickers (Cambridge: Cambridge University Press, 1984), 51–71; see also Stuart Clark, "The Scientific Status of Demonology," in Vickers, *Occult*, 351–74.
55. Crombie, *Styles*, 1:332.
56. Daston and Park, *Wonders*, 95.
57. Lindberg, "Introduction," xxiv.
58. Crombie, *Styles*, 1:380, 387, 1108.
59. Ibid., 1:324.

60. Analogously, astronomy was regarded as applied geometry, based on the geometry of the sphere.
61. Dirk Struik, *Geschiedenis van de wiskunde* (Amsterdam: Sua, 1977), 72.
62. Aristotle, *Metafysica* 1078a, 14–17.
63. Peter Dear, *Mersenne*, 64, 104.
64. Struik, *Geschiedenis*, 72.
65. Michael S. Mahoney, "Mathematics," in Lindberg, *Science in the Middle Ages*, 145–78; Alfred W. Crosby, *The Measure of Reality* (Cambridge: Cambridge University Press, 1997), 44, 111.
66. Mahoney, "Mathematics."
67. Rose, *Italian Renaissance*, 79, 82.
68. Struik, *Geschiedenis*, 100.
69. Crosby, *Measure*, 205.
70. Struik, *Geschiedenis*, 101.
71. Crombie, *Styles*, 1:419.
72. Ibid., 1:416.
73. Lindberg, *Beginnings*, chap. 13; Charles Talbot, "Medicine," in Lindberg, *Science in the Middle Ages*, 391–428.
74. Crombie, *Styles*, 1:318, 339–40.
75. Michael R. McVaugh, *Medicine Before the Plague: Practitioners and Their Patients in the Crown of Aragon* (Cambridge: Cambridge University Press, 1993), 153.
76. Nancy Siraisi, *The Clock and the Mirror: Girolamo Cardano and Renaissance Medicine* (Princeton: Princeton University Press, 1997).
77. Crombie, *Styles*, 1:343.
78. Ibid., 1:395, 399.
79. Carlo Ginzburg, "Clues: Roots of an Evidential Paradigm," in Ginzburg, *Clues, Myths*, 96–125.
80. Rose, *Italian Renaissance*, 84.
81. Ofer Gal and Raz Chen-Morris, "The Archeology of the Inverse-Square Law: (1) Metaphysical Images and Mathematical Practices," *History of Science* 43 (2005): 391–414.
82. Such scholars include Herbert Butterfield, A. Rupert Hall, Mary Boas Hall, Thomas Kuhn, and Floris Cohen.
83. Dijksterhuis, *De mechanisering*; Alistair Crombie, *Augustine to Galileo: The History of Science A.D. 400–1650* (London: Falcon Press), 1952.
84. Helge Kragh, *An Introduction to the Historiography of Science* (Cambridge: Cambridge University Press, 1987), 75–77.
85. David C. Lindberg, "Science as Handmaiden: Roger Bacon and the Patristic Tradition," *Isis* 78 (1987): 518–36.
86. John Henry, "Metaphysics and the Origins of Modern Science: Descartes and the Importance of Laws of Nature," *Early Science and Medicine* 9 (2004): 73–114, 88. Henry's central argument is that Descartes, drawing on his background as a mathematician, did refer back to the ideas of Roger Bacon (especially his conception of a natural law as an observed regularity), but that in his role as advocate of a new kind of physics he weighed down the concept of natural law with the heavier burden of physical causality. Incidentally, this can also be seen as a return to Aristotle—in particular, to the *Posterior Analytics*.

87. Steven Shapin, *The Scientific Revolution* (Chicago: The University of Chicago Press, 1996).

Chapter 4: From Scholar to Virtuoso

Epigraph: Johann Joachim Becher, *Methodus didactica*, Munich, 1668, quoted in Pamela H. Smith, *The Business of Alchemy* (Princeton: Princeton University Press, 1994), 56.

1. Galileo, *Il Saggiatore*; see Gary Hatfield, "Metaphysics and the New Science," *Reappraisals of the Scientific Revolution*, ed. David Lindberg and Robert S. Westman (Cambridge: Cambridge University Press, 1990), 93–166, 129.

2. Edgar Zilsel, "The Sociological Roots of Science," *American Journal of Sociology* 47 (1942): 245–79, reprinted in *Social Studies of Science* 30 (2000): 935–49; Edgar Zilsel, "The Origins of William Gilbert's Scientific Method," *Journal of the History of Ideas* 2 (1941): 1–32; Edgar Zilsel, *Die sozialen Ursprünge der neuzeitlichen Wissenschaft* (Frankfurt: Suhrkamp, 1976). See also Joseph Ben-David, *The Scientist's Role in Society: A Comparative Study* (Englewood Cliffs, N.J.: Prentice Hall, 1971).

3. Mario Biagioli, *Galileo, Courtier: The Practice of Science in the Culture of Absolutism* (Chicago: The University of Chicago Press, 1993).

4. A. Rupert Hall, "The Scholar and the Craftsman in the Scientific Revolution," and "Science, Technology and Warfare," in *Science and Society* (Aldershot: Variorum, 1994); John Henry, *The Scientific Revolution and the Origins of Modern Science*, 2nd ed. (Basingbroke: Palgrave, 2002), 35–36.

5. Alexandre Koyré, *Etudes Galiléennes* (Paris: Hermann, 1939).

6. The experiment in question involves dropping a stone on a moving ship. Galileo Galilei, *Galileo on the World Systems*, trans. and ed. Maurice Finocchiaro (Berkeley: University of California Press, 1977), 164.

7. Guidobaldo del Monte, quoted in Rose, *Italian Renaissance*, 231.

8. Pamela Long, "Power, Patronage, and the Authorship of *Ars*: From Mechanical Know-How to Mechanical Knowledge in the Last Scribal Age," *Isis* 88 (1997): 1–41; Henry, *Scientific Revolution*, 35–36; Lissa Roberts, Simon Schaffer, and Peter Dear, eds., *The Mindful Hand: Inquiry and Invention from the Late Renaissance to Early Industrialisation* (Amsterdam: Koninklijke Nederlandse Akademie van Wetenschappen, 2007).

9. Goodman and Russell, *Rise of Scientific Europe*, 129; Kristeller, *Medieval Aspects*, 52; Rienk Vermij, *The Calvinist Copernicans* (Amsterdam: Koninklijke Nederlandse Akademie van Wetenschappen, 2002), 23.

10. Jacob Burckhardt, *Die Kultur der Renaissance in Italien* (1860; repr., Stuttgart: Kröner, 1925).

11. Oliver Logan, *Culture and Society in Venice, 1470–1790* (New York: Scribners, 1972).

12. Kristeller, *Medieval Aspects*, 16; Ann Blair and Anthony Grafton, "Reassessing Humanism and Science," *Journal of the History of Ideas* 53 (1992): 535–40.

13. Lisa Jardine, *Erasmus, Man of Letters* (Princeton: Princeton University Press, 1993).

14. Rose, *Italian Renaissance*, 23, 44; James Hankins, *Plato in the Italian Renaissance*, vol. 1 (Leiden: Brill, 1990); Frances Yates, "The Italian Academies," in *Renais-*

sance and Reform: The Italian Contribution (1949; repr., London: Routledge & Kegan Paul, 1983), 6–29.

15. Rose, Italian Renaissance, 143.
16. Burckhardt, Kultur der Renaissance, 182.
17. Jardine, Erasmus, 83–98.
18. Goodman and Russell, Rise, 38.
19. Anthony Grafton, Defenders of the Text: The Traditions of Scholarship in an Age of Science (Cambridge, Mass.: Harvard University Press, 1991).
20. Lorenzo Valla, Discorsi sulla Donazione di Constantino [1440]. See also Crombie, Styles, 1:486.
21. Crombie, Styles, 1:426; Peter French, John Dee: The World of an Elizabethan Magus (London: Routledge and Kegan Paul, 1972), 31.
22. Schmitt, Aristotle and the Renaissance; on the Jesuits, see Rivka Feldhay, Galileo and the Church (Cambridge: Cambridge University Press, 1995), 157.
23. Crombie, Styles, 1:486, 490.
24. Mary Boas, The Scientific Renaissance, 1450–1630 (London: Collins, 1962), 23; Crombie, Styles, 1:432.
25. Elizabeth Eisenstein, The Printing Press as an Agent of Change, vols. 1 and 2 (Cambridge: Cambridge University Press, 1979); Roger Chartier, The Cultural Uses of Print in Early Modern France (Princeton: Princeton University Press, 1987).
26. On the recovery of lost proofs by Archimedes, see Rose, Italian Renaissance, 201.
27. Ibid., 81, 101, 175.
28. Ibid., 176.
29. Saliba, Islamic Science, 193 ff.
30. Ficino, quoted in Crombie, Styles, 1:469.
31. Regiomontanus, Oratorio, lectures delivered at the University of Padua, first published in 1537; Polydore Vergil, De Inventoribus Rerum, 1499 (see Rose, Italian Renaissance, 97, 256).
32. Mario Biagioli, "The Social Status of Italian Mathematicians, 1450–1600," History of Science 27 (1989): 41–95, 45.
33. Charles van den Heuvel, "De verspreiding van de Italiaanse vestingbouwkunde in de Nederlanden in de tweede helft van de zestiende eeuw," in Vesting. Vier eeuwen vestingbouw in Nederland, ed. J. Sneep, H. A. Treu, and M. Tydeman (The Hague: Stichting Menno van Coehoorn, 1982), 9–17.
34. Tartaglia, Quesiti et inventioni [1546], referred to by Bertrand Gille, Les ingénieurs de la Renaissance (Paris: Hermann, 1964), 206.
35. Robert Westman, "The Copernicans and the Churches," in God and Nature: Historical Essays on the Encounter between Christianity and Science, ed. David C. Lindberg and Ronald L. Numbers (Berkeley: California University Press, 1986), 76–82.
36. Dear, Mersenne, 37, 44; John Gascoigne, "The Role of the Universities," in Lindberg and Westman, Reappraisals, 207–23; Grafton, Defenders, 138.
37. Crombie, Styles, 1:492; Dear, Mersenne, 46.
38. Crombie, Styles, 1:488; Smith, Business of Alchemy, 36; Reyer Hooykaas, Humanisme, science et réforme. Pierre de la Ramée (Leiden: Brill, 1958).
39. Plato, Republic 525b, see Crosby, Measure of Reality, 179.
40. William Eamon, "From the Secrets of Nature to Public Knowledge," in Lindberg and Westman, Reappraisals, 333–65.

41. Burckhardt, *Kultur*, 130.
42. Ernst Gombrich, *The Story of Art* (London: Phaidon, 1950).
43. Crombie, *Styles*, 1:436; Pamela Long, "The Contribution of Architectural Writers to a 'Scientific' Outlook in the Fifteenth and Sixteenth Centuries," *Journal of Medieval and Renaissance Studies* 15 (1985): 265–98.
44. C. S. M. Rademaker, *Leven en werk van Gerardus Joannes Vossius (1577–1649)* (Hilversum: Verloren, 1999).
45. Leonardo, *Il Codice Atlantico*, quoted in Crombie, *Styles*, 1:471–73; Paolo Galluzzi, *Les ingénieurs de la Renaissance de Brunelleschi à Leónard de Vinci*, exhibition catalogue (Paris: Giunti, 1995).
46. Donald Sassoon, *Mona Lisa: The History of the World's Most Famous Painting* (London: HarperCollins, 2001); see the *London Review of Books*, April 4, 2002.
47. Gille, *Les ingénieurs*, 93–111, 116.
48. Ibid., 165, 143.
49. Crombie, *Styles*, 1:37–41, 484; Grafton, *Defenders*, 54; Lynn S. Joy, "Epicureanism in Renaissance Moral and Natural Philosophy," *Journal of the History of Ideas* 53 (1992): 573–83.
50. Crombie, *Styles*, 1:434; S. Y. Edgerton Jr., *The Heritage of Giotto's Geometry: Art and Science on the Eve of the Scientific Revolution* (Ithaca: Cornell University Press, 1991).
51. Caroline van Eck and Robert Zwijnenberg, "Inleiding," in *Over de schilderkunst, Leon Battista Alberti* (Amsterdam: Boom, 1996).
52. Crombie, *Styles*, 1:439.
53. Ibid., 1:461.
54. Stillman Drake, *Galileo at Work: His Scientific Biography* (Chicago: The University of Chicago Press, 1978), 165. The committee also reviewed the other discoveries that Galileo had made with his telescope: the moons of Jupiter, the oval shape of Saturn, and the phases of Venus. They confirmed the truth of all these discoveries (Feldhay, *Galileo*, 249).
55. Rose, *Italian Renaissance*, 5–6.
56. Baxandall, *Painting*, 87.
57. Gille, *Les ingénieurs*, 39; on the court of the Montefeltris, see B. Kempers, *Kunst, macht en mecenaat. Het beroep van schilder in sociale verhoudingen, 1250–1600* (Amsterdam: Arbeiderspers, 1987).
58. Crosby, *Measure*, 211.
59. Rose, *Italian Renaissance*, 144.
60. Biagioli, "Social Status," 43, 163.
61. Boas, *Scientific Renaissance*, 231.
62. Gille, *Les ingénieurs*, 119.
63. Biagioli, "Social Status," 55.
64. Tartaglia's chances to further his career were thwarted when he was defeated by a student of Cardano's in a mathematical contest.
65. Anthony Grafton, *Cardano's Cosmos: The Worlds and Works of a Renaissance Astrologer* (Cambridge, Mass.: Harvard University Press, 1999); Boas, *Scientific Renaissance*, 186; Charles Webster, *From Paracelsus to Newton: Magic and the Making of Modern Science* (Cambridge: Cambridge University Press, 1982).
66. E. J. Dijksterhuis, *Clio's stiefkind* (Amsterdam: Bert Bakker, 1990), 232; Vermij, *Calvinist Copernicans*, 58.

67. Dijksterhuis, *Simon Stevin (1548–1620)* (The Hague: Martinus Nijhoff, 1941); Crombie, *Styles*, 1:519.

68. The instructors in this program taught their courses in Dutch, while the rest of the university used Latin. This contrast reflects the military engineering program's unique emphasis on applied science within the largely theoretical world of the university.

69. Struik, *Geschiedenis*; Crombie, *Styles*; Michael Mahoney, "The Beginnings of Algebraic Thought in the Seventeenth Century," in *Descartes: Philosophy, Mathematics and Physics*, ed. Stephen Gaukroger (Brighton: The Harvester Press, 1980), 141–55.

70. Rose, *Italian Renaissance*, 146–47.

71. Westman, "Copernicans," 78.

72. Aquinas regarded mathematical entities as "quasi-substances," devoid of the properties that would make them part of physical reality. Aquinas, *Super Boethium De Trinitate (On The Trinity by Boethius)*, lectio 2, question 5, quoted in Feldhay, *Galileo and the Church*, 91.

73. See, for example, Gaukroger, *Descartes, Philosophy, Mathematics and Physics*, 57.

74. Rose, *Italian Renaissance*, 76.

75. Ibid., 82; Vincenzo di Grassi, a professor of philosophy at the University of Pisa, attacked Galileo in 1611 for using mathematics to describe "natural phenomena" (Biagioli, *Galileo*, 205).

76. Van Steenberghen, *Aristotle in the West*.

77. Biagioli, *Galileo, Courtier*.

78. Rose, *Italian Renaissance*, 283; see also Grafton, *Cardano's Cosmos*.

79. Biagioli, "Social Status," 45.

80. See Feldhay, *Galileo and the Church*, 146–49. The Jesuits also changed the organization of the educational system. They were the first to group students into homogeneous classes and to make sure that students internalized the objectives of their educational program through self-discipline.

81. Joseph E. Brown, "The Science of Weights," in Lindberg, *Science in the Middle Ages*, 179–205, 186; but see also Rose, *Italian Renaissance*, 11.

82. Brown, "Weights," 181–83.

83. Ibid., 190–201.

84. Rose, *Italian Renaissance*, 152.

85. Hall, "Science, Technology, and Warfare," 18.

86. Stillman Drake, *Galileo: Pioneer Scientist* (Toronto: University of Toronto Press, 1990), 48.

87. Rose, *Italian Renaissance*, 153.

88. Hall, "Science," 18.

89. Rose, *Italian Renaissance*, 155, 233.

90. Crombie, *Styles*, 1:482.

91. In particular by Josephus Blancanus, a student of Clavius's. See Feldhay, *Galileo*, 165–70. Galileo wanted to expand the domain of *scientia* by absorbing mathematics into physics.

92. Rose, *Italian Renaissance*, 227–29.

93. Biagioli, "Social Status," 65.

94. Rose, *Italian Renaissance*, 222–23; Biagioli, "Social Status," 60.

95. On the relationship between mathematics and reality in relation to machines, see Rose, *Italian Renaissance*, 234.

96. Rose, *Italian Renaissance*, 233.

97. The *Dialogue* is actually about a different topic, namely the accuracy of the Copernican view of the solar system. In this context, it explores whether the idea of free fall supports the hypothesis of a moving or a stationary Earth. Galileo's theory is not entirely modern in that it does not involve the concept of constant acceleration.

98. Galileo, quoted in Dijksterhuis, *De mechanisering*, 381. English translation from Maurice A. Finocchiaro, *The Galileo Affair: A Documentary History* (Berkeley: University of California Press, 1989), 154.

99. Ibid., 375, 380–81.

100. Dijksterhuis was undoubtedly influenced by Koyré's *Etudes galiléennes*.

101. Drake, *Galileo at Work*; Drake, *Pioneer*.

102. Drake, *Galileo at Work*, 29, 71, 68, 85.

103. Ibid., 60.

104. Leonardo had guessed that the time series 1, 2, 3, and 4 would correspond to successive distances of 1, 2, 3, and 4 (in other words, a cumulative distance of 1, 3, 6, and 10 units).

105. Drake, *Pioneer*, 111n; Drake, *Galileo at Work*, 91–95.

106. Drake, *Pioneer*, 15.

107. Drake, *Galileo at Work*, 17.

108. Vasari is the author of a famous collection of artists' biographies: *Le Vite delle più eccellenti pittori, scultori, e architettori* (*Lives of the Most Excellent Painters, Sculptors, and Architects*, 1550, 1568).

109. Rose, *Italian Renaissance*, 280.

110. Drake, *Galileo at Work*, 13–14.

111. Simon Schaffer, "The Charter'd Thames: Naval Architecture and Experimental Spaces in Georgian Britain," in Roberts, Schaffer, and Dear, *Mindful Hand*, 279–305.

112. Drake, *Galileo at Work*, 35, 38, 50.

113. Biagioli, *Galileo*, 7–8; Zilsel, *Ursprünge*, 60; Drake, *Galileo at Work*, 30.

114. Biagioli, *Galileo*, 162. Another scientist of the period who went through a similar series of patronage relationships was Pierre Gassendi. Gassendi was much more diplomatic than Galileo, however. See Lisa Sarasohn, *Gassendi's Ethics* (Ithaca: Cornell University Press, 1996), 5–12; see also Lisa Sarasohn, "Nicolas Claude Fabri de Peiresc and the Patronage of the New Science in the Seventeenth Century," *Isis* 84 (1993): 70–90.

115. Biagioli, *Galileo*, 276.

116. See William A. Wallace, *Galileo and His Sources* (Princeton: Princeton University Press, 1984).

117. The role of the Jesuits has been brought to light mainly by Pietro Redondi, in *Galileo Heretic* (Princeton: Princeton University Press, 1987), and by Feldhay in *Galileo*. Galileo's achievement in parlaying a mathematical background into a situation as a philosopher at an Italian court was not entirely unprecedented. More significantly, the Jesuits also united the humanistic tradition with the ideals of virtuosity and the *vita activa*. They put the full weight of their institutional authority behind reforms in education and scholarship. In one respect, in fact, they were more deeply influenced by the humanist tradition than Galileo was; their moderate skepticism with regard to science restrained them from throwing their support behind the Copernican system.

118. Biagioli, *Galileo*. Bagioli's interpretation of the trial of Galileo seems more plausible than Redondi's, which is that Galileo was "sacrificed" to the Jesuits by

Urbanus, who insisted on the condition that Galileo would not be convicted of any crime for which the penalty could be death.

119. Biagioli, *Galileo*, 348; Paula Findlen, *Possessing Nature: Museums, Collecting and Scientific Culture in Early Modern Italy* (Berkeley: University of California Press, 1994).

120. Biagioli, *Galileo*, 360.

Chapter 5: The Experimental Style II

Epigraph: Michel de Montaigne, *Essays*, trans. Charles Cotton, ed. William Carew Hazlitt, 1877, e-text produced by David Widger, Project Gutenberg.

1. Richard H. Popkin, *The History of Scepticism from Erasmus to Descartes* (Assen: Van Gorcum, 1960), 17.

2. These terms were so used by Philipp Melanchthon; see Nicholas Jardine, "The Forging of Modern Realism: Clavius and Kepler against the Sceptics," *Studies in the History and Philosophy of Science* 10 (1979): 141–73.

3. Popkin, *Scepticism*; Pieter Pekelharing, "Reflexiviteit, skepticism en vertrouwen," *Krisis* 52 (September 1993): 33–44.

4. Charles Taylor, *Sources of the Self* (Cambridge: Cambridge University Press, 1989), 178, 182.

5. Crombie, *Styles*, 1:503.

6. Alessandro Piccolomini in 1558, quoted in Crombie, *Styles*, 1:532. Piccolomini was arguing against the "reality" of epicycles and other mathematical tricks in Ptolemy's system; see below.

7. Popkin, *Scepticism*, 5–7.

8. Ibid., 8–9.

9. William Shea, "Galileo and the Church," in Lindberg and Numbers, *God and Nature*, 114–35, 121; Crombie, *Styles*, 1:597.

10. Dear, *Mersenne*, 28.

11. Hacking, *Emergence*, 18–30.

12. Dear, *Mersenne*, 29.

13. Blaise Pascal, *Les Provinciales*, letter 5, in *Oeuvres Complètes* (Paris: Gallimard [Pléiade], 1954), 706, 710ff.

14. John Heilbron, *Electricity in the 17th and 18th Century: A Study of Early Modern Physics* (Berkeley: University of California Press, 1979).

15. On Grassi, see Biagioli, *Galileo, Courtier*, 271.

16. Wallace, *Galileo and His Sources*, 347; Joella Yoder, *Unrolling Time: Christiaan Huygens and the Mathematization of Nature* (Cambridge: Cambridge University Press, 1988), 14; on Riccioli, see primarily Alexandre Koyré, *Metaphysics and Measurement: Essays in Scientific Revolutions* (London: Chapman and Hall, 1968).

17. William B. Ashworth, "Catholicism and Early Modern Science," in Lindberg and Numbers, *God and Nature*, 136–66.

18. Redondi, *Galileo: Heretic*.

19. David Noble, *A World without Women: The Christian Clerical Culture of Western Science* (New York: Knopf, 1992).

20. Duffy, *Saints and Sinners*.

21. Stephen Toulmin, *Cosmopolis* (New York: Free Press, 1990).

22. Averroës, *Commentary on Aristotle's De Caelo*, quoted in Crombie, *Styles*, 1:530; Aquinas, *Summa theologica* 1.q32.a1, quoted in Blumenberg, *Genesis*, 226.
23. Crombie, *Styles*, 1:343.
24. Biagioli, *Galileo, Courtier*, 9.
25. Blumenberg, *Genesis*, 253; Lloyd, *Greek Science*, 64, 119, 129.
26. Westman, "Copernicans," 79.
27. Westman, "Copernicans"; Blumenberg, *Genesis*, 151.
28. Blumenberg, *Genesis*, 250; Thomas Kuhn, *The Copernican Revolution* (Cambridge, Mass: Harvard University Press, 1957).
29. This was the main thrust of the attack on Copernicus by the astronomer Giovanni Maria Tolosani around 1547; see Feldhay, *Galileo and the Church*, 207–9. In Galileo's day, it was primarily the Dominicans who found fault with him on these grounds.
30. R. Hooykaas, *G. J. Rheticus' Treatise on Holy Scripture and the Motions of the Earth* (Amsterdam: North-Holland Publishing Company, 1984).
31. Robert S. Westman, "Magical Reform and Astronomical Reform: The Yates Thesis Revisited," in *Hermeticism and the Scientific Revolution*, ed. R. S. Westman and J. E. McGuire (Los Angeles: Williams Andrews Clark Memorial Library, 1977), 46. See also Vermij, *Calvinist Copernicans*, 57.
32. Westman, "Copernicans," 94.
33. Gascoigne, "Role of the Universities," 213; Westman, "Copernicans," 98; Crombie, *Styles*, 1:551.
34. Feldhay, *Galileo*, chap. 13. Scheiner arrived at this conclusion while researching sunspots. Galileo felt this approach was superfluous, because he could deduce Venus's orbit around the sun from its phases.
35. Shea, "Galileo," 119.
36. Westman, "Copernicans," 77.
37. Drake, *Galileo at Work*, 255.
38. Westman, "Copernicans," 102.
39. Ashworth, "Catholicism," 150; on Patrizi, see Yates, *Giordano Bruno*.
40. Biagioli, *Galileo, Courtier*, 286.
41. Crombie, *Styles*, 1:551.
42. Galileo had a reason for holding fast to the idea of circular planetary orbits, which was rooted in his beliefs about inertia.
43. Crombie, *Styles*, 1:543, 551; A. C. Crombie, "Sources of Galileo's Early Natural Philosophy," in *Science, Art and Nature in Medieval and Modern Thought* (London: The Hambledon Press), 155.
44. Taylor, *Sources*, 144; for a comparison with Aristotle's theory of knowledge, see 186.
45. Hatfield, "Metaphysics," 93–166.
46. Crombie, *Styles*, 1:585.
47. Galileo to Jean Baptiste Morin, quoted in Crombie, *Styles*, 1:565, 608.
48. Crombie, *Styles*, 1:552–53.
49. See Mario Biagioli, *Galileo's Instruments of Credit: Telescopes, Instruments, Secrecy* (Chicago: The University of Chicago Press, 2006), 178.
50. Ibid., 577, 581–88.
51. Jardine, "Forging of Modern Realism."
52. Ernan McMullin, "Conceptions of Science in the Scientific Revolution,"

in Lindberg and Westman, *Reappraisals*, 59. The first example has to do with the absence of parallax among distant stars, and the second with the configuration of the outermost planets in the solar system when they are closest to the earth.

53. Stephen Gaukroger, *Descartes: An Intellectual Biography* (Oxford: Oxford University Press, 1995), 24. Descartes arrived at La Flèche at the age of ten, and completed the equivalent of secondary school and the first years of general university studies there.

54. René Descartes, *Oeuvres philosophiques I*, ed. F. Alquié (Paris: Garnier, 1988), 641.

55. Quoted in Dear, *Mersenne*, 46.

56. Gaukroger, *Descartes: An Intellectual Biography*.

57. The characterization of this inference as inductive is borrowed from Ferdinand Alquié, see his *Descartes* (Paris: Hatier, 1956), 88. Descartes himself referred to his working method as analysis, in contrast to the synthesis of the classical geometers; "synthesis" should thus be taken as a synonym for "deduction," while "analysis" is a way of reasoning back to first principles. See *Réponses aux secondes objections*, which contains Descartes's response to Mersenne's objections to the *Meditations*.

58. Popkin, *History of Scepticism*, 196.

59. Richard Rorty, *Philosophy and the Mirror of Nature* (Princeton: Princeton University Press, 1979).

60. See also Martial Gueroult, "The Metaphysics and Physics of Force in Descartes," in Gaukroger, *Descartes: Philosophy, Mathematics, Physics*, 217.

61. Descartes, *Principles of Philosophy*, trans. Valentine Rodger Miller and Reese Miller (Dordrecht: D. Reidel Publishing Company, 1983), vol. 3, sec. 43, pp. 104–5.

62. Gaukroger, *Descartes: An Intellectual Biography*, 51, 54, 118. On the origin of this idea in Stoic philosophy, see Diogenes Laërtius, *Lives of the Philosophers*, bk. 7.

63. Aristotle, *De Anima*, bk. 3.

64. Gaukroger, *Descartes: An Intellectual Biography*, 121.

65. Quintilian, *Institutio Oratoria*, bk. 8, sec. 3.61.

66. Gaukroger, *Descartes: An Intellectual Biography*, 379.

67. Gassendi quoted in ibid., 343.

68. Ibid., 378, 386.

69. Toulmin, *Cosmopolis*.

70. Dear, *Mersenne*, 31.

71. These are, in fact, the tenets of the theological position called voluntarism.

72. Popkin, *Scepticism*, 134–35.

73. Mersenne quoted in Dear, *Mersenne*, 27.

74. Dear, *Mersenne*, 42, 73.

75. Crombie, *Styles*, 1:833, 836.

76. Dear, *Mersenne*, 43–45.

77. Crombie, *Styles*, 1:818–19.

78. H. F. Cohen, *Quantifying Music: The Science of Music at the First Stage of the Scientific Revolution, 1580–1650* (Dordrecht: Reidel, 1984). See also Dijksterhuis, *Simon Stevin*, chap. 13; on Beeckman, see Crombie, *Styles*, 1:808–10.

79. Cohen, *Music*, 4–7; Crombie, *Styles*, 1:787; Zarlino could not neatly fit the minor sixth (8:5) into this system. See also Crombie, *Styles*, 1:810.

80. Crombie, *Styles*, 1:834, 837.

81. Steven Shapin and Simon Schaffer, *Leviathan and the Air Pump* (Princeton: Princeton University Press, 1985), 113.
82. Steven Shapin, "The House of Experiment in Seventeenth-Century England," *Isis* 79 (1988): 373–404; Lorraine Daston, "The Moral Economy of Science," *Osiris* 10 (1995): 3–24.
83. Shapin and Shaffer, *Leviathan*, 213.
84. Robert Hooke, quoted in Eamon, "From the Secrets," 356.
85. Richard Popkin, "The Philosophy of the Royal Society of England," in *The Pimlico History of Western Philosophy* (London: Pimlico, 1999), 358–63.

Chapter 6: The Experimental Style III

Epigraphs: William Watts, quoted in J. A. Bennett, "The Mechanics' Philosophy and the Mechanical Philosophy," *History of Science* 24 (1986): 21; Francis Bacon, *Novum Organum* [1620], bk. 2, aphorism 52.

1. Kuhn, "Mathematical Versus Experimental," 31–65.
2. On Boyle's air pump, see Shapin and Shaffer, *Leviathan*; Frank Jr., *Harvey*, chap. 6.
3. The distinction between "classical" and "Baconian" sciences is not as great as Kuhn suggested in his well-known article. But it does not seem reasonable to me to do away with the distinction entirely, as J. A. Bennett has argued. The difference lies not in the use of mathematics per se (though the Baconian sciences hardly involved any mathematics at first and Bacon himself had little use for math) but in the extent to which the chosen mathematical methods yielded a deductively organized system. See Bennett, "Mechanics' Philosophy."
4. Webster, *From Paracelsus to Newton*; John Henry, "Occult Qualities and the Experimental Philosophy: Active Principles in Pre-Newtonian Matter Theory," *History of Science* 24 (1986): 335–81; Smith, *Business of Alchemy*; Ursula Klein and Wolfgang Lefèvre, *Materials in Eighteenth-Century Science: A Historical Ontology* (Cambridge, Mass.: MIT Press, 2007).
5. Paolo Rossi, *Francis Bacon: From Magic to Science* (London: Routledge and Kegan Paul, 1968), 18; Betty Jo Teeter Dobbs, *Foundations of Newton's Alchemy, or: "The Hunting of the Greene Lyon"* (Cambridge: Cambridge University Press, 1975), 59; William R. Newman, *Promethean Ambitions: Alchemy and the Quest to Perfect Nature* (Chicago: The University of Chicago Press, 2004).
6. This is according to Schweitzer in *Vitulus aureus* (*The Golden Calf*). See Wayne Shumaker, *The Occult Sciences in the Renaissance* (Berkeley: University of California Press, 1972), 163–65.
7. Smith, *Business of Alchemy*, 173. Becher unmasked a number of fraudulent gold-makers.
8. Johann Joachim Becher, *Trifolium Becherianum Hollandicum* (Frankfurt: Johann David Zunner, 1679), cited in Smith, *Business of Alchemy*, 249, 253.
9. Dobbs, *Foundations*, 44.
10. George Basalla, "Pop Science: The Depiction of Science in Popular Culture," in *Science and Its Public*, ed. G. A. Holton and W. Blanpied (Dordrecht: Reidel, 1976), 261–78.
11. Newman, *Promethean Ambitions*, 43.

12. Harold J. Cook, "Physicians and Natural History," in Jardine, Secord, and Spary, *Cultures of Natural History*, 91–105.

13. Michel Foucault, *Les mots et les choses: une archéologie des sciences humaines* (Paris: Gallimard, 1966), chap. 2. English edition: *The Order of Things: An Archeology of the Human Sciences* (New York: Random House, 1970).

14. Henry, "Occult Qualities."

15. David Teniers and Thomas van Wijck were two painters who depicted alchemists quite a bit more sympathetically. See Lawrence M. Principe and Lloyd DeWitt, *Transmutations: Alchemy in Art* (Philadelphia: Chemical Heritage Foundation, 2002).

16. On Marlowe and Shakespeare, see Frances Yates, *The Occult Philosophy in the Elizabethan Age* (London: Routledge, 1979).

17. William B. Ashworth, "Catholicism and Early Modern Science," in *God and Nature: Historical Essays on the Encounter between Christianity and Science*, ed. D. Lindberg and R. Numbers (Berkeley: University of California Press, 1986), 136–66, 149.

18. On Rudolf II, see Thomas DaCosta Kaufmann, *The Mastery of Nature: Aspects of Art, Science, and Humanism in the Renaissance* (Princeton: Princeton University Press, 1993); R. J. Evans, *Rudolf II and His World: A Study in Intellectual History, 1576–1612* (Oxford: Oxford University Press, 1973).

19. Smith, *Business of Alchemy*, 181.

20. Charles Webster, *The Great Instauration: Science, Medicine and Reform 1626–1660* (London: Duckworth, 1975), 327.

21. Dobbs, *Foundations*; Richard Westfall, "Newton and Alchemy," in *Occult and Scientific Mentalities*, ed. B. Vickers (Cambridge: Cambridge University Press, 1987), 315–35.

22. Henry, "Occult Qualities"; Brian Vickers, ed., *English Science, Bacon to Newton* (Cambridge: Cambridge University Press, 1987).

23. Brian Copenhaver, "Natural Magic, Hermetism, and Occultism in Early Modern Science," in Lindberg and Westman, *Reappraisals of the Scientific Revolution*, 261–301.

24. Edward Rosen, "Kepler's Attitude Towards Astrology and Mysticism," in Vickers, *Occult and Scientific Mentalities*, 253–72; Henry, *Scientific Revolution*, 61.

25. Smith, *Business of Alchemy*.

26. William Hine, "Marin Mersenne, Renaissance Naturalism and Renaissance Magic," in Vickers, *Occult and Scientific Mentalities*, 165–76.

27. Dobbs, *Foundations*, 57.

28. On Croll and Libavius, see Owen Hannaway, *The Chemists and the Word: The Didactic Origins of Chemistry* (Baltimore: Johns Hopkins University Press, 1975); Pagel, *Helmont*, 204.

29. *Encyclopédie*, vol. 18, 253–61, quoted in Copenhaver, "Natural Magic."

30. Newman, *Promethean Ambition*, 17; William R. Newman and Anthony Grafton, "Introduction: The Problematic Status of Astrology and Alchemy in Premodern Europe," in *Secrets of Nature: Astrology and Alchemy in Early Modern Europe*, ed. William R. Newman and Anthony Grafton (Cambridge, Mass.: The MIT Press, 2001), 1–38, 24.

31. Yates, *Giordano Bruno*, 15.

32. Daston and Park, *Wonders*, 95, 108. In the late seventeenth century, unicorn

horns were identified as coming from narwhals. By that time, they were no longer so rare.

33. "Geber" was the pseudonym of the anonymous author of *Summa perfectionis* ("The Sum of Perfection"), which sought a synthesis between alchemy and Aristotelianism. Geber may have been a man of medicine. Writing in the late thirteenth century, he probably took inspiration from the medical school in Salerno. What is exceptional about Geber is that he not only conducted experiments, but also interpreted their results with the aid of a kind of corpuscular theory. See William Newman, *Atoms and Alchemy: Chymistry and the Experimental Origin of the Scientific Revolution* (Chicago: The University of Chicago Press, 2006).

34. R. van den Broek and Gilles Quispel, "Inleiding," in *Corpus hermeticum* (Amsterdam: In de pelikaan, 1990), 13–28.

35. Yates, *Bruno*.

36. Yates, *Occult Philosophy*, 17 ff.

37. And with the reputation of other writings by Ficino, such as *De vita coelitus comparanda*. A number of historians have criticized Frances Yates with varying degrees of severity for her position that the magical and occult tradition which influenced science can be characterized as "hermetic," originating from the writings of Hermes Trismegistus. See, for example, Robert S. Westman and J. E. McGuire, *Hermeticism and the Scientific Revolution* (Los Angeles: Williams Andrews Clark Memorial Library, 1977). But after the smoke has cleared, this turns out to be largely a terminological issue: Yates's use of the adjective "hermetic" is somewhat unfortunate because it refers too exclusively to Hermes. "Occult" is better, according to Brian Vickers (*Occult and Scientific* Mentalities) and some other scholars. The slow acceptance of Yates's work is illustrated by Hall, *Scientific Renaissance*, which was once considered highly authoritative. The chapter on magic and astrology does not mention Ficino and Hermes, an omission that would be unthinkable today.

38. "Seminal words" might also be a reasonable translation, given the similarity to the first verse of the Gospel of John: "In the beginning was the Word . . ." See Frank Kermode, "John," in *The Literary Guide to the Bible*, ed. R. Alter and F. Kermode (Cambridge, Mass.: Harvard University Press, 1987), 440–66. The Neoplatonists borrowed the term *logoi spermatikoi* from the Stoics.

39. Yates, *Bruno*; Shumaker, *Occult Sciences*; Daston and Park, *Wonders*, 161; Newman and Grafton, "Introduction," 24.

40. Yates, *Occult Philosophy*, 1979.

41. Yates, *Bruno*, 165; Erasmus, *The Colloquies*, trans. Craig R. Thompson (Chicago: The University of Chicago Press, 1965), 238–45.

42. Webster, *Paracelsus to Newton*, 58.

43. Carolyn Merchant, *The Death of Nature* (San Francisco: Harper and Row, 1980).

44. Walter Pagel, *Paracelsus: An Introduction to Philosophical Medicine in the Era of the Renaissance*, 2nd ed. (Basel: Karger, 1982).

45. Pagel, *Paracelsus*, 142.

46. Ibid., 51, 61.

47. George Huppert, *The Idea of Perfect History: Historical Erudition and Historical Philosophy in Renaissance France* (Urbana: University of Illinois Press, 1970), 47.

48. Frank, *Harvey*, 119.

49. Webster, *Instauration*, 275; Norma E. Emerton, *The Scientific Reinterpretation of Form* (Ithaca: Cornell University Press, 1984).

50. Foucault, *Les mots et les choses*; Ian Maclean, "The Interpretation of Natural Signs," in Vickers, *Mentalities*, 231–52; Gerolamo Cardano, *De vita propria* (Paris, 1643), chap. 45.

51. Daston and Park, *Wonders*, 163–65; Siraisi, *Clock and Mirror*; on Cardano's personal relationships with demons, see Grafton, *Cardano's Cosmos*, 168.

52. Siraisi, *Clock and Mirror*.

53. Stillman Drake, "The Accademia dei Lincei," in *Galileo Studies* (Ann Arbor: The University of Michigan Press, 1970), 79–94; on the Lincei, see David Freedman, *The Eye of the Lynx: Galileo, His Friends, and the Beginnings of Modern Natural History* (Chicago: The University of Chicago Press, 2002).

54. William Eamon, "Secrets of Nature," 333–65.

55. Ginzburg, "Clues," 96–125.

56. Daston and Park, *Wonders*, 169.

57. Eamon, "Secrets of Nature"; Merchant, *Death of Nature*, 183; Gaukroger, *Descartes: An Intellectual Biography*, 59.

58. Webster, *Paracelsus to Newton*, 61.

59. William R. Newman and Lawrence M. Principe, *Alchemy Tried in the Fire: Starkey, Boyle, and the Fate of Helmontian Chymistry* (Chicago: The University of Chicago Press, 2002).

60. Pagel, *Helmont*, 13, 20, 28, 62, 64.

61. Allison Coudert, *The Impact of the Kabbalah in the Seventeenth Century: The Life and Thought of Francis Mercury van Helmont (1614–1698)* (Leiden: Brill, 1999).

62. Webster, *Instauration*, 276–79, 301.

63. Rossi, *Bacon*, 2; Merchant, *Death of Nature*, 220.

64. Webster, *Paracelsus to Newton*, 79.

65. Crombie, *Styles* 1:632; Gad Freudenthal, "Theory of Matter and Cosmology in William Gilbert's *De magnete*," *Isis* 74 (1983): 22–37; John Henry, "Animism and Empiricism: Copernican Physics and the Origin of William Gilbert's Experimental Method," *Journal of the History of Ideas* 62 (2001): 1–21.

66. In the terminology of the day, the type of mathematics applied to natural quantities was called mixed mathematics.

67. As quoted in Steven Shapin, "Rough Trade," *London Review of Books*, March 6, 2003. Shapin refers to works by Rob Iliffe and Adrian Johns.

68. Shapin, "Rough Trade."

69. Henry, "Occult Qualities"; on 1650 as the year that this view died out, see William Ashworth, "Natural History and the Emblematic World View," in Lindberg and Westman, *Reappraisals*, 303–32.

70. Rossi, *Bacon*, 13–14, 32, 104; Newman, *Promethean Ambitions*, 256–71.

71. Graham Rees, "Francis Bacon's Biological Ideas: A New Manuscript Source," in Vickers, *Occult and Scientific Mentalities*, 297–314.

72. Hutchison, "Occult Qualities," 233–53.

73. Ibid., 238.

74. The present scientific name is *Mandragora officinalis*. Not long after 1668, Abraham Bosse, an engraver who worked for the Académie des Sciences in Paris, made an engraving of a mandrake in which the root is strikingly similar to the lower part of a woman's body. Bosse was ridiculed for this work. See Alice Stroup, *A Company of Scientists* (Berkeley: University of California Press, 1990), 84.

75. Cellini relates in his *Vita* [1562] that he was five years old when his father called

him over to see the little creature in the fire. His father gave him a box on the ear so that he would never forget the occasion.

76. Daston and Park, *Wonders*, 36; Findlen, *Possessing Nature*, 225; Agnes Arber, *Herbals: Their Origin and Evolution: A Chapter in the History of Botany 1470–1670* (Cambridge: Cambridge University Press, 1938), 131.

77. Daston and Park, *Wonders*, 129.

78. Hutchison, "Occult Qualities," 239.

79. Daston and Park, *Wonders*, 311.

80. Feldhay, *Galileo*, 245–47, 273, 281.

81. Peter Dear, "Jesuit Mathematical Science and the Reconstruction of Experience in the Early Seventeenth Century," *Studies in History and Philosophy of Science* 18 (1987): 133–75.

82. Biagioli, *Galileo, Courtier*, 162.

83. Hutchison, "Occult Qualities," 236.

84. Carlo Ginzburg, "The High and the Low: The Theme of Forbidden Knowledge in the Sixteenth and Seventeenth Centuries," in Ginzburg, *Clues, Myths*, 60–76.

85. Augustine, *Confessions*, bk. 10, chap. 35. Both Augustine and Thomas offered certain escape routes, primarily through "study."

86. Eamon, "Secrets," 336.

87. On the genre of *problemata*, see Ann Blair, *The Theatre of Nature: Jean Bodin and Renaissance Science* (Princeton: Princeton University Press, 1997), 37.

88. Eamon, "Secrets."

89. Rossi, *Bacon*, 31.

90. Ginzburg, "High and the Low."

91. Daston and Park, *Wonders*, 223.

92. Clark, "Demonology," 355.

93. Hutchison, "Occult Qualities," 237; Newman, *Promethean Ambitions*, 45, 53. A number of theologians generally seen as Aristotelians, such as Thomas Aquinas, were in the Augustinian tradition, and they too express these views.

94. Carlo Ginzburg, *Ecstasies: Deciphering the Witches' Sabbath*, trans. Raymond Rosenthal (Chicago: The University of Chicago Press, 1991).

95. Clark, "Demonology," 359.

96. Nancy Siraisi, *Clock and Mirror*, 59.

97. Krzystof Pomian, *Collectors and Curiosities: Paris and Venice, 1500–1800* (Cambridge: Polity Press, 1990), 35.

98. Ashworth, "Natural History."

99. Blair, *Theatre*, 36; we can see a continuation of the *secreta* tradition in book titles such as *La minera del mondo* (*The Riches of the World*, 1589) by Giovanni Maria Bonardi.

100. Francis Bacon, "Of Tribute, Or Giving What Is Due" [1592], quoted in Julian Martin, *Francis Bacon, the State and the Reform of Natural Philosophy* (Cambridge: Cambridge University Press, 1992), 70.

101. Ashworth, "Natural History," 322.

102. Martin, *Bacon*, 153.

103. Rossi, *Bacon*, 237n90.

104. Findlen, *Possessing Nature*, 147, 149.

105. DaCosta Kaufmann, *Mastery*, 174.

106. Pomian, *Collectors*, 61, 63.

107. Barbara Shapiro, *Probability and Certainty in Seventeenth-Century England* (Princeton: Princeton University Press, 1983), 45, 66.
108. Webster, *Instauration*, 27, 339.
109. Popkin, *History of Scepticism*, 86.
110. "Lord Keeper" was a ceremonial title tied to the position of Lord Chancellor, the country's highest public official and judicial authority, answering only to the king.
111. Martin, *Bacon*, 113.
112. The Church of England was at least as much a product of the Reformation (with an iconoclasm and the like mandated by its leaders) as it was of Henry VIII's desire to divorce Catherine of Aragon despite the Pope's refusal. But one influential Anglican faction, known as the High Church, remained quite close to the Church of Rome in its beliefs about theology and liturgy.
113. Martin, *Bacon*, 73, 112. In 1621, however, he was removed from office by Parliament against the king's will on charges of corruption.
114. Ibid., 94.
115. John Langbein, *Prosecuting Crime in the Renaissance: England, Germany, France* (Cambridge, Mass.: Harvard University Press, 1974); Lorraine Daston, *Classical Probability in the Enlightenment* (Princeton: Princeton University Press, 1988), 42–43.
116. Martin, *Bacon*, 88.
117. McMullin, "Conceptions," 46; Martin, *Bacon*, 142.
118. Bacon, *Novum Organum*, quoted in Daston and Park, *Wonders*, 311.
119. Martin, *Bacon*, 77, 165.
120. Bacon, *The Masculine Birth of Time*, quoted in Merchant, *Death of Nature*, 170.
121. Martin, 82; Merchant, *Death of Nature*, 168–72; Merchant has been criticized by Peter Pesic for being too quick to equate *vexatio* with torture; see "Wrestling with Proteus: Francis Bacon and the 'Torture' of Nature," *Isis* 90 (1999): 81–94. Brian Vickers is one of the most outspoken critics of Merchant's thesis; Merchant responds to him in "Secrets of Nature: The Bacon Debates Revisited," *Journal of the History of Ideas* 69 (2008): 147–62.
122. Katherine Park, "Women, Gender, and Utopia: *The Death of Nature* and the Historiography of Early Modern Science," *Isis* 97 (2006): 487–95.
123. Martin, *Bacon*, 167; Crombie, *Styles*, 1:631.
124. Bacon, *Novum Organum*, bk. 2, aphorism 11; see Henry, *Scientific Revolution*, 65; McMullin, "Conceptions," 53.
125. Martin, *Bacon*, 144.
126. DaCosta Kaufmann, *Mastery*, 185.
127. Webster, *Instauration*, 333, 377–80.
128. Ibid., 222.
129. H. M. Solomon, *Public Welfare, Science and Propaganda in Seventeenth Century France* (Princeton: Princeton University Press, 1972); see also Mario Ambrosoli, *The Wild and the Sown: Botany and Agriculture in Western Europe, 1350–1850* (Cambridge: Cambridge University Press, 1997).
130. Webster, *Instauration*, 69, 76, 346, 375; Webster, *Paracelsus to Newton*, 62.
131. Webster, *Instauration*, 346.
132. Shapiro, *Probability*, 23.
133. Christiaan Huygens, letter to Philips Doublet, July 1666, quoted in C. D. Andriesse, *Titan kan niet slapen. Een biografie van Christiaan Huygens* (Amsterdam: Contact, 1993), 222; see also Stroup, *Company of Scientists*.

134. Shapin and Shaffer, *Leviathan*, 273.
135. Boas, *Scientific Renaissance*.
136. Newman, *Atoms and Alchemy*, 157 ff.
137. Boyle, quoted in Henry, "Occult Qualities," 344, 345.
138. Pagel, *Van Helmont*, 56n.
139. Shapin and Shaffer, *Leviathan*, 204.
140. Daston and Park, *Wonders*, 13, 311.
141. Ibid., 238–39.
142. Peter Dear, "Miracles, Experiments, and the Ordinary Course of Nature," *Isis* 81 (1990): 663–83.
143. Terry Shinn, *L'École Polytechnique 1794–1914. Savoir scientifique et pouvoir social* (Paris: Presses de la fondation nationale des sciences politiques, 1980); Eda Kranakis, *Constructing a Bridge: An Exploration of Engineering Culture, Design, and Research in Nineteenth-Century France and America* (Cambridge, Mass.: The MIT Press, 1996); Eda Kranakis, "Social Determinants of Engineering Practice," *Social Studies of Science* 19 (1989): 5–70.
144. Stroup, *Company of Scientists*.
145. Kuhn, "Mathematical versus Experimental Traditions."
146. Bellone, *World on Paper*, 32.
147. Janet Browne, *Charles Darwin: The Power of Place, Vol. 2 of a Biography* (London: Jonathan Cape, 2002), 260.
148. Browne, *Darwin Power*, 186.
149. Kuhn, "Mathematical versus Experimental Traditions." Ian Hacking, however, has argued that this development should be viewed partly in the context of the rise of statistics and the accompanying tendency to gather sets of data. The purely theoretical, deductive character that Kuhn attributes to the classical sciences is an oversimplification. See Hacking, *Taming*, 62.
150. Robert Merton, *Science, Technology, and Society in 17th Century England* (1938; repr., New York: Harper and Row, 1970).
151. Reyer Hooykaas, *Religion and the Rise of Modern Science* (Edinburgh: Scottish Academic Press, 1973); Lindberg and Numbers, *God and Nature*; Rivka Feldhay and Michael Heyd, "The Discourse of Pious Science," *Science in Context* 3 (1989): 109–42; Thomas F. Gieryn, "Distancing Science from Religion in Seventeenth-Century England," *Isis* 79 (1988): 582–93; Steven Shapin, "Understanding the Merton Thesis," *Isis* 79 (1988): 594–605.
152. Charles Taylor, *Sources of the Self* (Cambridge: Cambridge University Press, 1989).
153. Webster, *Instauration*, 421, 505.
154. Hooykaas, *Religion*, 114.
155. Webster, *Instauration*, 498–99.
156. Robert A. Schofield, *Mechanism and Materialism: British Natural Philosophy in an Age of Reason* (Princeton: Princeton University Press, 1970), 137.
157. Martin, *Bacon*, 58; Webster, *Instauration*, 274.
158. Pagel, *Helmont*, 199–200.
159. Charles Webster's monumental study *The Great Instauration* (1975) seems to have proved this once and for all. But his period of study ends with the Restoration and the end of the Republic. James Jacob and Margaret C. Jacob, in "The Anglican Origins

of Modern Science," *Isis* 71 (1980): 251–67, have shown that the same close relationship, albeit with differences in emphasis, still existed even after the Restoration, when Calvinist Puritanism lost the upper hand to moderate Anglicanism.

160. We would also hope to find such a relationship in Switzerland and a number of German countries. Chapter 8 discusses the new sciences in Basel, as well as in Montpellier in the period of Huguenot governance.

161. J. Huizinga, *Nederlandse beschaving in de zeventiende eeuw. Een schets* (Haarlem: Tjeenk Willink, 1941); see also Dirk Struik, "Further Thoughts on Merton in Context," *Science in Context* 3 (1989): 227–38.

162. Yoder, *Unrolling*, passim.

163. Stroup, *Company*, 70.

164. "Mais il faudroit raisonner avec méthode sur les experiences, et en amasser de nouvelles, à peu près suivant le projet de Verulamius." Christiaan Huygens, letter to G. W. Leibniz, November 16, 1691, quoted in Yoder, *Unrolling*, 225.

165. Yoder, *Unrolling Time*, 173.

166. Rademaker, *Leven en werk*, 54.

167. C. de Pater, "Petrus van Musschenbroek (1692–1761). Een Newtoniaans natuuronderzoeker" (PhD diss., University of Utrecht, 1979); Klaas van Berkel, *In het voetspoor van Stevin* (Meppel/Amsterdam: Boom, 1985); Rademaker, *Vossius*, 1999; Edward Ruestow, *Physics at Seventeenth- and Eighteenth-Century Leiden: Philosophy and the New Science in the University* (The Hague: Martinus Nijhoff, 1973).

168. On De Volder's Spinozism, see Wim Klever, *Mannen rond Spinoza* (Hilversum: Verloren, 1997).

169. This argument has been advanced by Harold J. Cook in *Matters of Exchange: Commerce, Medicine, and Science in the Dutch Golden Age* (New Haven: Yale University Press, 2007); see also Erik Jorink, *Het Boeck der Natuere. Nederlandse geleerden en de wonderen van Gods Schepping 1575–1715* (Leiden: Primavera Pers, 2006).

170. Simon Schaffer, "The Glorious Revolution and Medicine in Britain and the Netherlands," *Notes and Records of the Royal Society of London* 43 (1989): 167–90; Webster, *Instauration*, 144.

171. Rina Knoeff, *Herman Boerhaave (1668–1738): Calvinist Chemist and Physician* (Amsterdam: Koninklijke Nederlandse Akademie van Wetenschappen, 2002), 12.

172. Catherine Wilson, *The Invisible World: Early Modern Philosophy and the Invention of the Microscope* (Princeton: Princeton University Press, 1995).

173. Philip van Lansberge was a pastor and Copernican astronomer in the humanist tradition. Building on his work, his student Martinus Hortensius in 1632 calculated the diameters of a number of planets and fixed stars. See Albert van Helden, *Measuring the Universe* (Chicago: The University of Chicago Press, 1985), 101–4; see also Klaas van Berkel, *Citaten uit het boek der natuur* (Amsterdam: Bert Bakker, 1998), 64–68. Balthasar Bekker, a colorful and tragic figure, lost his pastorate in Amsterdam (though not his stipend) because his books opposing belief in witchcraft and demonology seemed too much like criticism of the Bible.

174. Boudewijn Bakker, *Landschap en wereldbeeld. Van Eyck tot Rembrandt* (Bussum: Thoth, 2004), 218–21, 383.

175. Ibid., 224.

176. Van Berkel, *Citaten*, 51.

177. Hooykaas, *Religion*, 130.

178. Gerhard Wiesenfeldt, *Leerer Raum in Minervas Haus. Experimentelle Naturlehre an der Universität Leiden, 1675–1715* (Amsterdam: Koninklijke Nederlandse Akademie van Wetenschappen, 2002), 41–44.

179. Jonathan Israel, *Radical Enlightenment: Philosophy and the Making of Modernity 1650–1750* (Oxford: Oxford University Press, 2001).

180. Vermij, *Calvinist Copernicans*, 320.

181. Wiesenfeldt, *Leerer Raum*, 51.

182. Vermij, *Calvinist Copernicans*, 336.

183. Robert A. Schofield, "An Evolutionary Taxonomy of Eighteenth-Century Newtonianisms," *Studies in 18th Century Culture* 7 (1978): 175–92.

184. A. Rupert Hall, *From Galileo to Newton 1630–1720* (London: Collins, 1963).

185. De Pater, *Van Musschenbroek*, 97.

186. Betty Jo T. Dobbs and Margaret C. Jacob, *Newton and the Culture of Newtonianism* (Atlantic Highlands, N.J.: Humanities Press, 1995), 83.

187. De Pater, *Van Musschenbroek*, 57; Schofield, *Mechanism*, 140–45.

188. Knoeff, *Boerhaave*, 95, 127, 141.

189. Rienk Vermij, *Secularisering en natuurwetenschap in de zeventiende en achttiende eeuw: Bernard Nieuwentijt* (Amsterdam: Rodopi, 1991).

190. De Pater, *Van Musschenbroek*; Clarence J. Glacken, *Traces on the Rhodian Shore: Nature and Culture in Western Thought from Ancient Times to the End of the Eighteenth Century* (Berkeley: University of California Press, 1967).

191. Knoeff, *Boerhaave*, 30–46.

192. Ibid., 59, 63; Simon Schama, *The Embarrassment of Riches: An Interpretation of Dutch Culture in the Golden Age* (New York: Knopf, 1987), 44.

193. "Hier lacht de goude tijd, in lieve lustpriëlen"; Joost van den Vondel, "De Beemster," in *Volledige dichtwerken en oorspronkelijk proza* (Amsterdam: Becht, 1937), 980. In Greek mythology, the golden age is the first age of human history, when mankind was not yet weighed down by misery, work, and illness (Hesiod, *Works and Days*, ll. 116–18). See Alette Fleischer, "The Beemster Polder: Conservative Invention and Holland's Great Pleasure Garden," in Roberts, Schaffer, and Dear, *Mindful Hand*, 145–68.

194. See, for example, Erik de Jong and M. Dominicus-van Soest, *Aardse paradijzen 1. De tuin in de Nederlandse kunst, 15^e tot 18^e eeuw* (Ghent: Snoeck-Ducaju and Zoon, 1996), 12.

195. Bennett, "Mechanics' Philosophy"; W. D. Hackmann, "Scientific Instruments: Models of Brass and Aids to Discovery," in *The Uses of Experiment: Studies in the Natural Sciences*, ed. David Gooding, Trevor Pinch, and Simon Schaffer (Cambridge: Cambridge University Press, 1989), 31–65.

196. Wilson, *Invisible World*, 84.

197. Fokko Jan Dijksterhuis, "Constructive Thinking: A Case for Dioptrics," in Roberts, Schaffer, and Dear, *Mindful Hand*, 59–82.

198. Simon Schaffer, "Experimenters' Techniques, Dyers' Hands, and the Electric Planetarium," *Isis* 88 (1997): 456–83, 460; Simon Schaffer, "Glass Works: Newton's Prisms and the Uses of Experiments," in Gooding, Pinch, Schaffer, *Uses of Experiment*, 67–104.

199. Klein and Lefèvre, *Materials*. In the system derived from Paracelsus, which remained in use among natural philosophers until the early eighteenth century,

mercury, sulfur, and salt were considered the elementary principles (somewhat like the Aristotelian *archai*) and therefore had a significance far beyond their material properties.

200. These included two Dutch universities in Franeker and Harderwijk.

201. Shinn, *L'Ecole Polytechnique*, 13; Robert Fox and George Weisz, eds., *The Organization in Science and Technology in France 1808–1914* (Cambridge: Cambridge University Press, 1980).

202. Georges Canguilhem, *La connaissance de la vie* (Paris: Vrin, 1980), 17–39; Frederic L. Holmes, *Claude Bernard and Animal Chemistry: The Emergence of a Scientist* (Cambridge, Mass.: Harvard University Press, 1974), 446.

203. Thomas Kuhn, "Energy Conservation as an Example of Simultaneous Discovery," in *The Essential Tension* (1959; repr., Chicago: Chicago University Press, 1977), 66–104.

204. Arnold Thackray, "Natural Knowledge in Cultural Context: The Manchester Model," *American Historical Review* 79 (1974): 672–709; Donald Cardwell, "Science and Technology: The Work of James Prescott Joule," *Technology and Culture* 17 (1976): 674–87.

205. Heinz Otto Sibum, "Reworking the Mechanical Value of Heat: Instruments of Precision and Gestures of Accuracy in Early Victorian England," *Studies in the History and Philosophy of Science* 26 (1995): 73–106.

206. Ibid.

207. Much later, in the 1920s, Frits Went isolated a chemical substance (a hormone) responsible for the movement of the tips of stems—in the well-equipped laboratory of the University of Utrecht. Soraya De Chaderevian, "Laboratory Science versus Country-House Experiments: The Controversy between Julius Sachs and Charles Darwin," *British Journal for the History of Science* 29 (1996): 17–41.

208. George McClelland, *State, Society, and University in Germany: 1700–1914* (Cambridge: Cambridge University Press, 1980), 113; Herbert Schnädelbach, *Philosophy in Germany 1831–1933* (Cambridge: Cambridge University Press, 1984), 21–27.

209. R. Steven Turner, "Justus Liebig versus Prussian Chemistry: Reflections on Early Institute Building in Germany," *Historical Studies in the Physical Sciences* 13 (1982): 129–62.

210. Frederic L. Holmes, "The Complementarity of Teaching and Research in Liebig's Laboratory," *Osiris* 5 (1989): 121–64.

211. Robert E. Kohler, "The Ph.D. Machine: Building on the Collegiate Base," *Isis* 81 (1990): 638–62.

212. Hacking, *Representing and Intervening*, 220–32.

213. Robert E. Kohler, *Landscapes and Labscapes: Exploring the Lab-Field Border in Biology* (Chicago: The University of Chicago Press, 2002), 212.

214. Hans-Jörg Rheinberger, *Toward a History of Epistemic Things: Synthesizing Proteins in the Test Tube* (Stanford: Stanford University Press, 1997).

215. Peter Galison, *How Experiments End* (Chicago: University of Chicago Press, 1987); Andrew Pickering, *The Mangle of Practice: Time, Agency and Science* (Chicago: Chicago University Press, 1995).

216. Hans Radder, *The World Observed/The World Conceived* (Pittsburgh: University of Pittsburgh Press, 2006), 19–32.

Chapter 7: The Hypothetical Style

Epigraphs: "Verum et factum convertuntur," Giambattista Vico, *De Antiquissima Italorum Sapientia*, 1710, as cited in Berlin, *Vico and Herder*, 15; "Was ist also Wahrheit? Ein beweglisches Heer von Metaphern, Metonymien, Anthropomorphismen, kurz eine Summe von menschlichen Relationen . . ." Friedrich Nietzsche, *Über Wahrheit und Lüge in aussermoralischen Sinn*, in *Werke in drei Bänden, dritter Band*, ed. Karl Schlechta (Munich: Carl Hanser Verlag, 1963), 314.

1. Nicholas of Cusa quoted in Crombie, *Styles*, 2:1097.
2. Crombie, *Styles*, 2:1090–92; Merchant, *Death of Nature*, 223.
3. Crosby, *Measure of Reality*.
4. Crombie, *Styles*, 2:1092; Otto Mayr, *Authority, Liberty, and Automatic Machinery in Early Modern Europe* (Baltimore: Johns Hopkins University Press, 1986), 7–13.
5. According to Cicero; see Dijksterhuis, *Archimedes*, 16.
6. Mayr, *Authority*; Crombie, *Styles*, 2:1101. The present-day clock was made between 1838 and 1842.
7. On Boyle, see Daston and Park, 298.
8. Robert Boyle, *A Free Inquiry into the Vulgarly Received Notion of Nature* [1682], as cited in Hooykaas, *Religion*, 14.
9. Bakker, *Landschap*, 57–65.
10. Merchant, *Death of Nature*, 223.
11. Cesare Ripa, *Baroque and Rococo Pictorial Imagery: The 1758–60 Hertel Edition of Ripa's Iconologia* (New York: Dover, 1971), image no. 106.
12. See, for example, Karl Pribram, "From Metaphors to Models: The Use of Analogy in Neuropsychology," in *Metaphors in the History of Psychology*, ed. David E. Leary (Cambridge: Cambridge University Press, 1990), 79–103.
13. Crombie, *Styles*, 2:1108–15.
14. Kepler, as quoted in Crombie, *Styles*, 2:1141.
15. Crombie, *Styles*, 2:1143.
16. Walter Pagel, *New Light on William Harvey* (Basel: S. Karger, 1976), 63; Stroup, *Company of Scientists*, 128.
17. Erich Nordenskiöld, in his classic *The History of Biology*, called Harvey's method "purely modern" (New York: Alfred A. Knopf, 1928; repr., New York: Tudor, 1948), 117.
18. For a fuller account of Servetus and Cesalpino, see Walter Pagel, *William Harvey's Biological Ideas* (Basel: S. Karger, 1967).
19. Pagel, *Harvey's Ideas*, 209.
20. Frank, 21.
21. Pagel, *New Light*, 168.
22. Ibid., 113–35; Frank, *Harvey*, 33; J. Schouten, *Johannes Walaeus: Zijn betekenis voor de verbreiding van de leer van de bloedsomloop* (Assen: Van Gorcum, 1972).
23. Pagel, *New Light*; Frank, *Harvey*.
24. Nordenskiöld, *History*, 117. The renowned historian Herbert Butterfield ignored Harvey's later work completely: see Butterfield, *Origins of Modern Science* (1949; repr., London: G. Bell and Sons, 1957), 37–54.
25. Thomas S. Hall, *History of General Physiology 600 B.C. to A.D. 1900*, vol. 2: *From the Enlightenment to the End of the Nineteenth Century* (Chicago: The University of Chicago Press, 1969); Pagel, *Harvey's Ideas*.

26. Pagel, *Harvey's Ideas*, 74.
27. Frank, *Harvey*, 38.
28. Pagel, *Harvey's Ideas*, 54, 83.
29. In other words, Harvey was a monist, not subscribing to any form of mind-body (or soul-body) dualism.
30. Crombie, *Styles*, 1:640; John Rogers, *The Power of Revolution* (Ithaca: Cornell University Press, 1996).
31. Jacob and Jacob, "Anglican Origins."
32. Zacharias Wood, quoted in Rogers, *Power*, 16.
33. Stroup, *Company*, 123.
34. Ibid., 139–42.
35. Crombie, *Styles*, 2:1171, 1193.
36. John Shearman, *Mannerism* (Harmondsworth: Penguin, 1967).
37. Gaukroger, *Descartes: An Intellectual Biography*, 62–63; Alfred Chapuis and Edmond Droz, *Automata* (London: Batsford, 1958); see also Simon Schaffer, "Enlightened Automata," in *Enlightened Europe*, ed. William Clark, Jan Golinski, and Simon Schaffer (Chicago: University of Chicago Press, 1999), 126–65.
38. Descartes to Mersenne, December 1629, as cited in Gaukroger, *Descartes: An Intellectual Biography*, 227; Descartes to Mersenne, June 1637, as cited in ibid., 270.
39. Gaukroger, *Descartes: An Intellectual Biography*, 271.
40. Ibid., 278.
41. McMullin, "Conceptions," 44.
42. Dijksterhuis, *De mechanisering*, 457.
43. McMullin, "Conceptions"; Gaukroger, *Descartes: An Intellectual Biography*, 71; it should be added that Descartes believed atoms would eventually prove to be divisible into even smaller parts.
44. Daston, *Classical Probability*, 248.
45. Dijksterhuis, *De mechanisering*, 546–47.
46. Descartes, *Le Monde*, chap. 5, in *Oeuvres*, 1:336–43. Descartes's use of the term "quality" harked back to the Aristotelian and Scholastic concept of qualities, such as cold, heat, wetness, and dryness. Descartes aimed to explain these properties of the corporeal world rather than using them as explanatory principles.
47. Gaukroger, *Descartes: An Intellectual Biography*, 237, 145–46.
48. "Ne désirant rien oublier de ce qu'il y a de plus général en cette Terre"; Descartes, *Principes*, 4, 133, in *Oeuvres*, 3:451.
49. "Permettez donc pour un peu de temps à votre pensée de sortir hors de ce Monde pour en venir voir un autre tout nouveau que je ferai naître en sa présence dans les espaces imaginaires"; Descartes, *Le Monde*, chap. 5, in *Oeuvres*, 1:343.
50. "Ainsi on peut aisément imaginer que toutes les mêmes choses arrivent aux planètes, et il ne faut que cela seul pour expliquer tous leurs phénomènes," Descartes, *Principes*, 3, 30, in *Oeuvres*, 3:238–39.
51. E. J. Aiton, *The Vortex Theory of Planetary Motion* (London: Macdonald, 1972).
52. Gaukroger, *Descartes: An Intellectual Biography*, 247.
53. Ibid.
54. On Kepler's sources and influence, see Ofer Gal and Raz Chen-Morris, "The Archeology of the Inverse-Square Law: (2) The Use and Non-Use of Mathematics," *History of Science* 44 (2006): 49–67.
55. Mary Hesse, *Forces and Fields* (New York: Philosophical Library: 1961), 131.

56. Ibid., 128, 132; Aiton, *Vortex Theory*, 58.
57. Aiton, *Vortex Theory*, 57.
58. Hesse, *Forces*, 108; Aiton, *Vortex Theory*, 78.
59. Andriesse, *Titan kan niet slapen*, 240–42.
60. On Thomas Hobbes, see Hesse, *Forces*, 113.
61. C. Huygens, *Traité de la lumière* [1690], preface, as cited in Andriesse, *Titan*, 351–52; Huygens, remarks on a biography of Descartes, as cited in ibid., 362.
62. Hesse, *Forces*, 157.
63. Vermij, *Calvinist Copernicans*, 336.
64. Newton, *Principia*, bk. 3.
65. Hesse, *Forces*, 144.
66. Hesse, *Forces*, 148.
67. Newton, in the second edition of the *Principia Mathematica* (General Scholium, 1719).
68. Gaukroger, *Descartes: An Intellectual Biography*, 375–77, 247. It is somewhat anachronistic to equate Descartes's concept of force (*vis*) and modern-day momentum, defined as mass times velocity. Descartes did not make a clear distinction between the concept of mass and that of extension. See Hesse, *Forces*, 111.
69. Newton, *Principia*, definition 4, as cited in Hesse, *Forces*, 135.
70. Newton, quoted in Hesse, *Forces*, 140.
71. Newton, quoted in Hesse, *Forces*, 150.
72. Hesse, *Forces*, 151.
73. Newton, *Opticks*, 2nd ed. [1717], queries 17–24, as quoted in Richard Westfall, *The Construction of Modern Science: Mechanisms and Mechanics* (1971; repr., Cambridge: Cambridge University Press, 1977), 157.
74. Ibid., query 31, 1717, as cited in Hesse, *Forces*, 153.
75. Newton, draft of query 31 [1705], as cited in Alan Gabby, "Force and Inertia in the Seventeenth Century: Descartes and Newton," in Gaukroger, *Descartes: Philosophy, Mathematics and Physics*, 230–320.
76. Newton, *Conclusio* [unpublished manuscript, 1686], as quoted in Westfall, "Newton and Alchemy," 328.
77. See Shapin, *Scientific Revolution*.
78. Henry, "Occult Qualities"; Margaret J. Osler, *Divine Will and the Mechanical Philosophy: Gassendi and Descartes on Contingency and Necessity in the Created World* (Cambridge: Cambridge University Press, 1994); Saul Fisher, "Gassendi's Atomist Account of Generation and Heredity in Plants and Animals," *Perspectives on Science* 11 (2003): 484–512.
79. Newton, "Vegetation of Metals" [undated manuscript, now dated at 1669], quoted in Henry, "Occult Qualities," 343.
80. Henry, "Occult Qualities," 342, 344.
81. Richard Westfall, *Never at Rest: A Biography of Isaac Newton* (Cambridge: Cambridge University Press, 1980), 331; Stephen D. Snobelen, "Isaac Newton, Heretic: The Strategies of a Nicodemite," *British Journal for the History of Science* 32 (1999): 381–419, see 394.
82. McMullin, "Conceptions"; Michael Ben-Chaim, "The Discovery of Natural Goods," *British Journal for the History of Science* 34 (2001): 395–416; Funkenstein, *Theology*.
83. In *Foundations of Newton's Alchemy*, Betty Jo Dobbs has argued that Newton's

use of the dichotomy "passive" and "active" comes from the Stoics, and that he adopted these terms so that he could avoid the more vitalistic and alchemical term "vegetable" (as opposed to "mechanical").

84. Mayr, *Authority*, 142.

85. Steven Shapin, "Of Gods and Kings: Natural Philosophy and Politics in the Leibniz-Clarke dispute," *Isis* 72 (1981): 187–215.

86. Shapin, "Gods and Kings," 201; similarly, Margaret Jacob, in *The Cultural Meaning of the Scientific Revolution* (New York: Knopf, 1988), 108, 119, emphasizes the strong ties between Newtonianism and constitutional monarchy, Cartesianism and absolutism, and Spinozism and freethinking.

87. Shapin, "Gods and Kings"; Jacob and Jacob, "Anglican Origins"; Margaret C. Jacob, *The Radical Enlightenment: Pantheists, Freemasons and Republicans* (London: Allen and Unwin, 1981). Toland was in contact with the Dutch Spinozist Tyssot de Patot, whose radical views led to his dismissal from his professorship at the Athenaeum in Deventer.

88. The appreciation was not entirely mutual. 'sGravesande believed that Desaguliers's book *A Course of Experimental Philosophy* took the right basic approach but was not mathematical enough. See C. de Pater, *Musschenbroek*, 80.

89. 'sGravesande, quoted in Dobbs and Jacob, 83.

90. Jacob, *Cultural*, 143–44.

91. Jean Desaguliers, quoted in Mayr, *Authority*, 153.

92. Dobbs and Jacob, *Newton*, 77.

93. Jacob, *Cultural Meaning*; see also Lissa Roberts, "Going Dutch: A Cultural History of Dutch Science in the Eighteenth Century," in *The Sciences in Enlightened Europe*, ed. W. Clark, J. Golinski, and S. Schaffer (Chicago: The University of Chicago Press, 1999), 350–88.

94. Margaret C. Jacob and Dorothée Sturkenboom, "A Women's Scientific Society in the West: The Late Eighteenth-Century Assimilation of Science," *Isis* 94 (2003): 217–52.

95. Margaret C. Jacob, "Radicalism in the Dutch Enlightenment," in *The Dutch Republic in the Eighteenth Century: Decline, Enlightenment, and Revolution*, ed. Margaret C. Jacob and Wijnand W. Mijnhardt (Ithaca: Cornell University Press, 1992), 224–40, see 238. Yet the question has been raised whether steam was really more than a symbol. Did it have a demonstrable economic impact during the first period of industrialization? On this issue, see the contributions by H. W. Lintsen in *Geschiedenis van de techniek in Nederland. De wording van een moderne samenleving 1800–1890*, vols. 4 and 6 (Zutphen: Walburg Pers, 1993).

96. In fact, the view held by the Cartesians was not a necessary consequence of Cartesian physics, but a result of their support for French geographers and mapmakers. See Mary Terrall, "Representing the Earth's Shape: The Polemics Surrounding Maupertuis' Expedition to Lapland," *Isis* 83 (1992): 218–37.

97. M. Norton Wise, "Work and Waste I," *History of Science* 27 (1989): 263–301. In the late nineteenth century, Henri Poincaré showed that Laplace's result was incorrect, and that every dynamic system composed of three or more bodies is unstable by its very nature.

98. On prophecy and natural science, see Webster, *From Paracelsus to Newton*; on Newton's belief in the predictions made in the Bible, see Snobelen, "Isaac Newton, Heretic."

99. Westfall, *Never at Rest*, 762; Derek Gjertsen, *The Newton Handbook* (London: Routledge and Kegan Paul), 1986.

100. Aiton, *Vortex Theory*, 211.

101. Frans van Lunteren, "Framing Hypotheses: Conceptions of Gravity in the 18th and 19th Centuries" (PhD diss., Utrecht University, 1991).

102. Euler to the Princess of Anhalt-Dessau, quoted in Aiton, *Vortex Theory*, 251.

103. C. Truesdell, *Essays in the History of Mechanics* (New York: Springer, 1968), 116.

104. Hesse, *Forces and Fields*, 192.

105. Truesdell, *Essays*, chap. 3.

106. The other was the Danish scientist Hans Christian Ørsted, who had hypothesized that there was a connection between electricity and magnetism in 1813, on the basis of his Romantic belief in the unity of the forces of nature. In 1820, he confirmed through observation that an electrical current generates a magnetic field.

107. Hesse, *Forces and Fields*, 206, 209; Van Lunteren, "Framing Hypotheses." See also Mary Jo Nye, *Before Big Science: The Pursuit of Modern Chemistry and Physics, 1800–1940* (1996; repr., Cambridge, Mass.: Harvard University Press, 1999).

108. William Thomson and P. G. Tait, *Treatise on Natural Philosophy* (Oxford: 1867).

109. Crosbie Smith and Norton Wise, *Energy and Empire: A Biographical Study of Lord Kelvin* (Cambridge: Cambridge University Press, 1989), 400, 423, 425, 441.

110. Robert M. Friedman, *Appropriating the Weather: Vilhelm Bjerknes and the Construction of a Modern Meteorology* (Ithaca: Cornell University Press, 1989), 11, 12; Van Lunteren, *Framing Hypotheses*.

111. Hacking, *Emergence*, 172.

112. Stroup, *Company*, 123.

113. Bellone, *World on Paper*, 52. Canguilhem, *La connaissance*, 1980.

114. Bellone, *World on Paper*, 32, 51.

115. Chunglin Kwa, "Radiation Ecology, Systems Ecology and the Management of the Environment," in *Science and Nature* (BSHS Monograph 8), ed. Michael Shortland (Stanford in the Vale: British Society for the History of Science, 1993), 213–49; Chunglin Kwa, "Modeling the Grasslands," *Historical Studies in the Physical and Biological Sciences* 24 (1993): 125–55, addendum in *HSPS* 25 (1994): 184–86.

116. Proposed by George Beadle and Arthur Tatum.

117. Lily E. Kay, *Who Wrote the Book of Life: A History of the Genetic Code* (Stanford: Stanford University Press, 2000), 54, 199, 217.

118. Evelyn Fox Keller, "Models, Simulation, and 'Computer Experiments,'" in *The Philosophy of Scientific Experimentation*, ed. Hans Radder (Pittsburgh: University of Pittsburgh Press, 2003), 198–215.

119. Michel Morange, *The Misunderstood Gene* (Cambridge, Mass.: Harvard University Press, 2001); Evelyn Fox Keller, *Refiguring Life, Metaphors of Twentieth-Century Biology* (New York: Columbia University Press, 1995).

120. Alfred I. Tauber and Leon Chernyak, *Metchnikoff and the Origins of Immunology: From Metaphor to Theory* (Oxford: Oxford University Press, 1991); Donna Haraway, *Simians, Cyborgs and Women: The Reinvention of Nature* (London: Free Association Books, 1991), 202–54.

121. Paul Forman, "Weimar Culture, Causality, and Quantum Theory, 1918–1927: Adaptation by German Physicists and Mathematicians to a Hostile Environment,"

Historical Studies in the Physical Sciences 3 (1971): 1–115; see also Hans Radder, "Kramers and the Forman Theses," *History of Science* 21 (1983): 165–82.

122. "Die Schicksalidee selbst verlangt Lebenserfahrung, nicht wissenschatliche Erfahrung, die Kraft des Schauens, nicht Berechnung, Tiefe, nicht Geist." Oswald Spengler, *Der Untergang des Abendlandes: Umrisse einer Morphologie der Weltgeschichte* (München: Beck, 1923), 152.

Chapter 8: The Taxonomic Style

Epigraphs: Virgil, *Georgics* 1.215–16 ("tum te quoque, Medica, putres, accipiunt sulci, et milio venit annua cura"), trans. James Rhoades, Internet Classics Archive, classics.mit.edu; Mary Douglas, *How Institutions Think* (Syracuse: Syracuse University Press, 1986), 55.

1. E. Durkheim and M. Mauss, "De quelques formes primitives de classification," *Année sociologique* (1903), reprinted in M. Mauss, *Essais de sociologie* (Paris: Minuit, 1968).

2. See, for example, Brent Berlin, Dennis Breedlove, and Peter H. Raven, "General Principles of Classification and Nomenclature in Folk Biology," *American Anthropologist* 75 (1973): 124–216; Scott Atran, *Cognitive Foundations of Natural History: Towards an Anthropology of Science* (Cambridge: Cambridge University Press, 1990).

3. Alain Desrosières, *La politique des grands nombres* (Paris: Editions La Découverte, 1993).

4. Geoffrey C. Bowker and Susan Leigh Star, *Sorting Things Out: Classification and Its Consequences* (Cambridge, Mass.: The MIT Press, 1999).

5. Desrosières, *Grand nombres*.

6. But see Bronislaw Malinowski, *Magic, Science, and Religion, and Other Essays* (New York: Doubleday, 1954); Marianne de Laet, *Vaders, sterren en hormonen* (PhD diss., Utrecht University, 1994).

7. Bernadette Bensaude-Vincent, "Mendeleev's Periodic System of Chemical Elements," *British Journal for the History of Science* 19 (1986): 3–17.

8. S. S. Schweber, "From 'Elementary' to 'Fundamental' Particles," in *Companion to Science in the Twentieth Century*, ed. John Krige and Dominique Pestre (1997; repr., London: Routledge, 2003), 599–616.

9. Mark Ridley, *Evolution and Classification: The Reformation of Cladism* (New York: Longman, 1986); Mark Ereshefsky, *The Poverty of the Linnean Hierarchy: A Philosophical Study of Biological Taxonomy* (Cambridge: Cambridge University Press, 2001).

10. See, in particular, Barney G. Glaser and Anselm L. Strauss, *The Discovery of Grounded Theory: Strategies for Qualitative Research* (Chicago: Aldine, 1967).

11. See also Arie Rip, "Science for the 21st Century," in *The Future of the Sciences and the Humanities*, ed. P. Tindemans, A. Verrijn-Stuart, and R. Visser (Amsterdam: Amsterdam University Press, 2002), 99–148.

12. Pomian, *Collectors and Curiosities*, 90; Thomas DaCosta Kaufmann, *Court, Cloister, City: The Art and Culture of Central Europe 1450–1800* (London: Weidenfeld and Nicolson, 1995), 167–83.

13. Lorraine Daston, "The Factual Sensibility," *Isis* 79 (1988): 452–67.

14. Klaas van Berkel, "Citaten uit het boek der natuur," in *De wereld binnen handbereik, Nederlandse kunst- en rariteitenverzamelingen*, ed. Ellinoor Bergvelt and

Renée Kistemaker (Zwolle/Amsterdam: Waanders/ Amsterdam Historical Museum, 1992), 169–91; Roelof van Gelder, "Liefhebbers en geleerde luiden," in ibid., 259–92.

15. Daston, however, in "Factual Sensibility," has pointed out that not all collections had an emblematic or encyclopedic orientation of this kind. Many focused on the rare, the bizarre, and the exceptional.

16. William Ashworth, "Natural History and the Emblematic World View," in Lindberg and Westman, *Reappraisals of the Scientific Revolution*, 314.

17. John Prest, *The Garden of Eden* (New Haven: Yale University Press, 1981).

18. Ashworth, "Natural History," 317; Foucault, *Les mots et les choses*, 141. The date Ashworth gives for the disappearance of the emblematic worldview is consistent with the aforementioned observation by Pomian, based on comparison of hundreds of art collections, that collections after 1650 were more specialized. See Pomian, *Collectors*, 90.

19. Ashworth, "Natural History."

20. Findlen, *Possessing Nature*, 73.

21. Rembert Dodoens, *Cruydeboeck* [1644], 1327. The text of this edition dates from 1608, and is based on notes made by Dodoens before his death in 1585.

22. Dodoens, *Cruydeboeck*, 18–20.

23. Crombie, *Styles*, 2:1263.

24. Findlen, *Possessing Nature*.

25. Karen Meier Reeds, *Botany in Medieval and Renaissance Universities* (New York: Garland, 1991), 4.

26. Findlen, *Possessing Nature*, 6, 166, and 180.

27. Ibid., 139; Aldrovandi received information about Paludanus's collection in 1596, see ibid., 367; Arber, *Herbals*, 102.

28. Ambrosoli, *Wild and Sown*, 181; Findlen, *Possessing Nature*.

29. Findlen, *Possessing Nature*, 262, 273. The same development took place in France, see Reeds, *Botany*, 25.

30. Wilfred Blunt, *The Compleat Naturalist: A Life of Linnaeus* (London: Collins, 1971).

31. David Elliston Allen, *The Naturalist in Britain: A Social History* (1976; repr., Princeton: Princeton University Press, 1994), 150, 167.

32. Ambrosoli, *Wild and Sown*, 206–7.

33. Rondelet was a model humanist. He had received his classical education at the Sorbonne and studied medicine in Montpellier at the same time as the writer François Rabelais. See Reeds, *Botany*.

34. Reeds, *Botany*, 116.

35. Findlen, *Possessing Nature*, 356.

36. Ibid.; Ambrosoli, *Wild and Sown*, 181; Dirk Imhof, "Het plantenrijk in beeld," *Wereldwijs. Wetenschappers rond Keizer Karel*, ed. Geert Vanpaemel and Tineke Padmos (Leuven: Davidsfonds, 2000), 64–73.

37. Arras remained part of the Habsburg Netherlands throughout Clusius's life.

38. The Collegium Trilingue was given its name because lectures there were given not only in Latin, but also in Greek and Hebrew.

39. J. Theunisz, *Carolus Clusius. Het merkwaardige leven van een pionier der wetenschap* (Amsterdam: van Kampen, 1939).

40. Ibid., 63.

41. Orta was the personal physician to the Portuguese viceroy in Goa. On Orta, see C. Boxer, *Two Pioneers of Tropical Medicine, Garcia d'Orta and Nicolás Monardes* (London: Wellcome, 1963).

42. Theunisz, *Clusius*, 55.

43. Monardes had not himself traveled to the Americas but relied on reports and materials sent to him. In several cases, botanical information had been extracted from indigenous informants by cunning and force. See José Pardo Tomas, "Two Glimpses of America from a Distance: Carolus Clusius and Nicolás Monardes," in *Carolus Clusius: Toward a Cultural History of a Renaissance Naturalist*, ed. Florike Egmond, Paul Hoftijzer, and Robert Visser (Amsterdam: Koninklijke Nederlandse Akademie van Wetenschappen, 2007), 173–94.

44. Anna Pavord, *The Tulip* (London: Bloomsbury, 1999); Theunisz, *Clusius*, 68; D. O. Wijnands, E. J. A. Zevenhuizen, J. Heniger, *Een sieraad voor de stad. De Amsterdamse Hortus Botanicus 1638–1993* (Amsterdam: Amsterdam University Press, 1994), 34.

45. Theunisz, *Clusius*, 108.

46. Ibid.; W. K. H. Karstens and H. Kleibrink, *De Leidse Hortus* (Zwolle: Waanders, 1982). The Clusius Garden, reconstructed on the basis of Clusius's and Clutius's inventory, has been open to visitors since 1933. It is not in precisely the same place as the original garden and is somewhat smaller.

47. On patronage of the sciences, particularly chemistry and alchemy, at the imperial court and various courts in Germany, see Smith, *Business of Alchemy*, 42.

48. Dodoens's *Cruydeboeck* [1554] included 715 woodcuts of plants; his Latin edition, *Stirpium* [1583], had 1,306. See Imhof, "Plantenrijk."

49. See Ambrosoli, *Wild and Sown*, for detailed analyses of the readership for various botanical works. On the importance of botanical illustrations in Italy in the early seventeenth century, see Freedman, *Eye of the Lynx*.

50. Oxford University's "Physick Garden," a collection of medicinal plants, was founded in 1621 with funding from a private donor. In 1669, a professor of botany was appointed to the medical faculty. For fifty years, Oxford's botanic garden remained the only one in England. See Frank, *Harvey*, 46, 49; Ambrosoli, *Wild and Sown*, 301; Webster, *Great Instauration*, 126.

51. E. C. Spary, *Utopia's Garden: French Natural History from Old Regime to Revolution* (Chicago: The University of Chicago Press, 2000).

52. On Van Reede, see J. Heniger, *Hendrik Adriaan van Reede tot Drakenstein (1636–1691) and Hortus Malabaricus* (Balkema: Rotterdam, 1986); Richard Grove, *Green Imperialism* (Cambridge: Cambridge University Press, 1995), 84–90. On Rumphius: G. Ballintijn, *Rumphius. De blinde ziener van Ambon* (Utrecht: De Haan, 1944). On the Dutch East India Company (VOC) as a flawed patron of both men: Klaas van Berkel, "Een onwillige mecenas? De VOC en het Indische natuuronderzoek," in van Berkel, *Citaten*, 131–46.

53. Clusius played a major role in the spread of the potato plant throughout Europe by way of the botanic garden in Leiden, but he was not the first European to cultivate this plant. See W. G. Burton, *The Potato* (Wageningen: Veenman en Zonen, 1966).

54. Wijnands et al., *Sieraad*, 88–89; Grove, *Green Imperialism*, 209.

55. Ambrosoli, *Wild and Sown*, 20; see also fig. 3. Pier de Crescenzi confused lucerne with *Melilotus* (melilot, also known as sweet clover).

56. See the epigraph that opens this chapter. See also *Georgics* 1.74: "there sow the golden grain where erst, luxuriant with its quivering pod, pulse, or the slender vetch-crop, thou hast cleared, and lupin sour" (trans. James Rhoades, Internet Classics Archive, classics.mit.edu).

57. Ambrosoli, *Wild and Sown*, III, 118, 173, 277.

58. The Dutch name that Dodoens used for *Medica* was *Boergoens hooi*, which is related to *foin de Bourgogne*, a name then used in France. Botanists were responsible for the spread of numerous varieties of clover and related leguminous plants throughout Europe. Lobelius, for instance, brought a variety of sainfoin that grew in Zeeland back to Montpellier with him (see Ambrosoli, *Wild and Sown*, 179).

59. Ambrosoli, *Wild and Sown*, 176, 180, 297.

60. Webster, *Instauration*, 469; Ambrosoli, *Wild and Sown*, chap. 6; B. H. Slicher van Bath, *De agrarische geschiedenis van West-Europa (500–1850)* (Utrecht: Spectrum, 1960).

61. Crombie, *Styles*, 2:1281; Lisbet Koerner, *Linnaeus: Nature and Nation* (Cambridge: Cambridge University Press, 1999).

62. Findlen, *Possessing Nature*, 167; Arber, *Herbals*, 221–26.

63. Arber, *Herbals*, 230.

64. Crombie, *Styles*, 2:1267; A. G. Morton, *History of Botanical Science* (London: Academic Press, 1981), 168–74; Frans Stafleu, *Linnaeus and the Linnaeans: The Spreading of Their Ideas in Systematic Botany, 1735–1789* (Utrecht: Oosthoek, 1971), 40.

65. Christien Brouwer, *Anatomische sekse als uitvinding in de botanie. Hoe stampers tot vrouwelijke en meeldraden tot mannelijke geslachtsorganen werden (1675–1735)* (PhD diss., Amsterdam University, 2004); Wilson, *Invisible World*, chap. 9.

66. This alphabetical system is found in books 2 to 5; the other twenty-five books of the *Cruydeboek* group plants by their shape and medical properties.

67. Dodoens, *Cruydeboeck*, 5, 13.

68. Rembertus Dodonaeus (Rembert Dodoens), *Stirpiae historiae. Pemptades sex* (Antwerp: Christopher Plantijn, 1583).

69. Theunisz, *Clusius*, 67; Reeds, *Botany*, 87.

70. Gerard J. van den Broek, *Baleful Weeds and Precious-Juiced Flowers: A Semiotic Approach to Botanical Practice* (PhD diss., Leiden University, 1986).

71. Arber, *Herbals*, 116.

72. This phrase is drawn from Bauhin's *Phytopinax* [1596], cited in Reeds, *Botany*, 122.

73. Reeds, *Botany*, 126; Arber, *Herbals*, 179.

74. Stroup, *Company*, 222.

75. James L. Larson, *Reason and Experience: The Representation of Natural Order in the Work of Carl von Linné* (Berkeley: University of California Press, 1971), 29. In modern biological terms, the plant parts that Cesalpino took to be equivalent are not now considered homologous.

76. In his writings on logic, Aristotle's rule was to divide groups analytically, on the basis of a single characteristic. But in his works about animals, he did not follow this rule, and instead defended a different approach, namely adopting a set of criteria, none of which is identified as essential. This inconsistency is closely connected to the discussion of Aristotle earlier in this volume. See also M. M. Slaughter, *Universal Languages and Scientific Taxonomy in the Seventeenth Century* (Cambridge: Cambridge University Press, 1982), 32, 33; Larson, *Reason*, 23.

77. Larson, *Reason*, 38.
78. Ibid., 29; Slaughter, *Universal Languages*, 27, 210.
79. Wilhelm Schmidt-Biggemann, *Topica universalis* (Hamburg: Felix Meiner Verlag, 1983); Jean-Marc Chatelain, "Du Parnasse à l'Amérique: l'imaginaire de l'encyclopédie à la Renaissance et à l'Age classique," *Tous les savoirs du monde*, ed. in Roland Schaer (Paris: Bibliothèque nationale de France, 1997), 156–63; Feldhay, *Galileo*, 150; Hannaway, *Chemists and the Word*.
80. Slaughter, *Universal Languages*, 62.
81. Stafleu, *Linnaeus*, 43.
82. Morton, *History*, 202, 228.
83. Ray was very interested in the philosophical systems developed by Epicurus and Gassendi, and was also in contact with the Cambridge Platonists. It would be reasonable to attribute the same sort of eclecticism to him that one finds among most British physicists of his day. See Morton, *History*, 212.
84. Slaughter, *Universal Languages*, 193.
85. Ibid., 212; Phillip R. Sloan, "The Buffon-Linnaeus Controversy," *Isis* 67 (1976): 356–75, at 365.
86. Larson, *Reason*, 41.
87. Wijnands et al., *Sieraad*, 39.
88. Stafleu, *Linnaeus*, 25, 39.
89. Morton, *History*, 213–14; Brouwer, *Anatomische sekse*.
90. Morton, *History*, 241–42.
91. Brouwer, *Anatomische sekse*, 92.
92. Ibid., 98.
93. Prest, *Eden*.
94. Brouwer, *Anatomische sekse*, 93, 95, 198.
95. There is an obvious connection to emblematic thinking here.
96. Thomas Laqueur, *Making Sex: Body and Gender from the Greeks to Freud* (Cambridge, Mass.: Harvard University Press, 1990); Londa Schiebinger, *The Mind Has No Sex? Women in the Origins of Modern Science* (Cambridge, Mass.: Harvard University Press, 1989); Thomas Laqueur, "Sex in the Flesh," *Isis* 94 (2003): 300–306; Londa Schiebinger, "Skelettestreit," *Isis* 94 (2003): 307–13; see also: Michael Stolberg, "A Woman Down to Her Bones: The Anatomy of Sexual Difference in the Sixteenth and Early Seventeenth Century," *Isis* 94 (2003): 274–99.
97. Linnaeus, quoted in Londa Schiebinger, *Nature's Body: Gender in the Making of Modern Science* (Boston: Beacon Press, 1993), 23.
98. Larson, *Reason*, 58.
99. Ibid., 57. Buffon is quoted on 59.
100. Stafleu, *Linnaeus*, 125.
101. Lisbet Koerner, "Carl Linnaeus in His Time and Place," in *Cultures of Natural History*, ed. N. Jardine, J. A. Secord, and E. C. Spary (Cambridge: Cambridge University Press, 1996), 145–62, 147.
102. Larson, *Reason*, 74, 83; Stafleu, *Linnaeus*, 70.
103. Slaughter, *Universal Languages*, 9.
104. Larson, *Reason*, 86; Crombie, *Styles*, 2:1281.
105. Linnaeus, *Genera plantarum*, quoted in Larson, *Reason*, 77.
106. Linnaeus, *Genera plantarum*, quoted in Slaughter, *Universal Languages*, 54.

107. Linnaeus, *Philosophia botanica*, section 167, 116–17, quoted in Larson, *Reason*, 91; see also Stafleu, *Linnaeus*, 69.

108. Crombie, *Styles*, 2:1281; Stafleu, *Linnaeus*, 42.

109. Linnaeus, *Philosophia botanica*, section 77, quoted in Stafleu, *Linnaeus*, 46, 133.

110. See, for example, E. Ashworth Underwood, *Boerhaave's Men at Leyden and After* (Edinburgh: Edinburgh University Press, 1977).

111. Michel Foucault, *La naissance de la clinique* (Paris: Presses universitaires de France, 1963).

112. Debora J. Meijers, *Kunst als natuur. De Habsburgse schilderijengalerij in Wenen omstreeks 1780* (Amsterdam: SUA, 1991).

113. Foucault, *Les mots et les choses*, 275–92.

114. According to Foucault, the classical episteme followed the Renaissance episteme. This use of the term "classical" stems from the French term *classique*, which is commonly used to describe a period of art history following the Baroque in the seventeenth century.

115. Linnaeus, quoted in Janet Browne, *The Secular Ark* (New Haven: Yale University Press, 1983), 17.

116. The other reason that Linnaeus stayed so long in the Netherlands was the large role of the Netherlands in book publishing, a situation of which he took full advantage.

117. Blunt, *Naturalist*, 85, 95.

118. van Gelder, "De wereld binnen handbereik," 27.

119. Daston, "Sensibility," 463.

120. Pomian, *Collectors*, 121, 138.

121. Ibid., 105, 219.

122. Koerner, "Linnaeus," 148.

123. Pomian, *Collectors*, 218, 228.

124. Florence Pieters, "Het schatrijke naturaliënkabinet van Stadhouder Willem V onder directoraat van topverzamelaar Arnout Vosmaer," in *Het verdwenen museum. Natuurhistorische verzamelingen 1750–1850*, ed. B. C. Sliggers and M. H. Besselink (Blaricum/Haarlem: V+K Publishing/Teylers Museum, 2002), 19–44.

125. See Sally Gregory Kohlstedt, "Curiosities and Cabinets: Natural History Museums and Education on the Antebellum Campus," *Isis* 79 (1988): 405–26; Pomian, *Collectors*, 191.

126. Stafleu, *Linnaeus*, 148.

127. Blunt, *Naturalist*, 184.

128. Ibid., 185, 188.

129. Bernard Smith, *European Vision and the South Pacific*, 2nd ed. (New Haven: Yale University Press, 1985).

130. Stafleu, *Linnaeus*, 154.

131. Ibid., 169; Wijnands et al., *Sieraad*.

132. Peter F. Stevens, *The Development of Biological Systematics: Antoine-Laurent de Jussieu, and the Natural System* (New York: Columbia University Press, 1994).

133. Stevens, *Development*, 24; Morton, *History*, 325.

134. Stevens, *Development*, 24.

135. In this respect, I disagree with Stevens.

136. Stevens, *Development*, 79–88.

137. Darwin's views about species were neither holistic nor essentialist.

138. Stevens, *Development*, 205; Harriet Ritvo, *The Platypus and the Mermaid and Other Figments of the Classifying Imagination* (Cambridge, Mass.: Harvard University Press, 1997).

139. On Schwann and Schleiden, see B. Theunissen and R. P. W. Visser, *De wetten van het leven: Historische grondslagen van de biologie 1750–1950* (Baarn: Ambo, 1996), 107.

140. Sharon E. Kingsland, "The Battling Botanist: Daniel Trembley MacDougal, Mutation Theory and the Rise of Experimental Evolutionary Biology in America, 1900–1912," *Isis* 82 (1991): 479–510, 494.

141. Charles Davenport, a contemporary of De Vries, made similar derogatory statements about taxonomy and taxonomists. See Kohler, *Landscapes and Labscapes*, 72.

142. Stevens, *Development*, 214; Rob Visser, "Het Rijksmuseum van Natuurlijke Historie in de 19de eeuw," in Sliggers and Besselink, *Het verdwenen museum*, 175–85; Lucile Brockway, *Science and Colonial Expansion: The Role of the British Royal Botanic Gardens* (New York: Academic Press, 1979).

143. Browne, *Ark*; Patrick Matagne, *Aux origines de l'écologie: Les naturalistes en France de 1800 à 1914* (Paris: Éditions du CHTS, 1999); Chunglin Kwa, "Alexander von Humboldt's Invention of the Natural Landscape," *European Legacy* 10 (2005): 149–62.

144. Ridley, *Evolution and Classification*; Ereshefsky, *Poverty of the Linnean Hierarchy*.

145. Stevens, *Development*.

146. David Takacs, *The Idea of Biodiversity: Philosophies of Paradise* (Baltimore: Johns Hopkins University Press, 1996).

Chapter 9: Statistical Analysis as a Style of Science

Epigraph: Francis Galton, as quoted in Nicholas Wright Gilham, *A Life of Sir Francis Galton: From African Exploration to the Birth of Eugenics* (Oxford: Oxford University Press, 2001).

1. Theodore Porter, *Trust in Numbers* (Princeton: Princeton University Press, 1995), 33–48; Alain Desrosières, "How to Make Things which Hold Together: Social Science, Statistics and the State," in *Discourses on Society: The Shaping of the Social Science Disciplines*, ed. Peter Wagner, Björn Wittrock, and Richard Whitley (Dordrecht: Kluwer, 1991), 195–218; Desrosières, *La politique*.

2. Hacking, *Taming*, 7.

3. Theodore Porter, *The Rise of Statistical Thinking, 1820–1900* (Princeton: Princeton University Press, 1986); Ida Stamhuis, *'Cijfers en Aequaties' en 'Kennis der Staatskrachten'. Statistiek in Nederland in de 19ᵉ eeuw* (Amsterdam: Rodopi, 1989), 56–62.

4. Hacking, *Emergence*, 102.

5. Ibid., 16–34; on the Netherlands, see Stamhuis, *Cijfers en Aequaties*, 140 ff.

6. Michel Foucault, *Histoire de la sexualité*, vol. 1: *La volonté de savoir* (Paris: Gallimard, 1976); see also Herbert Dreyfus and Paul Rabinow, *Michel Foucault: Beyond Structuralism and Hermeneutics* (Brighton: Harvester Press, 1982), 133 ff.

7. Marie-Noëlle Bourguet, "Décrire, Compter, Calculer: The Debate over Statistics during the Napoleonic Period," in *The Probabilistic Revolution*, vol. 1: *Ideas in History*,

ed. L. Krüger, L. Daston, and M. Heidelberger (Cambridge, Mass: The MIT Press, 1987), 305–16; Eileen Janes Yeo, "Social Surveys in Eighteenth and Nineteenth Centuries," in *The Modern Social Sciences*, ed. Theodore Porter and Dorothy Ross, vol. 7 of *History of Science* (Cambridge: Cambridge University Press, 2003), 83–99.

8. Desrosières, *Grands nombres*.

9. More specifically, the phenomenon in question is radioactive decay, expressed in half-lives. It was soon clear to researchers that there is no way of predicting *which* particle will decay.

10. Hacking, *Taming*, 1.

11. Desrosières, "Social Science," 203.

12. Aristotle, *Parts of Animals*, trans. E. S. Forster (Cambridge, Mass.: Harvard University Press, 1961), 56. At http://www.archive.org/stream/partsofanimals00arisuoft/partsofanimals00arisuoft_djvu.txt, accessed July 15, 2009.

13. Crombie, 2:1308–9.

14. Hacking, *Emergence*, 173, 183.

15. Shapiro, *Probability and Certainty*, 31.

16. Crombie, *Styles*, 2:1318–21.

17. Daston, *Classical Probability*, 20.

18. Gerolamo Cardano had written about this subject earlier, around 1530, but this work was not published until 1663.

19. Hacking, *Emergence*, 59, 93.

20. Christiaan Huygens, *Van rekeningh in spelen van geluck* [1660], new edition in modern Dutch spelling with notes by Wim Kleijne (Utrecht: Epsilon, 1998).

21. Daston, *Classical Probability*, 14.

22. Hacking, *Emergence*, 93.

23. Ibid., 111–21.

24. On Hudde as a mathematician, see Klever, *Mannen rond Spinoza*.

25. Hacking, *Emergence*, 116; Daston, *Probability*, 27; see also Stamhuis, *Cijfers en aequaties*, 26–33.

26. Daston, *Probability*, 139; see also Ida Stamhuis, "Radeloos, redeloos, noch reddeloos: Jan de Witts lijfrenteberekeningen rond het Rampjaar," *Nieuw Archief voor Wiskunde*, fourth series, vol. 17 (1999): 439–52.

27. Daston, *Probability*, 139, 168, 172, 176.

28. Ibid., 226–95.

29. Crombie, *Styles*, 2:1355–56; Daston, *Probability*, 239.

30. Pierre-Simon de Laplace, *Essai philosophique des probabilités* [1814].

31. Laplace, quoted in Lorenz Krüger, "The Slow Rise of Probabilism," in Krüger, Daston, and Heidelberger, *Probabilistic Revolution*, 1:59–89, 63.

32. David Hume, *A Treatise of Human Nature*, 1739, as quoted in Daston, *Probability*, 200.

33. Porter, *Rise*, 94; Hacking, *Taming*, 59n.

34. Porter, *Rise*, 95.

35. Hacking, *Taming*, 110–12.

36. "Statement of the Sizes of Men in Different Counties of Scotland, Taken from the Local Militia," *Edinburgh Medical and Surgical Journal* 13 (1917): 260–64. The data were gathered by an unnamed "gentleman of great observation and singular accuracy" (editorial note, 264).

37. Hacking, *Taming*.

38. Porter, *Rise*, 103–5.
39. Ibid., 110–28.
40. Hacking, *Taming*, 62.
41. Compare the term "price index." See also Porter, *Rise*, 52, 65.
42. Desrosières, *Grands nombres*.
43. Daston, *Classical Probability*, 187; Desrosières, *Grands nombres*, 124. In *Le suicide*, Durkheim moved toward a more Galtonian, individualistic stance.
44. Stamhuis, *Cijfers en Aequaties*; Ida Stamhuis, "An Unbridgeable Gap Between Two Approaches to Statistics," in *The Statistical Mind in a Pre-Statistical Era: The Netherlands 1750–1850*, ed. Paul Klep and Ida Stamhuis (Amsterdam: Aksant, 2002), 71–98.
45. Eddy Houwaart, *De hygiënisten: Artsen, staat & volksgezondheid in Nederland 1840–1890* (Groningen: Historische Uitgeverij, 1991). The movement of French hygienists was led by Louis Villermé, a good friend of Quételet's. See Desrosières, *Grands nombres*, 106.
46. Stamhuis, *Cijfers en Aequaties*, 161.
47. Ibid., 211–25.
48. Michael Dettelbach, "Humboldtian Science," in Jardine, Secord, and Spary, *Cultures of Natural History*, 287–304.
49. Humboldt is considered to have "discovered" Gauss, and the two were friends. In *Kosmos*, however, Humboldt's only mention of Gauss is in the context of his work on geomagnetism.
50. Kwa, "Alexander von Humboldt's Invention."
51. Susan Faye Cannon, *Science in Culture: The Early Victorian Period* (Folkestone: Dawson, 1978).
52. Browne, *Ark*, 53.
53. Ibid., 59–85.
54. Hacking, *Taming*, 180n.
55. Ibid., 181.
56. Galton probably borrowed this pair of terms from Shakespeare's description of Caliban in *The Tempest*.
57. That Galton was an anti-Semite is clear from his views about Jewish "parasitism" (though it should be added that he never actively promoted anti-Semitism). See Andrew Berry, "Whenever You Can, Count" *London Review of Books*, December 4, 2003. See also Porter, *Rise*, 129–31; Donald MacKenzie, *Statistics in Britain 1865–1930: The Social Construction of Scientific Knowledge* (Edinburgh: Edinburgh University Press, 1981); Daniel Kevles, *In the Name of Eugenics* (Chicago: The University of Chicago Press, 1984).
58. Porter, *Rise*, 140; Hacking, *Taming*, 186.
59. Porter, *Rise*, 291.
60. "Student" was the pseudonym of William S. Gossett, a student of Pearson's.
61. Gerd Gigerenzer, "Probabilistic Thinking and the Fight against Subjectivity," in *The Probabilistic Revolution*, vol. 2: *Ideas in the Sciences*, ed. Lorenz Krüger, Gerd Gigerenzer, and Mary Morgan (Cambridge, Mass.: The MIT Press), 1987.
62. John Turner, "Random Genetic Drift, R. A. Fisher, and the Oxford School of Ecological Genetics," in Krüger, Gigerenzer, and Morgan, *Probabilistic Revolution*, 2:313–54. See also the articles by John Beatty and M. J. Hodge in the same collection.
63. Hacking, *Taming*, 188.

64. MacKenzie, *Statistics*, 123; see also William Provine, *The Origins of Theoretical Population Genetics* (Chicago: The University of Chicago Press, 1971).
65. MacKenzie, *Statistics in Britain*, 211.
66. Gerd Gigerenzer, Zeno Swijtink, Theodore Porter, Lorraine Daston, John Beatty, and Lorenz Krüger, *The Empire of Chance: How Probability Changed Science and Everyday Life* (Cambridge: Cambridge University Press, 1989).
67. On this subject, see Harry Collins, *Changing Order: Replication and Induction in Scientific Practice* (London: Sage, 1985); Bruno Latour, *Science in Action* (Milton Keynes: Open University Press, 1987); Hans Radder, *The Material Realization of Science* (Assen, Netherlands: Van Gorcum, 1988).
68. This is therefore a form of the inverted Bernoulli theorem.
69. Gigerenzer et al., *Empire*, 262.
70. Ibid., 101–3, 210.
71. The technique involves counting the number of births in a limited area during a given period, along with the total population of that area. The number of births for the country as a whole could then be used to determine the country's population, within a certain margin.
72. Desrosières, *Grands nombres*, 37, 111, 258.
73. Ibid., 199, 263.
74. Ibid., 283.
75. Case studies can be placed in a typological, *pars pro toto* context, but they do not have to be, by any means.
76. Anne Beaulieu, "Voxels in the Brain: Neuroscience, Informatics and Changing Notions of Objectivity," *Social Studies of Science* 31 (2001): 635–80.
77. Kurt Danziger, "Statistical Method and the Historical Development of Research Practice in American Psychology," in Krüger, Gigerenzer, and Morgan, *Probabilistic Revolution*, 2:35–47.
78. Kurt Danziger, *Constructing the Subject* (Cambridge: Cambridge University Press, 1990), 80–83.
79. Ibid., 150.
80. Danziger, "Statistical Method."
81. Porter, *Trust*, 199.
82. Gigerenzer et al., *Empire*, 206, 209.
83. Gigerenzer, "Probabilistic Thinking."
84. In economics and econometrics, there is a sort of halfhearted acceptance of probabilism, while the use of statistics is mandatory. See the articles by Claude Ménard and Mary Morgan in Krüger, Gigerenzer, and Morgan, *Probabilistic Revolution*, vol. 2.
85. On the preoccupation with methodology, see also Trudy Dehue, *Changing the Rules: Psychology in the Netherlands, 1900–1985* (Cambridge: Cambridge University Press, 1995); Dehue also looks at the persistence of humanistic elements in psychological methodology.
86. Gigerenzer et al., *Empire*, 211, 228.
87. Fisher in 1955, as quoted in Gigerenzer et al., *Empire*, 211.
88. Porter, *Trust*, 148–89.
89. A similar dynamic is at work in numerous forms of "performance measurement." See Hans de Bruin, *Prestatiemeting in de publieke sector* (Utrecht: Lemma, 2001).

90. Porter, *Trust*; Lorraine Daston and Peter Galison, *Objectivity* (New York: Zone Books, 2007).

Chapter 10: The Evolutionary Style

Epigraph: Baudelaire, "La Géante," *Les fleurs du mal*, 19, trans. William Aggeler, *The Flowers of Evil* (Fresno, Calif.: Academy Library Guild, 1954) as quoted on http://fleursdumal.org/poem/118, accessed December 1, 2009.

1. Crombie, *Styles*, 3:1564.
2. Janet Browne, *Charles Darwin Voyaging: Vol. 1 of a Biography* (London: Jonathan Cape, 1995), 366.
3. Ibid., 448–72.
4. Ibid., 473–510.
5. Browne, *Darwin Power*.
6. Prest, *Eden*.
7. Browne, *Ark*.
8. Webster, *From Paracelsus to Newton*, 41.
9. Browne, *Ark*.
10. Crombie, *Styles*, 3:1634.
11. D'Arcy Wentworth Thompson, *On Growth and Form*, 2nd ed. (New York: Macmillan, 1942).
12. Stephen Jay Gould, *The Panda's Thumb: Further Reflections in Natural History* (New York: W. W. Norton and Co., 1980).
13. "Omnia in mensuro, et numero, et pondere disposuisti," Wisdom (of Solomon), 11:21. On this passage, see Blumenberg, *Legitimität der Neuzeit*, 407.
14. Crombie, *Styles*, 3:1638–58.
15. Northrop Frye, *The Great Code: The Bible and Literature* (New York: Harcourt Brace Jovanovich, 1982).
16. Clara Pinto-Correia, *The Ovary of Eve: Egg and Sperm and Preformation* (Chicago: The University of Chicago Press, 1997).
17. Ibid., 122.
18. Brouwer, *Anatomische*.
19. Barbara M. Stafford, *Body Criticism: Imaging the Unseen in Enlightenment Art and Medicine* (Cambridge, Mass.: The MIT Press, 1991), 249.
20. Crombie, *Styles*, 3:1682.
21. Arthur Lovejoy, *The Great Chain of Being* (Cambridge, Mass.: Harvard University Press, 1936).
22. Pinto-Correia, *Ovary of Eve*.
23. Glacken, *Traces*.
24. Maupertuis, *Lettre sur le progrès des sciences* [1752], sec. 6, as quoted in Blumenberg, *Legitimität der Neuzeit*, 478.
25. Jacques Roger, *Les sciences de la vie dans la pensée française du xviiie siècle* (Paris: Amand Colin, 1971), 477–87. See also Crombie, *Styles*, 3:1731; and Peter Bowler, *The Mendelian Revolution: The Emergence of Heriditarian Concepts in Modern Science and Society* (London: Athlone, 1989), 33.
26. Maupertuis, *Système de la nature*, sec. 45, as quoted in Roger, *Sciences de la vie*, 484.

27. The use of the term "Baroque" is inspired by Gilles Deleuze, *Le pli: Leibniz et le baroque* (Paris: Minuit, 1988). See also Chunglin Kwa, "Romantic and Baroque Conceptions of Complex Wholes in the Sciences," in *Complexities: Social Studies of Knowledge Practices*, ed. John Law and Annemarie Mol (Durham, NC: Duke University Press, 2002), 23–52.

28. See Foucault, *Les mots et les choses*, 166–67. There is greater similarity between Maupertuis and Linnaeus than between either of them and Darwin.

29. Kant, "Von den verschiedenen Racen der Menschen" [1775], www.ikuni-bonn.de/kant/aa02/, accessed March 27, 2008.

30. Crombie, *Styles* 3:1707.

31. Foucault, *Les mots et les choses*, 1966.

32. Daston and Galison, *Objectivity*.

33. Immanuel Kant, *Kritik der Urteilskraft*, sec. 80. See also Wolf Lepenies, *Das Ende der Naturgeschichte* (1976; repr., Frankfurt: Suhrkamp Verlag, 1978), 37–38.

34. Crombie, *Styles*, 3:1707, 1713; Yves Laissus, *Le Muséum national d'histoire naturelle* (Paris: Gallimard, 1995); Dorinda Outram, "New Spaces in Natural History," in Jardine, Secord, and Spary, *Cultures of Natural History*, 249–65.

35. The great chain of being nonetheless had many years of life left in it, and reemerged in various forms throughout the nineteenth century. One typical family tree drawn by Ernst Haeckel, Darwin's German advocate, gave the impression that only one line of development really mattered: the one from the amoeba to the human. Human beings were portrayed as the crowning glory of evolution, with sponges, insects, and reptiles banished to unimportant side branches. See Peter J. Bowler, *Life's Splendid Drama* (Chicago: The University of Chicago Press, 1996).

36. Alexander von Humboldt, *Voyage de Humboldt et Bonpland*, vol. 1 (Paris, 1814), as quoted in Anthony Pagden, *European Encounters with the New World* (New Haven: Yale University Press, 1993), 48. See also Kwa, "Alexander von Humboldt's Invention."

37. The phrase is actually the title of a book: G. Evelyn Hutchinson, *The Ecological Theater and the Evolutionary Play* (New Haven: Yale University Press, 1965).

38. Malthus, *Essay*, as quoted in Crombie, *Styles*, 3:1442.

39. Young, "Malthus and the Evolutionists."

40. Browne, *Darwin Voyaging*, 338.

41. Redmond O'Hanlon, *In Trouble Again* (London: Hamish Hamilton, 1988).

42. Browne, *Darwin Voyaging*, 385.

43. Note that Darwin, however much he may have borrowed from Quételet, was more of a precursor of Galton and Pearson's individualistic thinking and showed little affinity with Quételet's higher levels of aggregation, conceived as truly existing entities.

44. Crombie, *Styles*, 3:1741, 1758; Browne, *Darwin Power*, 56.

45. Worster, *Nature's Economy*, 156.

46. Robert M. Young, *Darwin's Metaphor: Nature's Place in Victorian Culture* (Cambridge: Cambridge University Press, 1985); Gillian Beer, "'The Face of Nature': Anthropomorphic Elements in the Language of the Origin of Species," in *Languages of Nature*, ed. L. Jordanova (London: Free Association Books, 1986), 207–43.

47. Browne, *Darwin Voyaging*, 516.

48. Robert Olby, *Origins of Mendelism* (Chicago: The University of Chicago Press, 1986), 45–47.

49. Browne, *Darwin Voyaging*, 517.

50. Garland Allen, *Life Science in the Twentieth Century* (1975; repr., Cambridge: Cambridge University Press 1978), 14.

51. Peter J. Bowler, *The Eclipse of Darwinism: Anti-Darwinian Evolution Theories in the Decades Around 1900* (Baltimore: Johns Hopkins University Press, 1983).

52. Olby, *Mendelism*, 7–14.

53. Ibid., 23–31; Browne, *Darwin Voyaging*.

54. R. Olby, "Mendel no Mendelian?" *History of Science* 17 (1979): 53–72, reprinted as an appendix in Olby, *Origins of Mendelism*.

55. Ibid.

56. Olby, *Origins of Mendelism* 103.

57. Eugene Cittadino, *Nature as the Laboratory: Darwinian Plant Ecology in the German Empire, 1880–1900* (New York: Cambridge University Press, 1990), 21, 122; Bowler, *Mendelian Revolution*.

58. Olby, *Origins of Mendelism*, 229.

59. Lindley Darden, *Theory Change in Science: Strategies from Mendelian Genetics* (New York: Oxford University Press), 40. Johannsen had originally come up with the term "genotype" to refer to a population of highly similar individuals. See Kingsland, "Battling Botanist."

60. Olby, "Mendel." The historians of science Onno Meijer and Bert Theunissen argue, however, that Mendel did have some insight into the possibility that each trait had a material carrier. See Meijer, "Hugo de Vries no Mendelian?" *Annals of Science* 42 (1985): 189–232; Theunissen, "Closing the Door on Hugo de Vries' Mendelism," *Annals of Science* 51 (1994): 225–48. See also Nordenskiöld, *History of Biology*, 533–44.

61. Erik Zevenhuizen, *Vast in het spoor van Darwin: Biografie van Hugo de Vries* (Amsterdam: Atlas, 2008).

62. Browne, *Darwin Power*, 465; Theunissen, "Closing," 245.

63. Browne, *Darwin Power*, 288.

64. For many years, the case for De Vries's precedence was viewed with great skepticism by historians, but it was recently shown to be correct. See Ida Stamhuis, Onno Meijer, and Erik Zevenhuizen, "Hugo de Vries on Heredity: Statistics, Mendelian Laws, Pangenes, Mutations," *Isis* 90 (1999): 238–67.

65. Theunissen, "Closing," 232–34.

66. Kingsland, "Battling Botanist."

67. Robert Kohler, *Lords of the Fly:* Drosophila *Genetics and the Experimental Life* (Chicago: University of Chicago Press, 1994).

68. Garland Allen, "Hugo de Vries and the Reception of Mutation Theory," *Journal for the History of Biology* 2 (1969): 55–87.

69. Kingsland, "Battling Botanist."

70. See Theunissen and Visser, *Wetten*, 241–66, for a survey of this period.

71. Ibid.

72. Karl Popper, "Darwinism as a Metaphysical Research Programme," in *The Philosophy of Karl Popper*, ed. P. A. Schillp (LaSalle, Il.: Open Court, 1974), 133–43.

73. Mary Williams, "Deducing the Consequences of Evolution," *Journal of Theoretical Biology* 29 (1970): 343–85.

74. Alexander Rosenberg, *The Structure of Biological Science* (New York: Cambridge University Press, 1985), 141–43.

75. Rosenberg, *Structure*, 216, 125; Ernst Mayr, *The Growth of Biological Thought* (Cambridge, Mass.: Harvard University Press, 1982).

76. Nancy Cartwright, *How the Laws of Physics Lie* (Oxford: Oxford University Press, 1983).
77. Peter Taylor, "Historical Versus Selectionist Explanations in Evolutionary Biology," *Cladistics* 3 (1987): 1–13.
78. F. John Odling-Smee, Kevin N. Laland, and Marcus W. Feldman, *Niche Construction* (Princeton: Princeton University Press, 2003); Rachel L. Day, Kevin L. Laland, and John Odling-Smee, *Perspectives in Biology and Medicine* 46 (2003): 80–95. The idea of niche construction goes back to an article by Richard Lewontin, "Gene, Organism, and Environment," in *Evolution from Molecules to Men*, ed. D. S. Bendall (Cambridge: Cambridge University Press, 1983).
79. Bellone, *World on Paper*, 44.
80. Wise, "Work and Waste I."
81. Philip Mirowski, *More Heat than Light* (Cambridge: Cambridge University Press, 1989), 44.
82. Ibid., 6.
83. Donald S. L. Cardwell, *From Watt to Clausius: The Rise of Thermodynamics in the Early Industrial Age* (Ithaca: Cornell University Press, 1971).
84. M. Norton Wise, "Work and Waste II," *History of Science* 27 (1989): 391–449.
85. Smith and Wise, *Energy and Empire*, 291.
86. Bellone, *World on Paper*, 48.
87. Wise, "Work and Waste II." But Kelvin always strove to resist Lucretius's "monstrous hypothesis" (namely, *clinamen*; see Bellone, *World on Paper*).
88. William Thomson, "The Age of the Sun's Heat," *Macmillan's Magazine* [1862] as quoted in Gillian Beer, *Open Fields: Science in Cultural Encounter* (Oxford: Oxford University Press, 1996), 229.
89. Although Thomson did discuss Clausius's version of the Second Law in *Macmillan's Magazine*, this apparently made no impression on Darwin. See also Janet Browne, *Charles Darwin*, vols. 1 and 2.
90. Kingsland, "Battling Botanist."
91. Charles Darwin, quoted in Beer, *Open Fields*, 220; Browne, *Darwin Power*, 315.
92. Beer, *Open Fields*, 235; Crombie and Smith, *Energy and Empire*, 549, 608–11.
93. L. Brillouin, "Life, Thermodynamics, and Cybernetics," *American Scientist* 37 (1949): 544–68.
94. J. M. Burgers, "Over de verhouding tussen de entropie en de levensfuncties," *Verhandelingen Nederlandse Akademie van Wetenschappen, Afdeling Natuurkunde, 1^e sectie XVIII* (1943): 1–39; see Mechtild de Jong, "Scheidslijnen in het denken over natuurbeheer in Nederland" (PhD diss., Delft University of Technology, 2002).
95. J. M. Burgers, *Ervaring en conceptie* (Arnhem: Van Loghum Slaterus, 1956), 93.
96. Erwin Schrödinger, *What Is Life?* [1944], reprinted in *What Is Life and Mind and Matter* (Cambridge: Cambridge University Press, 1967).
97. Keller, *Refiguring Life*.
98. See Kwa, "Romantic and Baroque."
99. N. Katherine Hayles, *Chaos Bound: Orderly Disorder in Contemporary Literature and Science* (Ithaca: Cornell University Press, 1990).
100. Ilya Prigogine and Isabelle Stengers, *Entre le temps et l'éternité* (Paris: Fayard, 1988), 46.
101. Lily E. Kay, *The Molecular Vision of Life* (New York: Oxford University Press, 1993; see also Kay, *Who Wrote the Book of Life*.

102. Paul Rabinow, *Making PCR: A Story of Biotechnology* (Chicago: The University of Chicago Press, 1996); on Kary Mullis, who won the Nobel Prize in 1993 for his contribution to the invention of PCR, see Steven Shapin, *The Scientific Life: A Moral History of a Late Modern Vocation* (Chicago: The University of Chicago Press, 2008), 226.

103. Morange, *Misunderstood Gene.*

104. C. Woese, as quoted in Michel Morange, "What History Tells Us XIII: Fifty Years of the Central Dogma," *Journal of Bioscience* 33 (2008): 171–75, 174; see also Keller, *Reconfiguring Life.* Morange supports the new theory that RNA was the original self-reproducing macromolecule; only later in the history of life did DNA and proteins emerge and take over some of the functions of the various RNA molecules.

Chapter 11: Science in the Twentieth Century

Epigraph: Letter from Ernest Rutherford to his mother [1900], quoted in Schweber, "From 'Elementary,'" 603.

1. Ideas presented in this chapter were developed earlier in Chunglin Kwa, "Interdisciplinarity and Postmodernity in the Environmental Sciences," *History and Technology* 21 (2005): 331–44; Chunglin Kwa, "Shifts in Dominance of Scientific Styles: From Modernity to Postmodernity," *History and Technology* 23 (2007): 166–75; Chunglin Kwa, "Onderzoek aan de universiteiten. Een geschiedenis van de academische autonomie," *Krisis* 8, no. 4 (2007): 36–51.

2. Ben de Pater and Herman van der Wusten, *Het geografische huis. De opbouw van een wetenschap* (Muiderberg: Coutinho, 1991), 132–56.

3. Rip, "Science for the 21st Century."

4. See, for example, Hacking, *Representing,* and Rheinberger, *Toward a History.*

5. Karl Popper, *The Logic of Scientific Discovery* (London: Hutchinson, 1959), 107.

6. Paul van der Vet, *The Aborted Takeover of Chemistry by Physics* (PhD diss., University of Amsterdam, 1987).

7. Cartwright, *How the Laws of Physics Lie.*

8. Paul Forman, "The Primacy of Science in Modernity, of Technology in Postmodernity, and of Ideology in the History of Technology," *History and Technology* 23 (2007): 1–152.

9. J. D. Bernal, *The Social Function of Science* (London: George Routledge, 1939), 29, 95.

10. Gerard Alberts, *Jaren van berekening. Toepassingsgerichte initiatieven in de Nederlandse wiskundebeoefening 1945–1960* (Amsterdam: Amsterdam University Press, 1998).

11. Mary Jo Nye, *From Chemical Philosophy to Theoretical Chemistry: Dynamics of Matter and Dynamics of Disciplines 1800–1950* (Berkeley: University of California Press, 1993), 34–40.

12. Trevor H. Levere, *Poetry Realized in Nature: Samuel Taylor Coleridge and Early Nineteenth-Century Science* (Cambridge: Cambridge University Press, 1981), 73.

13. Schnädelbach, *Philosophy in Germany,* 33–40.

14. Harry W. Paul, *From Knowledge to Power: The Rise of the Science Empire in France, 1860–1939* (Cambridge: Cambridge University Press, 1985), 61.

15. Nye, *Chemical Philosophy,* 41, 54.

16. Peter Scott, *The Crisis of the University* (London: CroomHelm, 1984).

17. Ernst Homburg, *Van beroep "Chemiker'. De opkomst van de industriële chemicus en het polytechnische onderwijs in Duitsland (1790–1850)* (Delft: Delftse universitaire pers, 1993).

18. Donald Cardwell, Peter Harman, Crosbie Smith, and David Knight, quoted in Nye, *Chemical Philosophy*, 37.

19. Christa Jungnickel and Russell McCormach, *Intellectual Mastery of Nature: Theoretical Physics from Ohm to Einstein*, vol. 2: *The Now Mighty Theoretical Physics, 1870–1925* (Chicago: The University of Chicago Press, 1986), 220.

20. Daston and Galison, *Objectivity*, 301–2, 261.

21. Einstein noted dryly: "For velocities higher than that of light our considerations cease to make sense" (Für Überlichtgeschwindigkeiten werden unsere Überlegungen sinnlos), Albert Einstein, "Zur Elektrodynamik bewegter Körper," *Annalen der Physik* IV, folge 17 (1905): 891–921, 903.

22. Albert Einstein, *Mein Weltbild* (Amsterdam: Querido, 1934), quoted in Gerald Holton, *Thematic Origins of Scientific Thought: Kepler to Einstein* (Cambridge, Mass.: Harvard University Press, 1973), 234.

23. Holton, *Thematic Origins*, 235–37. See also James W. McAllister, *Beauty and Revolution in Science* (Ithaca: Cornell University Press, 1996).

24. Daston and Galison, *Objectivity*, 305.

25. McClelland, *State, Society*, 244, 254.

26. W. Bernard Carlson, "Innovation and the Modern Corporation: From Heroic Invention to Industrial Science," in Krige and Pestre, *Companion to Science*, 203–26.

27. David F. Noble, *America by Design: Science, Technology and the Rise of Corporate Capitalism* (New York: Knopf, 1977).

28. Dominique Pestre, "Science, Political Power, and the State," in Krige and Pestre, *Companion to Science*, 61–76; B. Theunissen, "Natuursport en levensgeluk. Hugo de Vries, Eli Heimans en Jac. Thijsse," *Gewina* 16 (1993), 287–307.

29. Anne Kox, "Hendrik A. Lorentz 1853–1928," in *Van Stevin tot Lorentz. Portretten van achttien Nederlandse natuurwetenschappers*, ed. Anne Kox (Amsterdam: Bert Bakker, 1990), 226–42.

30. Nye, *Before Big Science*.

31. On Max Planck, see John Heilbron, *The Dilemmas of an Upright Man: Max Planck as a Spokesman for German Science* (Berkeley: California University Press, 1986).

32. Zygmunt Bauman, *Legislators and Interpreters: On Modernity, Post-Modernity, and Intellectuals* (Cambridge: Polity Press, 1987), 75; the phrase "speaking truth to power," while not literally employed by Don K. Price, expresses his view on the desired formal separation of the institutions concerned with truth and institutions exercising power. See Don K. Price, *The Scientific Estate* (Cambridge, Mass.: Harvard University Press, 1965), 191.

33. Jean-François Lyotard, *La condition postmoderne: Rapport sur le savoir* (Paris: Minuit, 1979).

34. Shapin, *Scientific Life*.

35. Daniel Kevles, *The Physicists: The History of a Scientific Community in America* (1978; repr., New York: Vintage, 1979), 100; John Servos, "The Industrial Relations of Science: Chemical Engineering at MIT, 1900–1939," *Isis* 71 (1980): 531–49.

36. Servos, "Industrial Relations."

37. Pnina Abir-Am, "The Discourse of Physical Power in the 1930s: A Reappraisal of the Rockefeller Foundation's 'Policy' in Molecular Biology," *Social Studies of Science* 12 (1982): 341–82; Robert E. Kohler, *Partners in Science: Foundations and Natural Scientists 1900–1945* (Chicago: The University of Chicago Press, 1991), 331, 354.

38. Kohler, *Partners*, 241.

39. Ibid., 396.

40. J. Merton England, *A Patron for Pure Science: The National Science Foundation's Formative Years, 1945–57* (Washington D.C.: The National Science Foundation, 1982).

41. Report of the Advisory Committee to the Prime Minister, 1946, quoted in *Nederlandse Organisatie voor Zuiver-Wetenschappelijk Onderzoek: Voorbereiding en werkzaamheden in de oprichtingsperiode 1945–1949* (Report of the Provisional Board of the Dutch Organization for Pure Scientific Research, ZWO) (The Hague, 1950), 8, 30. See also Alberts, *Jaren van berekening*.

42. Kevles, *Physicists*.

43. Paul Forman, "Behind Quantum Electronics: National Security as Basis for Physical Research in the United States, 1940–1960," *Historical Studies in the Physical and Biological Sciences* 18 (1987): 149–229; Daniel Kevles, "Cold War and Hot Physics: Science, Security and the American State, 1945–1956," *Historical Studies in the Physical and Biological Sciences* 20 (1990): 239–64; Stuart W. Leslie, *The Cold War and American Science: The Military-Industrial-Academic Complex at MIT and Stanford* (New York: Columbia University Press, 1993); Ronald Doel, "Constituting the Postwar Earth Sciences: The Military's Influence on the Environmental Sciences in the USA after 1945," *Social Studies of Science* 33 (2003): 635–66; Jessica Wang, *American Science in an Age of Anxiety: Scientists, Anti-Communism, & the Cold War* (Chapel Hill: University of North Carolina Press, 1999).

44. Leslie, *Cold War*, 30.

45. Daniel Kevles, "The National Science Foundation and the Debate over Postwar Research Policy," 1942–1945, *Isis* 68 (1977): 5–26.

46. Personal communication, Arthur Hasler to the author, University of Wisconsin–Madison, 1985; on ONR, see Roger L. Geiger, "Science, Universities, and National Defense, 1945–1970," *Osiris* 7 (1992): 26–48.

47. James H. Capshew and Karen A. Rader, "Big Science: Price to the Present," *Osiris* 7 (1992): 3–25.

48. Leslie, *Cold War*.

49. Forman, "Behind Quantum Electronics," 150.

50. D. E. H. Edgerton, "British Scientific Intellectuals and the Relations of Science, Technology and War," in *National Military Establishments and the Advancement of Science and Technology*, ed. Paul Forman and José Sánchez-Ron (Dordrecht: Kluwer, 1996), 1–35.

51. Scott, *Crisis*, 31.

52. John Krige and Dominique Pestre, "The How and Why of the Birth of CERN," in *History of CERN*, vol. 1: *Launching the European Organization for Nuclear Research*, ed. A. Hermann, J. Krige. U. Mersits, and D. Pestre (Amsterdam: North-Holland, 1987), 523–44.

53. Chunglin Kwa, "Modelling Technologies of Control," *Science as Culture* 4, no. 20 (1994): 363–91.

54. Forman, "Quantum Electronics," 150.

55. In 1998, Seitz made the news with a frontal assault on the conclusion of the United Nations Intergovernmental Panel on Climate Change that industrial greenhouse gas emissions play a role in climate change. He made his own report look as though it came from the National Academy of Sciences, but the NAS distanced itself from him. See Myanna Lahsen, "Technocracy, Democracy, and U.S. Climate Politics: The Need for Demarcations," *Science, Technology and Human Values* 30 (2005): 137–69.

56. Polanyi, quoted in Steve Fuller, *The Governance of Science: Ideology and the Future of the Open Society* (Buckingham: Open University Press, 2000). Let there be no misunderstanding: Polanyi was utterly convinced of the value of autonomous, fundamental science.

57. Daniel Greenberg, *The Politics of Pure Science* (New York: New American Library, 1967).

58. One example was the book written by science journalists William Broad and Nicholas Wade, *Betrayers of the Truth: Fraud and Deceit in the Halls of Science* (New York: Simon and Schuster, 1982).

59. Daniel Kevles, *The Baltimore Case: A Trial of Politics, Science and Character* (New York: W. W. Norton, 1998).

60. Björn Wittrock, "Dinosaurs or Dolphins? Rise and Resurgence of the Research-Oriented University," in *The University Research System: The Public Policy of the Home of the Scientists*, edited by B. Wittrock and A. Elzinga (Stockholm: Almqvist and Wiksell International, 1985), 13–37; see also David Dickson, *The New Politics of Science* (New York: Pantheon, 1984).

61. On RANN, see Toby A. Appel, *Shaping Biology: The National Science Foundation and American Biological Research, 1945–1975* (Baltimore: Johns Hopkins University Press, 2000), 239.

62. Chunglin Kwa, "Representations of Nature Mediating between Ecology and Science Policy: The Case of the International Biological Programme," *Social Studies of Science* 17 (1987): 413–42.

63. Kwa, "Modeling the Grasslands."

64. Chunglin Kwa, "Local Ecologies, Global Science: Discourses and Strategies of the International Geosphere-Biosphere Programme," *Social Studies of Science* 35 (2005): 923–50; Kwa, "Interdisciplinary Research."

65. Chunglin Kwa and René Rector, "A Data Bias in Interdisciplinary Cooperation in the Sciences: Ecology in Climate Change Research," in *Collaboration in the New Life Sciences*, ed. John Parker, Niki Vermeulen, Bart Penders (London: Ashgate, 2010), 161–76.

66. Erik M. Conway, "Drowning in Data: Satellite Oceanography and Information Overload in the Earth Sciences," *Historical Studies in the Physical and Biological Sciences* 37 (2006): 127–51.

67. James Collins, "Foreword," in Joyce E. Canaan and Wesley Shumar, *Structure and Agency in the Neoliberal University* (New York: Routledge, 2008), xiv.

68. Shapin, *Scientific Life*, 228.

69. Sheila Jasanoff, *Designs on Nature: Science and Democracy in Europe and the United States* (Princeton: Princeton University Press, 2005), 238.

70. Hans Radder, "Exploiting Abstract Possibilities: A Critique of the Concept and Practice of Product Patenting," *Journal of Agricultural and Environmental Ethics* 17 (2004): 275–91.

71. Sheila Slaughter and Gary Rhoades, "The Emergence of a Competitive Research and Development Policy Coalition and the Commercialization of Academic Science and Technology," *Science, Technology & Human Values* 21 (1996): 303–40.

72. Daniel S. Greenberg, *Science for Sale: The Perils, Rewards, and Delusions of Campus Capitalism* (Chicago: University of Chicago Press, 2007).

73. Jasanoff, *Designs*, 236.

74. Philip Mirowski and Robert Van Horn, "The Contract Research Organization and the Commercialization of Scientific Research," *Social Studies of Science* 35 (2005): 503–48, at 520, 528.

75. The principles of the neoliberal society were worked out, starting in 1932, by a group of German economists that included Walter Eucken and Wilhelm Röpke. They presented themselves as critics of John Maynard Keynes, and their critique of his strategy for combating the Great Depression of the 1930s showed striking similarities to the views of a group of Austrian economists including F. A. von Hayek and Joseph Schumpeter. Michel Foucault, *Naissance de la biopolitique: Cours au Collège de France, 1978–1979* (Paris: Gallimard/Seuil, 2004).

76. Stefan Collini, "HiEdBiz," *London Review of Books*, November 6, 2003.

77. Lyotard, *La condition*, 16.

78. Slaughter and Rhoades, "Competitiveness Coalition," 311.

79. Jasanoff, *Designs*, 240, 243.

80. Van den Hoeven, quoted in Michael Persson, "Geef onderzoekers rust," *De Volkskrant*, May 27, 2006.

81. Lyotard, *La condition*.

82. Michael Gibbons et al., *The New Production of Knowledge: The Dynamics of Science and Technology in Contemporary Societies* (London: Sage, 1994); Slaughter and Rhoades, "Competitiveness Coalition"; Donna Haraway, *Modest_Witness@Second_Millennium. FemaleMan^C_Meets_Oncomouse^{TM}: Feminism and Technoscience* (New York: Routledge, 1997), 85–94; Steven Shapin, "Ivory Trade," *London Review of Books*, September 11, 2003; Philip Mirowski and Esther-Mirjam Sent, eds., *Science Bought and Sold: Essays in the Economics of Science* (Chicago: The University of Chicago Press, 2002).

83. Daniel Kevles, "Big Science and Big Politics in the United States: Reflections on the Death of the SSC and the Life of the Human Genome Project," *Historical Studies in the Physical and Biological Sciences* 27 (1997): 269–99. In the past decade, new supercollider projects have been launched, but they have had to go to great lengths to demonstrate their potential usefulness in fields other than physics.

84. Capshew and Rader, "Big Science."

85. Pierre Bourdieu, *La distinction: critique sociale du jugement* (Paris: Minuit, 1979).

86. Bill, quoted in *Nederlandse Organisatie voor Zuiver-Wetenschappelijk Onderzoek: Voorbereiding en werkzaamheden in de oprichtingsperiode 1945–1949* (Report of the Provisional Board of the Dutch Organization for Pure Scientific Research, ZWO) (The Hague, 1950), 26.

87. Harvey Brooks, "Research Universities and the Social Contract for Science," in *Empowering Technology: Implementing a US Strategy*, ed. Lewis Branscomb (Cambridge, Mass.: MIT Press, 1993), 202–34.

88. Shapin, "Ivory Tower."

89. Dominique Pestre, *Science, argent et politique: un essai d'interprétation* (Paris: Institut National de la Recherche Agronomique, 2003).

90. Sheila Jasanoff, "(No?) Accounting for Expertise," *Science and Public Policy* 30 (2003), 157–62; Shapin, "Ivory Tower"; Jasanoff, *Designs*; Lahsen, "Technocracy."

BIBLIOGRAPHY

Abir-Am, Pnina. "The Discourse of Physical Power in the 1930s: A Reappraisal of the Rockefeller Foundation's 'Policy' in Molecular Biology." *Social Studies of Science* 12 (1982): 341–82.
Acot, Pascal. *Histoire de l'écologie,* Que sais je 2870. Paris: Presses Universitaires de France, 1994.
Aiton, E. J. *The Vortex Theory of Planetary Motion.* London: Macdonald, 1972.
Alberts, Gerard. *Jaren van berekening. Toepassingsgerichte initiatieven in de Nederlandse wiskundebeoefening 1945–1960.* Amsterdam: Amsterdam University Press, 1998.
Allen, David Elliston. *The Naturalist in Britain: A Social History.* 1976. Reprint, Princeton: Princeton University Press, 1994.
Allen, Garland. "Hugo de Vries and the Reception of Mutation Theory." *Journal for the History of Biology* 2 (1969): 55–87.
———. *Life Science in the Twentieth Century.* 1975. Reprint, Cambridge: Cambridge University Press, 1978.
Alquié, Ferdinand. *Descartes.* Paris: Hatier, 1956.
Ambrosoli, Mario. *The Wild and the Sown: Botany and Agriculture in Western Europe, 1350–1850.* Cambridge: Cambridge University Press, 1997.
Andriesse, C. D. *Titan kan niet slapen. Een biografie van Christiaan Huygens.* Amsterdam: Contact, 1993.
Appel, Toby A. *Shaping Biology: The National Science Foundation and American Biological Research, 1945–1975.* Baltimore: Johns Hopkins University Press, 2000.
Arber, Agnes. *Herbals: Their Origin and Evolution: A Chapter in the History of Botany 1470–1670.* Cambridge: Cambridge University Press, 1938.
Ashworth, William B. "Catholicism and Early Modern Science." In Lindberg and Numbers, *God and Nature,* 136–66.
———. "Natural History and the Emblematic World View." In Lindberg and Westman, *Reappraisals of the Scientific Revolution,* 303–32.
Atran, Scott. *Cognitive Foundations of Natural History: Towards an Anthropology of Science.* Cambridge: Cambridge University Press, 1990.
Bakker, Boudewijn. *Landschap en wereldbeeld. Van Van Eyck tot Rembrandt.* Bussum: Toth, 2004.

Ballintijn, G. *Rumphius. De blinde ziener van Ambon.* Utrecht: de Haan, 1944.
Basalla, George. "Pop Science: The Depiction of Science in Popular Culture." In *Science and Its Public,* edited by G. A. Holton and W. Blanpied, 261–78. Dordrecht: Reidel, 1976.
Bauman, Zygmunt. *Legislators and Interpreters: On Modernity, Post-Modernity, and Intellectuals.* Cambridge: Polity Press, 1987.
Beaulieu, Anne. "Voxels in the Brain: Neuroscience, Informatics and Changing Notions of Objectivity." *Social Studies of Science* 31 (2001): 635–80.
Baxandall, Michael. *Painting and Experience in Fifteenth-Century Italy: A Primer in the Social History of Pictorial Style.* Oxford: Oxford University Press, 1972.
Beer, Gillian. "'The Face of Nature': Anthropomorphic Elements in the Language of the Origin of Species." In *Languages of Nature,* edited by L. Jordanova, 207–43. London: Free Association Books, 1986.
———. *Open Fields: Science in Cultural Encounter.* Oxford: Oxford University Press, 1996.
Bellone, Enrico. *A World on Paper: Studies on the Second Scientific Revolution.* Cambridge, Mass.: The MIT Press, 1980.
Ben-Chaim, Michael. "The Discovery of Natural Goods." *British Journal for the History of Science* 34 (2001): 395–416.
Ben-David, Joseph. *The Scientist's Role in Society: A Comparative Study.* Englewood Cliffs, N.J.: Prentice Hall, 1971.
Bennett, J. A. "The Mechanics' Philosophy and the Mechanical Philosophy." *History of Science* 24 (1986): 1–28.
Bensaude-Vincent, Bernadette. "Mendeleev's Periodic System of Chemical Elements." *British Journal for the History of Science* 19 (1986): 3–17.
Bergvelt, Ellinoor, and Renée Kistemaker, eds. *De wereld binnen handbereik, Nederlandse kunst- en rariteitenverzamelingen.* Zwolle/Amsterdam: Waanders/Amsterdams Historisch Museum, 1992.
Berkel, Klaas van. "Citaten uit het boek der natuur." In Bergvelt and Kistemaker, *De wereld binnen handbereik,* 169–91.
———. *Citaten uit het boek der natuur. Opstellen over Nederlandse wetenschapsgeschiedenis.* Amsterdam: Bert Bakker, 1998.
———. *In het voetspoor van Stevin.* Meppel/Amsterdam: Boom, 1985.
Berlin, Brent, Dennis Breedlove, and Peter H. Raven. "General Principles of Classification and Nomenclature in Folk Biology." *American Anthropologist* 75 (1973): 124–216.
Berlin, Isaiah. *Vico and Herder: Two Studies in the History of Ideas.* London: Chatto and Windus, 1976.
Bernal, J. D. *The Social Function of Science.* London: George Routledge, 1939.
Berry, Andrew. "Whenever You Can, Count." *London Review of Books,* December 4, 2003.
Biagioli, Mario. *Galileo, Courtier: The Practice of Science in the Culture of Absolutism.* Chicago: The University of Chicago Press, 1993.
———. *Galileo's Instruments of Credit: Telescopes, Instruments, Secrecy.* Chicago: The University of Chicago Press, 2006.
———. "The Social Status of Italian Mathematicians, 1450–1600." *History of Science* 27 (1989): 41–95.

Blair, Ann. *The Theatre of Nature: Jean Bodin and Renaissance Science*. Princeton: Princeton University Press, 1997.
Blair, Ann, and Anthony Grafton. "Reassessing Humanism and Science." *Journal of the History of Ideas* 53 (1992): 535–40.
Blumenberg, Hans. *Arbeit am Mythos*. Frankfurt: Suhrkamp, 1979.
———. *Die Genesis der kopernikanischen Welt*. Frankfurt: Suhrkamp, 1975.
———. *Die Legitimität der Neuzeit*. Frankfurt: Suhrkamp, 1988.
Blunt, Wilfred. *The Compleat Naturalist: A Life of Linnaeus*. London: Collins, 1971.
Boas Hall, Mary. *The Scientific Renaissance, 1450–1630*. London: Collins, 1962.
Böhme, Gernot. *Alternativen der Wissenschaft*. Frankfurt: Suhrkamp (stw334), 1980.
Bourdieu, Pierre. *La distinction: critique sociale du jugement*. Paris: Minuit, 1979.
———. "Postface." In Erwin Panovsky, *Architecture gothique et pensée scholastique*, 133–67. 1946. Reprint, Paris: Minuit, 1979.
Bourguet, Marie-Noëlle. "Décrire, Compter, Calculer: The Debate over Statistics during the Napoleonic Period." In Krüger, Daston, and Heidelberger, *Probabilistic Revolution*, 1:305–16.
Bowker, Geoffrey C., and Susan Leigh Star. *Sorting Things Out: Classification and Its Consequences*. Cambridge, Mass.: The MIT Press, 1999.
Bowler, Peter J. *The Eclipse of Darwinism: Anti-Darwinian Evolution Theories in the Decades Around 1900*. Baltimore: Johns Hopkins University Press, 1983.
———. *Life's Splendid Drama*. Chicago: The University of Chicago Press, 1996.
———. *The Mendelian Revolution: The Emergence of Heriditarian Concepts in Modern Science and Society*. London: Athlone, 1989.
Boxer, C. *Two Pioneers of Tropical Medicine, Garcia d'Orta and Nicolás Monardes*. London: Wellcome, 1963.
Brillouin, L. "Thermodynamics and Cybernetics." *American Scientist* 37 (1949): 544–68.
Broad, William, and Nicholas Wade. *Betrayers of the Truth: Fraud and Deceit in the Halls of Science*. New York: Simon and Schuster, 1982.
Brockway, Lucile H. *Science and Colonial Expansion: The Role of the British Royal Botanic Gardens*. New York: Academic Press, 1979.
Broek, Gerard J. van den. *Baleful Weeds and Precious-Juiced Flowers: A Semiotic Approach to Botanical Practice*. PhD dissertation, Leiden University, 1986.
Broek, R. van den, and Gilles Quispel. "Inleiding." In *Corpus hermeticum*. Amsterdam: In de pelikaan, 1990.
Brooks, Harvey. "Research Universities and the Social Contract for Science." In *Empowering Technology: Implementing a US Strategy*, edited by Lewis Branscomb, 202–34. Cambridge, Mass.: The MIT Press, 1993.
Brouwer, Christien. *Anatomische sekse als uitvinding in de botanie. Hoe stampers tot vrouwelijke en meeldraden tot mannelijke geslachtsorganen werden (1675–1735)*. PhD dissertation, Amsterdam University, 2004.
Brown, Joseph E. "The Science of Weights." In Lindberg, *Science in the Middle Ages*, 179–205.
Browne, Janet. *Charles Darwin: The Power of Place. Vol. 2 of a Biography*. London: Jonathan Cape, 2002.
———. *Charles Darwin Voyaging: Vol. 1 of a Biography*. London: Jonathan Cape, 1995.
———. *The Secular Ark*. New Haven: Yale University Press, 1983.

Bruin, Hans de. *Prestatiemeting in de publieke sector*. Utrecht: Lemma, 2001.
Burgers, J. M. *Ervaring en conceptie*. Arnhem: van Loghum Slaterus, 1956.
———. "Over de verhouding tussen de entropie en de levensfuncties." *Verhandelingen Nederlandse Akademie van Wetenschappen, Afdeling Natuurkunde*, 1^e sectie XVIII 3 (1943): 1–39.
Burckhardt, J. *Die Kultur der Renaissance in Italien. Ein Versuch*. 1860. Reprint, Stuttgart: Alfred Kröner Verlag, 1925.
Burkert, Walter. *Lore and Science in Ancient Pythagoreanism*. Cambridge, Mass.: Harvard University Press, 1972.
Butterfield, Herbert. *The Origins of Modern Science 1300–1800*. 1949. Reprint, London: G. Bell and Sons, 1957.
Burton, W. G. *The Potato*. Wageningen: Veenman en Zonen, 1966.
Canaan, Joyce E., and Wesley Shumar. *Structure and Agency in the Neoliberal University*. New York: Routledge, 2008.
Canfora, Luciano. *The Vanished Library: A Wonder of the Ancient World*. Berkeley: University of California Press, 1990.
Canguilhem, Georges. *La connaissance de la vie*. Paris: Vrin, 1980.
Cannon, Susan Faye. *Science in Culture: The Early Victorian Period*. Folkestone: Dawson, 1978.
Capshew, James H., and Karen A. Rader. "Big Science: Price to the Present." *Osiris* 7 (1992): 3–25.
Cardano, Gerolamo. *De vita propria*. Paris, 1643.
Cardwell, Donald S. L. *From Watt to Clausius: The Rise of Thermodynamics in the Early Industrial Age*. Ithaca: Cornell University Press, 1971.
———. "Science and Technology: The Work of James Prescott Joule." *Technology and Culture* 17 (1976): 674–87.
Carlson, W. Bernard. "Innovation and the Modern Corporation: From Heroic Invention to Industrial Science." In Krige and Pestre, *Companion to Science*, 203–26.
Cartwright, Nancy. *How the Laws of Physics Lie*. Oxford: Oxford University Press, 1983.
Chapuis, Alfred, and Edmond Droz. *Automata*. London: Batsford, 1958.
Chartier, Roger. *The Cultural Uses of Print in Early Modern France*. Princeton: Princeton University Press, 1987.
———. "Texts, Printing, Readings." In Hunt, *New Cultural History*, 1989, 154–75.
Chatelain, Jean-Marc. "Du Parnasse à l'Amérique: l'imaginaire de l'encyclopédie à la Renaissance et à l'Age classique." In *Tous les savoirs du monde*, edited by Roland Schaer, 156–63. Paris: Bibliothèque nationale de France, 1997.
Cittadino, Eugene. *Nature as the Laboratory: Darwinian Plant Ecology in the German Empire, 1880–1900*. New York: Cambridge University Press, 1990.
Clagett, Marshall. "The Influence of Archimedes on Medieval Science." *Isis* 50 (1959): 419–29.
Clark, Stuart. "The Scientific Status of Demonology." In Vickers, *Occult and Scientific Mentalities*, 351–74.
Clulee, Nicholas H. "At the Crossroads of Magic and Science: John Dee's Archemastrie." In Vickers, *Occult and Scientific Mentalities*, 57–71.
Cohen, H. F. *Quantifying Music: The Science of Music at the First Stage of the Scientific Revolution, 1580–1650*. Dordrecht: Reidel, 1984.
Collini, Stefan. "HiEdBiz." *London Review of Books*, November 6, 2003.

Collins, Harry. *Changing Order: Replication and Induction in Scientific Practice.* London: Sage, 1985.
Collins, James. "Foreword." In Canaan and Shumar, *Structure and Agency,* xiii–xviii.
Conway, Erik M. "Drowning in Data: Satellite Oceanography and Information Overload in the Earth Sciences." *Historical Studies in the Physical and Biological Sciences* 37 (2006): 127–51.
Cook, Harold J. *Matters of Exchange: Commerce, Medicine, and Science in the Dutch Golden Age.* New Haven: Yale University Press, 2007.
———."Physicians and Natural History." In Jardine, Secord, and Spary, *Cultures of Natural History,* 91–105.
Copenhaver, Brian. "Natural Magic, Hermetism, and Occultism in Early Modern Science." In Lindberg and Westman, *Reappraisals of the Scientific Revolution,* 261–301.
Coudert, Allison. *The Impact of the Kabbalah in the Seventeenth Century: The Life and Thought of Francis Mercury van Helmont (1614–1698).* Leiden: Brill, 1999.
Crombie, Alistair Cameron. *Augustine to Galileo: The History of Science A.D. 400–1650.* London: Falcon Press, 1952.
———. *Styles of Scientific Thinking in the European Tradition: The History of Argument and Explanation Especially in the Mathematical and Biomedical Sciences and Arts.* 3 vols. London: Duckworth, 1994.
———. "Sources of Galileo's Early Natural Philosophy." In *Science, Art and Nature in Medieval and Modern Thought,* 149–63. London: The Hambledon Press, 1996.
Crosby, Alfred W. *The Measure of Reality.* Cambridge: Cambridge University Press, 1997.
DaCosta Kaufmann, Thomas. *Court, Cloister, City: The Art and Culture of Central Europe 1450–1800.* London: Weidenfeld and Nicolson, 1995.
———. *The Mastery of Nature: Aspects of Art, Science, and Humanism in the Renaissance.* Princeton: Princeton University Press, 1993.
Danziger, Kurt. *Constructing the Subject.* Cambridge: Cambridge University Press, 1990.
———. "Statistical Method and the Historical Development of Research Practice in American Psychology." In Krüger, Gigerenzer, and Morgan, *Probabilistic Revolution,* vol. 2, 35–47.
Darden, Lindley. *Theory Change in Science: Strategies from Mendelian Genetics.* New York: Oxford University Press, 1991.
Darwin, John. *After Tamerlane: The Rise and Fall of Global Empires 1400–2000.* London: Allen Lane, 2007.
Daston, Lorraine. *Classical Probability in the Enlightenment.* Princeton: Princeton University Press, 1988.
———. "The Factual Sensibility." *Isis* 79 (1988): 452–67.
———. "The Moral Economy of Science." *Osiris* 10 (1995): 3–24.
Daston, Lorraine, and Peter Galison. *Objectivity.* New York: Zone Books, 2007.
Daston, L., and K. Park. *Wonders and the Order of Nature.* New York: Zone Books, 1998.
Davidson, Arnold. "Styles of Reasoning, Conceptual History, and the Emergence of Psychiatry." In *The Disunity of Science: Boundaries, Contexts and Power,* edited by Peter Galison and David J. Stump, 75–100. Stanford: Stanford University Press, 1996.

Day, Rachel L., Kevin L. Laland, and John Odling-Smee. *Perspectives in Biology and Medicine* 46 (2003): 80–95.

Dear, Peter. "Jesuit Mathematical Science and the Reconstruction of Experience in the Early Seventeenth Century." *Studies in History and Philosophy of Science* 18 (1987): 1331–75.

———. *Mersenne and the Learning of the Schools*. Ithaca: Cornell University Press, 1988.

———. "Miracles, Experiments, and the Ordinary Course of Nature." *Isis* 81 (1990): 663–83.

De Chaderevian, Soraya. "Laboratory Science versus Country-House Experiments: The Controversy between Julius Sachs and Charles Darwin." *British Journal for the History of Science* 29 (1996): 17–41.

Dehue, Trudy. *Changing the Rules: Psychology in the Netherlands, 1900–1985*. Cambridge: Cambridge University Press, 1995.

Deleuze, Gilles. *Le pli: Leibniz et le baroque*. Paris: Minuit, 1988.

Descartes, René. *Oeuvres philosophiques*. Edited by F. Alquié. Paris: Garnier, 1988.

———. *Principles of Philosophy*. Translated by Valentine Rodger Miller and Reese Miller. Dordrecht: D. Reidel Publishing Company, 1983.

Desrosières, Alain. "How to Make Things which Hold Together: Social Science, Statistics and the State." In *Discourses on Society: The Shaping of the Social Science Disciplines*, edited by Peter Wagner, Björn Wittrock, and Richard Whitley, 195–218. Dordrecht: Kluwer, 1991.

———. *La politique des grands nombres*. Paris: Editions La Découverte, 1993.

Dettelbach, Michael. "Humboldtian Science." In Jardine, Secord, and Spary, *Cultures of Natural History*, 287–304.

Dickson, David. *The New Politics of Science*. New York: Pantheon, 1984.

Dijksterhuis, E. J. *Archimedes*. Groningen: Noordhoff, 1938.

———. *Clio's stiefkind*. Amsterdam: Bert Bakker, 1990.

———. *De mechanisering van het wereldbeeld*. 1950. Reprint, Amsterdam: Meulenhoff, 1975.

———. *Simon Stevin (1548–1620)*. The Hague: Martinus Nijhoff, 1941.

Dijksterhuis, Fokko Jan. "Constructive Thinking: A Case for Dioptrics." In Roberts, Schaffer, and Dear, *Mindful Hand*, 59–82.

Dobbs, Betty Jo T. *Foundations of Newton's Alchemy, or: "The Hunting of the Greene Lyon."* Cambridge: Cambridge University Press, 1975.

Dobbs, Betty Jo T., and Margaret C. Jacob. *Newton and the Culture of Newtonianism*. Atlantic Highlands, N.J.: Humanities Press, 1995.

Doel, Ronald P. "Constituting the Postwar Earth Sciences: The Military's Influence on the Environmental Sciences in the USA after 1945." *Social Studies of Science* 33 (2003): 635–66.

Dodonaeus, Rembertus (Rembert Dodoens). *Stirpiae historiae. Pemptades sex.* Antwerp: Christopher Plantijn, 1583.

Douglas, Mary, *How Institutions Think*. Syracuse: Syracuse University Press, 1986.

Drake, Stillman. "The Accademia dei Lincei." In *Galileo Studies*, 79–94. Ann Arbor: The University of Michigan Press, 1970.

———. *Galileo at Work: His Scientific Biography*. Chicago: The University of Chicago Press, 1978.

———. *Galileo: Pioneer Scientist*, Toronto: University of Toronto Press, 1990.

Dreyfus, Herbert, and Paul Rabinow. *Michel Foucault: Beyond Structuralism and Hermeneutics*. Brighton: Harvester Press, 1982.
Duffy, Eamon. *Saints and Sinners: A History of the Popes*. New Haven: Yale University Press, 1997.
Durkheim, E., and M. Mauss. "De quelques formes primitives de classification." *Année sociologique*. 1903. Reprinted in M. Mauss. *Essais de sociologie*. Paris: Minuit, 1968.
Eamon, William. "From the Secrets of Nature to Public Knowledge." In Lindberg and Westman, *Reappraisals*, 333–65.
Eck, Caroline van, and Robert Zwijnenberg. "Inleiding." In *Over de schilderkunst*, by Leon Battista Alberti. Meppel/Amsterdam: Boom, 1996.
Edgerton, D. E. H. "British Scientific Intellectuals and the Relations of Science, Technology and War." In *National Military Establishments and the Advancement of Science and Technology*, edited by Paul Forman and José Sánchez-Ron, 1–35. Dordrecht: Kluwer, 1996.
Edgerton Jr., S. Y. *The Heritage of Giotto's Geometry: Art and Science on the Eve of the Scientific Revolution*. Ithaca: Cornell University Press, 1991.
Einstein, Albert. "What Is the Theory of Relativity?" *London Times*, November 28, 1919.
———. "Zur Elektrodynamik bewegter Körper." *Annalen der Physik* IV, folge 17 (1905): 891–921.
Eisenstein, E. *The Printing Press as an Agent of Change*. Vols. 1 and 2. Cambridge: Cambridge University Press, 1979.
Emerton, Norma E. *The Scientific Reinterpretation of Form*. Ithaca: Cornell University Press, 1984.
England, J. Merton. *A Patron for Pure Science: The National Science Foundation's Formative Years, 1945–57*. Washington D.C.: The National Science Foundation, 1982.
Erasmus. *The Colloquies*. Translated by Craig R. Thompson. Chicago: The University of Chicago Press, 1965.
Ereshefsky, Mark. *The Poverty of the Linnean Hierarchy: A Philosophical Study of Biological Taxonomy*. Cambridge: Cambridge University Press, 2001.
Evans, R. J. *Rudolf II and His World: A Study in Intellectual History, 1576–1612*. Oxford: Oxford University Press, 1973.
Feldhay, Rivka. *Galileo and the Church*. Cambridge: Cambridge University Press, 1995.
Feldhay, Rivka, and Michael Heyd. "The Discourse of Pious Science." *Science in Context* 3 (1989): 109–42.
Ferguson, Eugene. *Engineering and the Mind's Eye*. Cambridge, Mass.: The MIT Press, 1992.
Findlen, Paula. *Possessing Nature: Museums, Collecting and Scientific Culture in Early Modern Italy*. Berkeley: University of California Press, 1994.
Finocchiaro, Maurice A. *The Galileo Affair: A Documentary History*. Berkeley: University of California Press, 1989.
Fisher, Saul. "Gassendi's Atomist Account of Generation and Heredity in Plants and Animals." *Perspectives on Science* 11 (2003): 484–512.
Fleck, Ludwik. *Entstehung und Entwicklung einer wissenschaftlicher Tatsache: Einführung in die Lehre vom Denkkollektiv und Denkstil*. 1935. Reprint, Frankfurt am Main: Suhrkamp, 1980.

Fleischer, Alette. "The Beemster Polder: Conservative Invention and Holland's Great Pleasure Garden." In Roberts, Schaffer and Dear, *Mindful Hand*, 145–68.
Forman, Paul. "Behind Quantum Electronics: National Security as Basis for Physical Research in the United States, 1940–1960." *Historical Studies in the Physical and Biological Sciences* 18 (1987): 149–229.
———. "The Primacy of Science in Modernity, of Technology in Postmodernity, and of Ideology in the History of Technology." *History and Technology* 23 (2007): 1–152.
———. "Weimar Culture, Causality, and Quantum Theory, 1918–1927: Adaptation by German Physicists and Mathematicians to a Hostile Environment." *Historical Studies in the Physical Sciences* 3 (1971): 1–115.
Foucault, Michel. *Histoire de la sexualité vol. 1: La volonté de savoir.* Paris: Gallimard, 1976.
———. *La naissance de la clinique.* Paris: Presses Universitaires de France, 1963.
———. *Les mots et les choses: une archéologie des sciences humaines.* Paris: Gallimard, 1966.
———. *Naissance de la biopolitique: Cours au Collège de France, 1978–1979.* Paris: Gallimard/Seuil, 2004.
Fox, Robert, and George Weisz, eds. *The Organization in Science and Technology in France 1808–1914.* Cambridge: Cambridge University Press, 1980.
Frank Jr., Robert G. *Harvey and the Oxford Physiologists.* Berkeley: University of California Press, 1980.
Freedman, David. *The Eye of the Lynx: Galileo, His Friends, and the Beginnings of Modern Natural History.* Chicago: The University of Chicago Press, 2002.
French, Peter. *John Dee: The World of an Elizabethan Magus.* London: Routledge and Kegan Paul, 1972.
Freudenthal, Gad. "Theory of Matter and Cosmology in William Gilbert's *De magnete*." *Isis* 74 (1983): 22–37.
Friedman, Robert M. *Appropriating the Weather: Vilhelm Bjerknes and the Construction of a Modern Meteorology.* Ithaca: Cornell University Press, 1989.
Frye, Northrop. *The Great Code: The Bible and Literature.* New York: Harcourt Brace Jovanovich, 1981.
Fuller, Steve. *The Governance of Science: Ideology and the Future of the Open Society.* Buckingham: Open University Press, 2000.
Funkenstein, Amos. *Theology and the Scientific Imagination.* Princeton: Princeton University Press, 1986.
Gabby, Alan. "Force and Inertia in the Seventeenth Century: Descartes and Newton." In Gaukroger, *Descartes: Philosophy, Mathematics and Physics*, 230–320.
Gal, Ofer, and Raz Chen-Morris. "The Archeology of the Inverse-Square Law: (1) Metaphysical Images and Mathematical Practices." *History of Science* 43 (2005): 391–414.
———. "The Archeology of the Inverse-Square Law: (2) The Use and Non-Use of Mathematics." *History of Science* 44 (2006): 49–67.
Galilei, Galileo. *Galileo on the World Systems.* Translated and edited by Maurice Finocchiaro. Berkeley: University of California Press, 1977.
Galison, Peter. *How Experiments End.* Chicago: The University of Chicago Press, 1987.
Galluzzi, Paolo. *Les ingénieurs de la Renaissance de Brunelleschi à Leónard de Vinci.* Exhibition Catalogue. Paris, Florence: Giunti, 1995.

Gascoigne, John. "The Role of the Universities." In Lindberg and Westman, *Reappraisals*, 207–60.
Gaukroger, Stephen. *Descartes: An Intellectual Biography*. Oxford: Oxford University Press, 1995.
———, ed. *Descartes: Philosophy, Mathematics and Physics*. Brighton: The Harvester Press, 1980.
Geertz, Clifford. *The Interpretation of Cultures*. New York: Basic Books, 1973.
Geiger, Roger L. "Science, Universities, and National Defense, 1945–1970." *Osiris* 7 (1992): 26–48.
Gelder, Roelof van. "De wereld binnen handbereik. Nederlandse kunst- en rariteitenverzamelingen, 1585–1735." In Bergvelt and Kistemaker, *De wereld binnen handbereik*, 15–38.
———. "Liefhebbers en geleerde luiden. Nederlandse kabinetten en hun bezoekers." In Bergvelt and Kistemaker, *De wereld binnen handbereik*, 259–92.
Geschiedenis van de techniek in Nederland. De wording van een moderne samenleving 1800–1890. Parts IV and VI. Walburg Pers, 1993.
Gibbons, Michael, et al. *The New Production of Knowledge: The Dynamics of Science and Technology in Contemporary Societies*. London: Sage, 1994.
Gieryn, Thomas F. "Distancing Science from Religion in Seventeenth-Century England." *Isis* 79 (1988): 5825–93.
Gigerenzer, Gerd. "Probabilistic Thinking and the Fight against Subjectivity." In Krüger, Gigerenzer, and Morgan, *Probabilistic Revolution*, vol. 2, 11–33.
Gigerenzer, Gerd, Zeno Swijtink, Theodore Porter, Lorraine Daston, John Beatty, and Lorenz Krüger. *The Empire of Chance: How Probability Changed Science and Everyday Life*. Cambridge: Cambridge University Press, 1989.
Gilham, Nicholas Wright. *A Life of Sir Francis Galton: From African Exploration to the Birth of Eugenics*. Oxford: Oxford University Press, 2001.
Gille, Bertrand. *Les ingénieurs de la Renaissance*. Paris: Hermann, 1964.
Ginzburg, Carlo. "Clues: Roots of an Evidential Paradigm." In *Clues, Myths, and the Historical Method*, 96–125. Baltimore: Johns Hopkins University Press, 1989.
———. *Ecstasies: Deciphering the Witches' Sabbath*. Translated by Raymond Rosenthal. Chicago: The University of Chicago Press, 1991.
———. "From Aby Warburg to Ernst Hans Gombrich: A Problem of Method." In *Clues, Myths, and Historical Method*, 17–59. Baltimore: Johns Hopkins University Press, 1989.
———. "The High and the Low: The Theme of Forbidden Knowledge in the Sixteenth and Seventeenth Centuries." In *Clues, Myths, and the Historical Method*, 60–76. Baltimore: Johns Hopkins University Press, 1989.
Gjertsen, Derek. *The Newton Handbook*. London: Routledge and Kegan Paul, 1986.
Glacken, Clarence J. *Traces on the Rhodian Shore: Nature and Culture in Western Thought from Ancient Times to the End of the Eighteenth Century*. Berkeley: University of California Press, 1967.
Glockner, Hermann. *Die europäische Philosophie von der Anfängen bis zur Gegenwart*, 1958. Reprint, Stuttgart: Reclam, 1980.
Goodman, David, and Colin A. Russell. *The Rise of Scientific Europe, 1500–1800*. Sevenoaks: Hodder and Stoughton, 1991.
Gombrich, Ernst. *Art and Illusion*. London: Phaidon, 1960.

———. *Meditations on a Hobby Horse*. London: Phaidon, 1963.
———. *The Story of Art*. London: Phaidon, 1950.
Gould, Stephen Jay. *The Panda's Thumb: Further Reflections in Natural History*. New York: W. W. Norton and Co., 1980.
Grafton, Anthony. *Cardano's Cosmos: The Worlds and Works of a Renaissance Astrologer*. Cambridge, Mass.: Harvard University Press, 1999.
———. *Defenders of the Text: The Traditions of Scholarship in an Age of Science*. Cambridge, Mass.: Harvard University Press, 1991.
Greenberg, Daniel. *The Politics of Pure Science*. New York: New American Library, 1967.
———. *Science for Sale: The Perils, Rewards, and Delusions of Campus Capitalism*. Chicago: University of Chicago Press, 2007.
Grove, Richard. *Green Imperialism: Colonial Expansion, Tropical Island Edens and the Origins of Environmentalism, 1600–1860*. Cambridge: Cambridge University Press, 1995.
Gueroult, Martial. "The Metaphysics and Physics of Force in Descartes." In Gaukroger, *Descartes: Philosophy, Mathematics, and Physics*, 196–225.
Hacking, Ian, *The Emergence of Probability*. Cambridge: Cambridge University Press, 1975.
———. "Raison et véracité, les choses, les gens, la raison, cours 2005/2006." Lecture, Collège de France. www.college-de-france.fr/default/EN/all/historique/ian_hacking.htm. Accessed December 27, 2010.
———. *Representing and Intervening: Introductory Topics in the Philosophy of Natural Science*. Cambridge: Cambridge University Press, 1983.
———. "Style for Historians and Philosophers." *Studies in the History and Philosophy of Science* 23 (1993): 1–20.
———. *The Taming of Chance*. Cambridge: Cambridge University Press, 1990.
Hackmann, W. D. "Scientific Instruments: Models of Brass and Aids to Discovery." In *The Uses of Experiment: Studies in the Natural Sciences*, edited by David Gooding, Trevor Pinch, and Simon Schaffer, 31–65. Cambridge: Cambridge University Press, 1989.
Hadot, Pierre. *Le voile d'Isis: Essai sur l'histoire de l'idée de Nature*. Paris: Gallimard, 2004.
Hall, A. Rupert. *From Galileo to Newton 1630–1720*. London: Collins, 1963.
———. *Science and Society*. Aldershot: Variorum, 1994.
Hall, Thomas S. *History of General Physiology 600 B.C. to A.D. 1900*. Vol. 2: *From the Enlightenment to the End of the Nineteenth Century*. Chicago: The University of Chicago Press, 1969.
Hankins, James. *Plato in the Italian Renaissance*. Vol. 1. Leiden: Brill, 1990.
Hannaway, Owen. *The Chemists and the Word: The Didactic Origins of Chemistry*. Baltimore: Johns Hopkins University Press, 1975.
Haraway, Donna. *Modest_Witness@Second_Millennium.FemaleMan^C_Meets_Oncomouse^{TM}: Feminism and Technoscience*. New York: Routledge, 1997.
———. *Simians, Cyborgs and Women: The Reinvention of Nature*. London: Free Association Books, 1991.
Hård, Mikael, and Andrew Jamison. *Hubris and Hybrids: A Cultural History of Technology and Science*. New York: Routledge, 2005.

Hatfield, Gary. "Metaphysics and the New Science." In Lindberg and Westman, *Reappraisals*, 93–166.
Hayles, N. Katherine. *Chaos Bound: Orderly Disorder in Contemporary Literature and Science*. Ithaca: Cornell University Press, 1990.
Helden, Albert van. *Measuring the Universe*. Chicago: The University of Chicago Press, 1985.
Heilbron, John. *The Dilemmas of an Upright Man: Max Planck as a Spokesman for German Science*. Berkeley: California University Press, 1986.
———. *Electricity in the 17th and 18th Century: A Study of Early Modern Physics*. Berkeley: University of California Press, 1979.
Heniger, J. *Hendrik Adriaan van Reede tot Drakenstein (1636–1691) and Hortus Malabaricus*. Rotterdam: Balkema, 1986.
Henry, John. "Animism and Empiricism: Copernican Physics and the Origin of William Gilbert's Experimental Method." *Journal of the History of Ideas* 62 (2001): 1–21.
———. "Metaphysics and the Origins of Modern Science: Descartes and the Importance of Laws of Nature." *Early Science and Medicine* 9 (2004): 73–114.
———. "Occult Qualities and the Experimental Philosophy: Active Principles in Pre-Newtonian Matter Theory." *History of Science* 24 (1986): 335–81.
———. *The Scientific Revolution and the Origins of Modern Science*. Basingbroke: Palgrave, 2002.
Hesse, Mary. *Forces and Fields: The Concept of Action at a Distance in the History of Physics*. New York: Philosophical Library, 1961.
———. "Rationality and the Generalization of Scientific Style." In *The Light of Nature*, edited by J. D. North and J. D. Roche, 365–81. Dordrecht: Martinus Nijhoff, 1985.
Heuvel, Charles van den. "De verspreiding van de Italiaanse vestingbouwkunde in de Nederlanden in de tweede helft van de zestiende eeuw." In *Vesting. Vier eeuwen vestingbouw in Nederland*, edited by J. Sneep, H. A. Treu, and M. Tydeman, 9–17. The Hague: Stichting Menno van Coehoorn, 1982.
Hine, William. "Marin Mersenne, Renaissance Naturalism and Renaissance Magic." In Vickers, *Occult and Scientific Mentalities*, 165–76.
Hogendijk, Jan P., and Abdelhamid I. Sabra. *The Enterprise of Science in Islam: New Perspectives*. Cambridge, Mass.: The MIT Press, 2003.
Holmes, Frederic L. *Claude Bernard and Animal Chemistry: The Emergence of a Scientist*. Cambridge, Mass.: Harvard University Press, 1974.
———. "The Complementarity of Teaching and Research in Liebig's Laboratory." *Osiris* 5 (1989): 121–64.
Holton, Gerald. *Thematic Origins of Scientific Thought: Kepler to Einstein*. Cambridge, Mass.: Harvard University Press, 1973.
Homburg, Ernst. *Van beroep "Chemiker": De opkomst van de industriële chemicus en het polytechnische onderwijs in Duitsland (1790–1850)*. Delft: Delftse universitaire pers, 1993.
Hooykaas, Reyer. *G. J. Rheticus' Treatise on Holy Scripture and the Motions of the Earth*. Amsterdam: North-Holland Publishing Company, 1984.
———. *Humanisme, science et réforme. Pierre de la Ramée (1515–1572)*. Leiden: Brill, 1958.
———. *Religion and the Rise of Modern Science*. Edinburgh: Scottish Academic Press, 1973.

Houwaart, Eddy. *De hygiënisten: Artsen, staat & volksgezondheid in Nederland 1840–1890*. Groningen: Historische Uitgeverij, 1991.
Huizinga, J. *Herfsttij der Middeleeuwen*. 1919. Reprint, Haarlem: Tjeenk Willink, 1952.
———. *Nederlandse beschaving in de zeventiende eeuw. Een schets*. Haarlem: Tjeenk Willink, 1941.
Hunt, Lynn, ed. *The New Cultural History*. Berkeley: University of California Press, 1989.
Huppert, George. *The Idea of Perfect History: Historical Erudition and Historical Philosophy in Renaissance France*. Urbana: University of Illinois Press, 1970.
Hutchinson, G. Evelyn. *The Ecological Theater and the Evolutionary Play*. New Haven: Yale University Press, 1965.
Hutchison, Keith. "What Happened to Occult Qualities in the Scientific Revolution?" *Isis* 73 (1982): 233–53.
Huygens, Christiaan. *Van rekeningh in spelen van geluck*. New edition in modern Dutch spelling with notes by Wim Kleijne. Utrecht: Epsilon, 1998.
Imhof, Dirk. "Het plantenrijk in beeld." In *Wereldwijs. Wetenschappers rond Keizer Karel*, edited by Geert Vanpaemel and Tineke Padmos, 64–73. Leuven: Davidsfonds, 2000.
Israel, Jonathan. *Radical Enlightenment: Philosophy and the Making of Modernity 1650–1750*. Oxford: Oxford University Press, 2001.
Jacob, James R., and Margaret C. Jacob. "The Anglican Origins of Modern Science," *Isis* 71 (1980): 251–67.
Jacob, Margaret C. *The Cultural Meaning of the Scientific Revolution*. New York: Knopf, 1988.
———. *The Radical Enlightenment: Pantheists, Freemasons, and Republicans*. London: Allen and Unwin, 1981.
———. "Radicalism in the Dutch Enlightenment." In *The Dutch Republic in the Eighteenth Century: Decline, Enlightenment, and Revolution*, edited by Margaret C. Jacob and Wijnand W. Mijnhardt, 224–40. Ithaca: Cornell University Press, 1992.
Jacob, Margaret C., and Dorothée Sturkenboom. "A Women's Scientific Society in the West: The Late Eighteenth-Century Assimilation of Science." *Isis* 94 (2003): 217–52.
Jaeger, Werner, *Aristoteles. Grundlegung einer Geschichte seiner Entwicklung*, Berlijn: Weidmannsche Buchhandlung, 1923.
Jardine, Lisa. *Erasmus, Man of Letters*. Princeton: Princeton University Press, 1993.
Jardine, Nicholas. "The Forging of Modern Realism: Clavius and Kepler against the Sceptics." *Studies in the History and Philosophy of Science* 10 (1979): 141–73.
Jardine, N., J. A. Secord, and E. C. Spary, eds. *Cultures of Natural History*. Cambridge: Cambridge University Press, 1996.
Jasanoff, Sheila. *Designs on Nature: Science and Democracy in Europe and the United States*. Princeton: Princeton University Press, 2005.
———. "(No?) Accounting for Expertise." *Science and Public Policy* 30 (2003): 157–62.
Jong, Erik de, and M. Dominicus-van Soest. *Aardse paradijzen 1. De tuin in de Nederlandse kunst, 15ᵉ tot 18ᵉ eeuw*. Ghent: Snoeck-Ducaju and Zoon, 1996.
Jong, Mechtild de. "Scheidslijnen in het denken over natuurbeheer in Nederland." PhD dissertation, Delft University of Technology, 2002.
Jorink, Erik. *Het Boeck der Natuere. Nederlandse geleerden en de wonderen van Gods Schepping 1575–1715*. Leiden: Primavera Pers, 2006.

Joy, Lynn S. "Epicureanism in Renaissance Moral and Natural Philosophy." *Journal of the History of Ideas* 53 (1992): 573–83.
Jungnickel, Christa, and Russell McCormmach. *Intellectual Mastery of Nature: Theoretical Physics from Ohm to Einstein.* Vol 2: *The Now Mighty Theoretical Physics, 1870–1925.* Chicago: The University of Chicago Press, 1986.
Kal, Victor. *On Intuition and Discursive Reasoning in Aristotle.* Leiden: Brill, 1988.
Karstens, W. K. H., and H. Kleibrink. *De Leidse Hortus.* Zwolle: Waanders, 1982.
Kay, Lily E. *The Molecular Vision of Life.* New York: Oxford University Press, 1993.
———. *Who Wrote the Book of Life: A History of the Genetic Code.* Stanford: Stanford University Press, 2000.
Keller, Evelyn Fox. "Models, Simulation, and 'Computer Experiments.'" In *The Philosophy of Scientific Experimentation,* edited by Hans Radder, 198–215. Pittsburgh: University of Pittsburgh Press, 2003.
———. *Refiguring Life: Metaphors of Twentieth-Century Biology.* New York: Columbia University Press, 1995.
Kempers, B. *Kunst, macht en mecenaat. Het beroep van schilder in sociale verhoudingen, 1250–1600.* Amsterdam: Arbeiderspers, 1987.
Kermode, Frank. "John." In *The Literary Guide to the Bible,* edited by R. Alter and F. Kermode, 440–66. Cambridge, Mass.: Harvard University Press, 1987.
Kevles, Daniel. *The Baltimore Case: A Trial of Politics, Science and Character.* New York: W. W. Norton, 1998.
———. "Big Science and Big Politics in the United States: Reflections on the Death of the SSC and the Life of the Human Genome Project." *Historical Studies in the Physical and Biological Sciences* 27 (1997): 269–99.
———. "Cold War and Hot Physics: Science, Security and the American State, 1945–1956." *Historical Studies in the Physical and Biological Sciences* 20 (1990): 239–64.
———. *In the Name of Eugenics.* Chicago: The University of Chicago Press, 1984.
———. "The National Science Foundation and the Debate over Postwar Research Policy, 1942–1945." *Isis* 68 (1977): 5–26.
———. *The Physicists: The History of a Scientific Community in America.* 1978. Reprint, New York: Vintage, 1979.
Kibre, P., and N. G. Siraisi. "The Institutional Setting: the Universities." In Lindberg, *Science in the Middle Ages,* 1978, 120–44.
Kingsland, Sharon E. "The Battling Botanist: Daniel Trembley MacDougal, Mutation Theory and the Rise of Experimental Evolutionary Biology in America, 1900–1912." *Isis* 82 (1991): 479–510.
Klein, Ursula, and Wolfgang Lefèvre. *Materials in Eighteenth-Century Science: A Historical Ontology.* Cambridge, Mass.: The MIT Press, 2007.
Klever, Wim. *Mannen rond Spinoza.* Hilversum: Verloren, 1997.
Knoeff, Rina. *Herman Boerhaave (1668–1738): Calvinist Chemist and Physician.* Amsterdam: Koninklijke Nederlandse Akademie van Wetenschappen, 2002.
Koerner, Lisbet. "Carl Linnaeus in His Time and Place." In Jardine, Secord, and Spary, *Cultures of Natural History,* 1996, 145–62.
———. *Linnaeus: Nature and Nation.* Cambridge: Cambridge University Press, 1999.
Kohler, Robert E. *Landscapes and Labscapes: Exploring the Lab-Field Border in Biology.* Chicago: The University of Chicago Press, 2002.

———. *Lords of the Fly: Drosophila Genetics and the Experimental Life*. Chicago: University of Chicago Press, 1994.

———. *Partners in Science: Foundations and Natural Scientists 1900–1945*. Chicago: The University of Chicago Press, 1991.

———. "The Ph.D. Machine: Building on the Collegiate Base." *Isis* 81 (1990): 638–62.

Kohlstedt, Sally Gregory. "Curiosities and Cabinets: Natural History Museums and Education on the Antebellum Campus." *Isis* 79 (1988): 405–26.

Kox, Anne. "Hendrik A. Lorentz, 1853–1928." In *Van Stevin tot Lorentz. Portretten van achttien Nederlandse natuurwetenschappers*, edited by Anne Kox, 226–42. Amsterdam: Bert Bakker, 1990.

Koyré, Alexandre. *Etudes Galiléennes*. Paris: Hermann, 1939.

———. *Metaphysics and Measurement: Essays in Scientific Revolution*. London: Chapman and Hall, 1968.

Kragh, Helge. *An Introduction to the Historiography of Science*. Cambridge: Cambridge University Press, 1987.

Kranakis, Eda. *Constructing a Bridge: An Exploration of Engineering Culture, Design, and Research in Nineteenth-Century France and America*. Cambridge, Mass.: The MIT Press, 1996.

———. "Social Determinants of Engineering Practice." *Social Studies of Science* 19 (1989): 5–70.

Krige, John, and Dominique Pestre, eds. *Companion to Science in the Twentieth Century*. 1997. Reprint, London: Routledge, 2003.

Krige, John, and Dominique Pestre. "The How and Why of the Birth of CERN." In *History of CERN*. Vol. 1: *Launching the European Organization for Nuclear Research*, edited by A. Hermann, J. Krige, U. Mersits, and D. Pestre, 523–44. Amsterdam: North-Holland, 1987.

Kristeller, Paul Oskar. *Medieval Aspects of Renaissance Learning*. 1974. Reprint, New York: Columbia University Press, 1992.

Krüger, Lorenz. "The Slow Rise of Probabilism." In Krüger, Daston, and Heidelberger, *Probabilistic Revolution*, vol. 1, 59–89.

Krüger, L., L. Daston, and M. Heidelberger, eds. *The Probabilistic Revolution*. Vol. 1: *Ideas in History*. Cambridge, Mass.: The MIT Press, 1987.

Krüger, Lorenz, Gerd Gigerenzer, and Mary Morgan, eds. *The Probabilistic Revolution*. Volume 2: *Ideas in the Sciences*. Cambridge, Mass.: The MIT Press, 1987.

Kuhn, Thomas. *The Copernican Revolution*. Cambridge, Mass.: Harvard University Press, 1957.

———. "Energy Conservation as an Example of Simultaneous Discovery." In *The Essential Tension*, 66–104. 1959. Reprint, Chicago: Chicago University Press, 1977.

———. "Mathematical Versus Experimental Traditions in the Development of Physical Science." In *The Essential Tension*, 31-65. Chicago: The University of Chicago Press, 1977.

———. *The Structure of Scientific Revolutions*. Chicago: The University of Chicago Press, 1970.

Kwa, Chunglin. "Alexander von Humboldt's Invention of the Natural Landscape." *European Legacy* 10 (2005): 149–62.

———. "Alliances between Styles: A New Model for the Interaction Between Science and Technology." In *Science and Its Recent History: Epochal Break or Business*

as Usual?, edited by Alfred Nordmann, Hans Radder, and Gregor Schiemann, Pittsburgh: University of Pittsburgh Press, 2011.
———. "De retorische structuur van geschiedkundige teksten." *Krisis* 37 (1989): 54–60.
———. "Interdisciplinary Research through Informal Science-Policy Interactions." *Science and Public Policy* 10 (2006): 457–67.
———. "Local Ecologies, Global Science: Discourses and Strategies of the International Geosphere-Biosphere Programme." *Social Studies of Science* 35 (2005): 923–50.
———. "Modeling the Grasslands." *Historical Studies in the Physical and Biological Sciences* 24 (1993): 125–55. Addendum in *HSPS* 25 (1994): 184–86.
———. "Modelling Technologies of Control." *Science as Culture* 4, no. 20 (1994): 363–91.
———."Onderzoek aan de universiteiten. Een geschiedenis van de academische autonomie." *Krisis* 8, no. 4 (2007): 36–51.
———. "Radiation Ecology, Systems Ecology and the Management of the Environment." In *Science and Nature* (BSHS Monograph 8), edited by Michael Shortland, 213–49. Stanford in the Vale: British Society for the History of Science, 1993.
———. "Representations of Nature Mediating between Ecology and Science Policy: The Case of the International Biological Programme." *Social Studies of Science* 17 (1987): 413–42.
———. "Romantic and Baroque Conceptions of Complex Wholes in the Sciences." In *Complexities: Social Studies of Knowledge Practices*, edited by John Law and Annemarie Mol, 23–52. Durham: Duke University Press, 2002.
———. "Shifts in Dominance of Scientific Styles: From Modernity to Postmodernity." *History and Technology* 23 (2007): 166–75.
Kwa, Chunglin, and René Rector. "A Data Bias in Interdisciplinary Cooperation in the Sciences: Ecology in Climate Change Research." In *Collaboration in the New Life Sciences*, edited by John Parker, Niki Vermeulen, and Bart Penders, 161–76. London: Ashgate, 2010.
LaCapra, Dominick. *Rethinking Intellectual History: Texts, Contexts, Language*. Ithaca: Cornell University Press, 1983.
Laet, Marianne de. "Vaders, sterren en hormonen." PhD dissertation, Universiteit Utrecht, 1994.
Lahsen, Myanna. "Technocracy, Democracy, and U.S. Climate Politics: The Need for Demarcations." *Science, Technology and Human Values* 30 (2005): 137–69.
Laissus, Yves. *Le Muséum national d'histoire naturelle*. Paris: Gallimard, 1995.
Langbein, John. *Prosecuting Crime in the Renaissance: England, Germany, France*. Cambridge, Mass.: Harvard University Press, 1974.
Laqueur, Thomas. *Making Sex: Body and Gender from the Greeks to Freud*. Cambridge, Mass.: Harvard University Press, 1990.
———. "Sex in the Flesh." *Isis* 94 (2003): 300–306.
Larson, James L. *Reason and Experience: The Representation of Natural Order in the Work of Carl von Linné*. Berkeley: University of California Press, 1971.
Latour, Bruno. *Science in Action*. Milton Keynes: Open University Press, 1987.
LeGoff, Jacques. *Les intellectuels au Moyen Age*. 1956. Reprint, Paris: Le Seuil, 1976.
Lepenies, Wolf. *Das Ende der Naturgeschichte*. 1976. Reprint, Frankfurt: Suhrkamp Verlag, 1978.

Leslie, Stuart W. *The Cold War and American Science: The Military-Industrial-Academic Complex at MIT and Stanford*. New York: Columbia University Press, 1993.
Levere, Trevor H. *Poetry Realized in Nature: Samuel Taylor Coleridge and Early Nineteenth-Century Science*. Cambridge: Cambridge University Press, 1981.
Lewontin, Richard. "Gene, Organism, and Environment." In *Evolution from Molecules to Men*, edited by D. S. Bendall, 151–70. Cambridge: Cambridge University Press, 1983.
Lindberg, David C. *The Beginnings of Western Science*. Chicago: The University of Chicago Press, 1992.
———, ed. *Roger Bacon's Philosophy of Nature*. Oxford: Oxford University Press, 1983.
———. "Science as Handmaiden: Roger Bacon and the Patristic Tradition." *Isis* 78 (1987): 518–36.
———, ed. *Science in the Middle Ages*. Chicago: The University of Chicago Press, 1978.
———. "Science of Optics." In Lindberg, *Science in the Middle Ages*, 338–68.
———. "The Transmission of Greek and Arabic Learning to the West." In Lindberg, *Science in the Middle Ages*, 1978, 52–90.
Lindberg David C., and Ronald L. Numbers, eds. *God and Nature: Historical Essays on the Encounter between Christianity and Science*. Berkeley: University of California Press, 1986.
Lindberg, David C., and Robert S. Westman, eds. *Reappraisals of the Scientific Revolution*. Cambridge: Cambridge University Press, 1990.
Lintsen, H. W., ed. *Geschiedenis van de techniek in Nederland. De wording van een moderne samenleving 1800–1890*. Vols. 4 and 6. Zutphen: Walburg Pers, 1993.
Lloyd, A. C. *The Anatomy of Neoplatonism*. Oxford: Oxford University Press, 1990.
Lloyd, G. E. R. *The Ambitions of Curiosity: Understanding the World in Ancient Greece and China*. Cambridge: Cambridge University Press, 2002.
———. *Greek Science after Aristotle*. New York: Norton, 1973.
Logan, Oliver. *Culture and Society in Venice, 1470–1790: The Renaissance and Its Heritage*. New York: Scribners, 1972.
Long, Pamela. "The Contribution of Architectural Writers to a 'Scientific' Outlook in the Fifteenth and Sixteenth Centuries." *Journal of Medieval and Renaissance Studies* 15 (1985): 265–98.
———. "Power, Patronage, and the Authorship of Ars: From Mechanical Know-How to Mechanical Knowledge in the Last Scribal Age." *Isis* 88 (1997): 1–41.
Lovejoy, Arthur. *The Great Chain of Being*. Cambridge, Mass.: Harvard University Press, 1936.
Low, Morris, ed. *Beyond Joseph Needham: Science, Technology, and Medicine in East and South East Asia*. *Osiris* 13, special issue (1998).
Lunteren, Frans van. "Framing Hypotheses: Conceptions of Gravity in the 18th and 19th Centuries." PhD dissertation, Rijksuniversiteit Utrecht, 1991.
Lyotard, Jean-François. *La condition postmoderne: Rapport sur le savoir*. Paris: Minuit, 1979.
MacKenzie, Donald. *Statistics in Britain 1865–1930: The Social Construction of Scientific Knowledge*. Edinburgh: Edinburgh University Press, 1981.
Maclean, Ian. "The Interpretation of Natural Signs." In Vickers, *Occult and Scientific Mentalities in the Renaissance*, 231–52.
Mahoney, Michael S. "The Beginnings of Algebraic Thought in the Seventeenth Century." In Gaukroger, *Descartes: Philosophy, Mathematics, and Physics*, 141–55.

———. "Mathematics." In Lindberg, *Science in the Middle Ages*, 145–78.
Malinowski, Bronislaw. *Magic, Science, and Religion, and Other Essays*. New York: Doubleday, 1954.
Mannheim, Karl. *Ideology and Utopia: An Introduction to the Sociology of Knowledge*. 1936. Reprint, London: Routledge and Kegan Paul, 1976.
Martin, Julian. *Francis Bacon, the State and the Reform of Natural Philosophy*. Cambridge: Cambridge University Press, 1992.
Mason, Stephen S. *Main Currents of Scientific Thought*. New York: Henry Schuman, 1953.
Matagne, Patrick. *Aux origines de l'écologie: Les naturalistes en France de 1800 à 1914*. Paris: Éditions du CHTS, 1999.
Mayr, Ernst. *The Growth of Biological Thought*. Cambridge, Mass.: Harvard University Press, 1982.
Mayr, Otto. *Authority, Liberty, and Automatic Machinery in Early Modern Europe*. Baltimore: Johns Hopkins University Press, 1986.
McAllister, James W. *Beauty and Revolution in Science*. Ithaca: Cornell University Press, 1996.
McClelland, Charles. *State, Society, and University in Germany 1700–1914*. Cambridge: Cambridge University Press, 1980.
McEvoy, James. *The Philosophy of Robert Grosseteste*. Oxford: Oxford University Press, 1982.
McMullin, Ernan. "Conceptions of Science in the Scientific Revolution." In Lindberg and Westman, *Reappraisals of the Scientific Revolution*, 27–92.
McVaugh, Michael R. *Medicine before the Plague: Practitioners and Their Patients in the Crown of Aragon*. Cambridge: Cambridge University Press, 1993.
Meijer, Onno. "Hugo de Vries no Mendelian?" *Annals of Science* 42 (1985): 189–232.
Meijers, Debora J. *Kunst als natuur. De Habsburgse schilderijengalerij in Wenen omstreeks 1780*. Amsterdam: SUA, 1991.
Merchant, Carolyn. *The Death of Nature*. San Francisco: Harper and Row, 1980.
———. "Secrets of Nature: The Bacon Debates Revisited." *Journal of the History of Ideas* 69 (2008): 147–62.
Merton, Robert. *Science, Technology, and Society in Seventeenth Century England*. 1938. Reprint, New York: Harper and Row, 1970.
Mirowski, Philip. *More Heat than Light: Economics as Social Physics, Physics as Nature's Economics*. Cambridge: Cambridge University Press, 1989.
Mirowski, Philip, and Esther-Mirjam Sent, eds. *Science Bought and Sold: Essays in the Economics of Science*. Chicago: The University of Chicago Press, 2002.
Mirowski, Philip, and Robert Van Horn. "The Contract Research Organization and the Commercialization of Scientific Research." *Social Studies of Science* 35 (2005): 503–48.
Misa, Thomas. *Leonardo to the Internet: Technology and Culture from the Renaissance to the Present*. Baltimore: Johns Hopkins University Press, 2004.
Morange, Michel. *The Misunderstood Gene*. Cambridge, Mass.: Harvard University Press, 2001.
———. "What History Tells Us XIII: Fifty Years of the Central Dogma." *Journal of Bioscience* 33 (2008): 171–75.
Morton, A. G. *History of Botanical Science*. London: Academic Press, 1981.

Murdoch, John D., and Edith D. Sylla. "The Science of Motion." In Lindberg, *Science in the Middle Ages*, 206–64.

Needham, Joseph. *Clerks and Craftsmen in China and the West*. Cambridge: Cambridge University Press, 1970.

———. *Human Law and the Laws of Nature in China and the West*. London: Oxford University Press, 1951.

Newman, William R. *Atoms and Alchemy: Chymistry and the Experimental Origin of the Scientific Revolution*. Chicago: The University of Chicago Press, 2006.

———. *Promethean Ambitions: Alchemy and the Quest to Perfect Nature*. Chicago: The University of Chicago Press, 2004.

Newman, William R., and Anthony Grafton. "Introduction: The Problematic Status of Astrology and Alchemy in Premodern Europe." In *Secrets of Nature: Astrology and Alchemy in Early Modern Europe*, edited by William R. Newman and Anthony Grafton, 1–38. Cambridge, Mass.: The MIT Press, 2001.

Newman, William R., and Lawrence M. Principe. *Alchemy Tried in the Fire: Starkey, Boyle, and the Fate of Helmontian Chymistry*. Chicago: The University of Chicago Press, 2002.

Nietzsche, Friedrich. Über Wahrheit und Lüge in aussermoralischen Sinn. In *Werke in drei Bänden, dritter Band*, edited by Karl Schlechta. Munich: Carl Hanser Verlag, 1963.

Noble, David F. *America by Design: Science, Technology and the Rise of Corporate Capitalism*. New York: Knopf, 1977.

———. *A World Without Women: The Christian Clerical Culture of Western Science*. New York: Knopf, 1992.

Nordenskiöld, Erich. *The History of Biology*. New York: Alfred A. Knopf, 1928. Reprint, New York: Tudor, 1946.

Nussbaum, M. *The Fragility of Goodness: Luck and Ethics in Greek Tragedy and Philosophy*. New York: Cambridge University Press, 1986.

Nye, Mary Jo. *Before Big Science: The Pursuit of Modern Chemistry and Physics, 1800–1940*. 1996. Reprint, Cambridge, Mass.: Harvard University Press, 1999.

———. *From Chemical Philosophy to Theoretical Chemistry: Dynamics of Matter and Dynamics of Disciplines 1800–1950*. Berkeley: University of California Press, 1993.

Odling-Smee, F. John, Kevin N. Laland, and Marcus W. Feldman. *Niche Construction*. Princeton: Princeton University Press, 2003.

O'Hanlon, Redmond. *In Trouble Again*. London: Hamish Hamilton, 1988.

Olby, R. "Mendel no Mendelian?" *History of Science* 17 (1979): 53–72, reprinted as appendix in Olby, *Origins of Mendelism*.

———. *Origins of Mendelism*. Chicago: The University of Chicago Press, 1986.

Osler, Margaret J. *Divine Will and the Mechanical Philosophy: Gassendi and Descartes on Contingency and Necessity in the Created World*. Cambridge: Cambridge University Press, 1994.

Outram, Dorinda. "New Spaces in Natural History." In Jardine, Secord, and Spary, *Cultures of Natural History*, 249–65.

Pagden, Anthony. *European Encounters with the New World*. New Haven: Yale University Press, 1993.

Pagel, Walter. *Joan Baptista van Helmont: Reformer of Science and Medicine*. Cambridge: Cambridge University Press, 1982.

———. *New Light on William Harvey*. Basel: S. Karger, 1976.

———. *Paracelsus: An Introduction to Philosophical Medicine in the Era of the Renaissance*. 2nd ed. Basel: S. Karger, 1982.
———. *William Harvey's Biological Ideas*. Basel: S. Karger, 1967.
Panofsky, Erwin. *Gothic Architecture and Scholasticism*. Latrobe, Pa.: The Archabbey Press, 1951.
Pardo Tomas, José. "Two Glimpses from America from a Distance: Carolus Clusius and Nicolás Monardes." In *Carolus Clusius: Toward a Cultural History of a Renaissance Naturalist*, edited by Florike Egmond, Paul Hoftijzer, and Robert Visser, 173–94. Amsterdam: Koninklijke Nederlandse Akademie van Wetenschappen, 2007.
Park, Katherine. "Women, Gender, and Utopia: *The Death of Nature* and the Historiography of Early Modern Science." *Isis* 97 (2006): 487–95.
Pascal, Blaise. *Oeuvres Complètes*. Paris: Gallimard (Pléiade), 1954.
Pater, Ben de, and Herman van der Wusten. *Het geografische huis. De opbouw van een wetenschap*. Muiderberg: Coutinho, 1991.
Pater, C. De. "Petrus van Musschenbroek (1692–1761). Een Newtoniaans natuuronderzoeker." PhD dissertation, Rijksuniversiteit Utrecht, 1979.
Paul, Harry W. *From Knowledge to Power: The Rise of the Science Empire in France, 1860–1939*. Cambridge: Cambridge University Press, 1985.
Pavord, Anna. *The Tulip*. London: Bloomsbury, 1999.
Pekelharing, Pieter. "Reflexiviteit, scepticisme en vertrouwen." *Krisis* 52 (September 1993): 33–44.
Pels, Dick. "Karl Mannheim and the Sociology of Scientific Knowledge: Toward a New Agenda." *Sociological Theory* 14 (1996): 30–48.
Pesic, Peter. "Wrestling with Proteus: Francis Bacon and the 'Torture' of Nature." *Isis* 90 (1999): 81–94.
Pestre, Dominique. *Science, argent et politique: un essai d'interprétation*. Paris: Institut National de la Recherche Agronomique, 2003.
———. "Science, Political Power, and the State." In Krige and Pestre, *Companion to Science*, 61–76.
Pickering, Andrew. *The Mangle of Practice: Time, Agency and Science*. Chicago: Chicago University Press, 1995.
Pickstone, John. *Ways of Knowing: A New History of Science, Technology and Medicine*. Manchester: Manchester University Press, 2000.
Pieters, Florence. "Het schatrijke naturaliënkabinet van Stadhouder Willem V onder directoraat van topverzamelaar Arnout Vosmaer." In Sliggers and Besselink, *Het verdwenen museum*, 19–44.
Pinto-Correia, Clara. *The Ovary of Eve: Egg and Sperm and Preformation*. Chicago: The University of Chicago Press, 1997.
Pomian, Krzystof. *Collectors and Curiosities: Paris and Venice, 1500–1800*. Cambridge: Polity Press, 1990.
Popkin, Richard H. *The History of Scepticism from Erasmus to Descartes*. Assen: Van Gorcum, 1960.
———. "The Philosophy of the Royal Society of England." In *The Pimlico History of Western Philosophy*, 358–63. London: Pimlico, 1999.
Popper, Karl. "Darwinism as a Metaphysical Research Programme." In *The Philosophy of Karl Popper*, edited by P. A. Schillp, 133–43. LaSalle, Il.: Open Court, 1974.
———. *The Logic of Scientific Discovery*. London: Hutchinson, 1959.

Porter, Theodore. *The Rise of Statistical Thinking, 1820–1900*. Princeton: Princeton University Press, 1986.

———. *Trust in Numbers: The Pursuit of Objectivity in Science and Public Life*. Princeton: Princeton University Press, 1995.

Prest, John. *The Garden of Eden*. New Haven: Yale University Press, 1981.

Pribram, Karl. "From Metaphors to Models: The Use of Analogy in Neuropsychology." In *Metaphors in the History of Psychology*, edited by David E. Leary, 79–103. Cambridge: Cambridge University Press, 1990.

Price, Don K. *The Scientific Estate*. Cambridge, Mass.: Harvard University Press, 1965.

Prigogine, Ilya, and Isabelle Stengers. *Entre le temps et l'éternité*. Paris: Fayard, 1988.

Principe, Lawrence M., and Lloyd DeWitt. *Transmutations: Alchemy in Art*. Philadelphia: Chemical Heritage Foundation, 2002.

Provine, William. *The Origins of Theoretical Population Genetics*. Chicago: The University of Chicago Press, 1971.

Quispel, Gilles. *Valentinus de gnosticus en zijn Evangelie der Waarheid*. Amsterdam: In de Pelikaan, 2003.

Rabinow, Paul. *Making PCR: A Story of Biotechnology*. Chicago: The University of Chicago Press, 1996.

Radder, Hans. "Exploiting Abstract Possibilities: A Critique of the Concept and Practice of Product Patenting." *Journal of Agricultural and Environmental Ethics* 17 (2004): 275–91.

———. "Kramers and the Forman Theses." *History of Science* 21 (1983): 165–82.

———. *The Material Realization of Science*. Assen: Van Gorcum, 1988.

———. *The World Observed/ The World Conceived*. Pittsburgh: University of Pittsburgh Press, 2006.

Rademaker, C. S. M. *Leven en werk van Gerardus Johannes Vossius (1577–1649)*. Hilversum: Verloren, 1999.

Raphael, Lutz. *Die Erben von Bloch und Febvre: Annales-Geschichtsschreibung und nouvelle histoire in Frankreich 1945–1980*. Stuttgart: Klett-Cotta, 1994.

Redondi, Pietro. *Galileo Heretic*. Princeton: Princeton University Press, 1987.

Reeds, Karen Meier. *Botany in Medieval and Renaissance Universities*. New York: Garland, 1991.

Rees, Graham. "Francis Bacon's Biological Ideas: A New Manuscript Source." In Vickers, *Occult and Scientific Mentalities in the Renaissance*, 297–314.

Rheinberger, Hans-Jörg. *Toward a History of Epistemic Things: Synthesizing Proteins in the Test Tube*. Stanford: Stanford University Press, 1997.

Ridley, Mark. *Evolution and Classification: The Reformation of Cladism*. New York: Longman, 1986.

Rip, Arie. "Science for the 21st Century." In *The Future of the Sciences and the Humanities*, edited by P. Tindemans, A. Verrijn-Stuart, and R. Visser, 99–148. Amsterdam: Amsterdam University Press, 2002.

Ripa, Cesare. *Baroque and Rococo Pictorial Imagery: The 1758–60 Hertel Edition of Ripa's Iconologia*. New York: Dover, 1971.

Rist, John M. *The Mind of Aristotle: A Study in Philosophical Growth*, Toronto: University of Toronto Press, 1989.

Ritvo, Harriet. *The Platypus and the Mermaid and Other Figments of the Classifying Imagination*. Cambridge, Mass.: Harvard University Press, 1997.

Roberts, Lissa. "Going Dutch: A Cultural History of Dutch Science in the Eighteenth Century." In *The Sciences in Enlightened Europe*, edited by William Clark, Jan Golinski, and Simon Schaffer, 350–88. Chicago: The University of Chicago Press, 1999.

Roberts, Lissa, Simon Schaffer, and Peter Dear, eds. *The Mindful Hand: Inquiry and Invention from the Late Renaissance to Early Industrialisation*. Amsterdam: Koninklijke Nederlandse Akademie van Wetenschappen, 2007.

Roger, Jacques. *Les sciences de la vie dans la pensée française du xviiie siècle*. Paris: Armand Colin, 1971.

Rogers, John. *The Power of Revolution*. Ithaca: Cornell University Press, 1996.

Ronan, Colin A. *The Shorter Science and Civilization in China*. Vols. 1–5. Cambridge: Cambridge University Press, 1978.

Rorty, Richard. *Philosophy and the Mirror of Nature*. Princeton: Princeton University Press, 1979.

Rose, Paul. *The Italian Renaissance of Mathematics*. Geneva: Librairie Droz, 1975.

Rosen, Edward. "Kepler's Attitude Towards Astrology and Mysticism." In Vickers, *Occult and Scientific Mentalities*, 253–72.

Rosenberg, Alexander. *The Structure of Biological Science*. New York: Cambridge University Press, 1985.

Rosenberg, Charles. "Towards an Ecology of Knowledge." In *The Organization of Knowledge in Modern America, 1860–1920*, edited by Alexandra Oleson and John Voss, 440–55. Baltimore: Johns Hopkins University Press, 1979.

Rossi, Paolo. *Francis Bacon: From Magic to Science*. London: Routledge and Kegan Paul, 1968.

Ruestow, Edward. *Physics at Seventeenth- and Eighteenth-Century Leiden: Philosophy and the New Science in the University*. The Hague: Martinus Nijhoff, 1973.

Saliba, George. *Islamic Science and the Making of the European Renaissance*. Cambridge, Mass.: The MIT Press, 2007.

Sambursky, S. *The Physical World of the Greeks*. London: Routledge & Kegan Paul, 1956.

Sarasohn, Lisa. *Gassendi's Ethics*. Ithaca: Cornell University Press, 1996.

———. "Nicolas Claude Fabri de Peiresc and the Patronage of the New Science in the Seventeenth Century." *Isis* 84 (1993): 70–90.

Sassoon, Donald. *Mona Lisa: The History of the World's Most Famous Painting*, London: HarperCollins, 2001.

Schaffer, Simon. "'The Charter'd Thames': Naval Architecture and Experimental Spaces in Georgian Britain." In Roberts, Schaffer, and Dear, *The Mindful Hand*, 279–305.

———. "Enlightened Automata." In *Enlightened Europe*, edited by William Clark, Jan Golinski, and Simon Schaffer, 126–65. Chicago: The University of Chicago Press, 1999.

———. "Experimenters' Techniques, Dyers' Hands, and the Electric Planetarium." *Isis* 88 (1997): 456–83.

———. "Glass Works: Newton's Prisms and the Uses of Experiments." In *The Uses of Experiment: Studies in the Natural Sciences*, edited by David Gooding, Trevor Pinch, and Simon Schaffer, 67–104. Cambridge: Cambridge University Press, 1989.

———. "The Glorious Revolution and Medicine in Britain and the Netherlands." *Notes and Records of the Royal Society of London* 43 (1989): 167–90.

Schama, Simon. *The Embarrassment of Riches: An Interpretation of Dutch Culture in the Golden Age*. New York: Knopf, 1987.
Schiebinger, Londa. *The Mind Has No Sex? Women in the Origins of Modern Science*. Cambridge, Mass.: Harvard University Press, 1989.
———. *Nature's Body: Gender in the Making of Modern Science*. Boston: Beacon Press, 1993.
———. "Skelettestreit." *Isis* 94 (2003): 307–13.
Schmidt-Biggemann, Wilhelm. *Topica universalis*. Hamburg: Felix Meiner Verlag, 1983.
Schmitt, Charles B. *Aristotle in the Renaissance*. Cambridge, Mass.: Harvard University Press, 1983.
Schnädelbach, Herbert. *Philosophy in Germany 1831–1933*. Cambridge: Cambridge University Press, 1984.
Schofield, Robert A. "An Evolutionary Taxonomy of Eighteenth-Century Newtonianisms." *Studies in 18th Century Culture* 7 (1978): 175–92.
———. *Mechanism and Materialism: British Natural Philosophy in an Age of Reason*. Princeton: Princeton University Press, 1970.
Schouten, J. *Johannes Walaeus: Zijn betekenis voor de verbreiding van de leer van de bloedsomloop*. Assen: Van Gorcum, 1972.
Schrödinger, Erwin. *What Is Life?* 1944. Reprinted in *What Is Life and Mind and Matter*. Cambridge: Cambridge University Press, 1967.
Schweber, S. S. "From 'Elementary' to 'Fundamental' Particles." In Krige and Pestre, *Companion to Science*, 599–616.
Scott, Peter. *The Crisis of the University*. London: Croom Helm, 1984.
Serres, Michel. *La naissance de la physique dans le texte de Lucrèce*. Paris: Minuit, 1977.
Servos, John. "The Industrial Relations of Science: Chemical Engineering at MIT, 1900–1939." *Isis* 71 (1980): 531–49.
Shapin, Steven. "The House of Experiment in Seventeenth-Century England." *Isis* 79 (1988): 373–404.
———. "Ivory Trade." *London Review of Books*, September 11, 2003.
———. "Of Gods and Kings: Natural Philosophy and Politics in the Leibniz-Clarke Dispute." *Isis* 72 (1981): 187–215.
———. "Rough Trade." *London Review of Books*, March 6, 2003.
———. *The Scientific Life: A Moral History of a Late Modern Vocation*. Chicago: The University of Chicago Press, 2008.
———. *The Scientific Revolution*, Chicago: The University of Chicago Press, 1996.
———. "Understanding the Merton Thesis." *Isis* 79 (1988): 594–605.
Shapin, Steven, and Simon Schaffer. *Leviathan and the Air Pump*. Princeton: Princeton University Press, 1985.
Shapiro, Barbara. *Probability and Certainty in Seventeenth-Century England*. Princeton: Princeton University Press, 1983.
Shea, William. "Galileo and the Church." In Lindberg and Numbers, *God and Nature*, 114–35.
Shearman, John. *Mannerism*. Harmondsworth: Penguin, 1967.
Shinn, Terry. *L'École Polytechnique 1794–1914. Savoir scientifique et pouvoir social*. Paris: Presses de la fondation nationale des sciences politiques, 1980.
Shumaker, Wayne. *The Occult Sciences in the Renaissance*. Berkeley: University of California Press, 1972.

Sibum, Heinz Otto. "Reworking the Mechanical Value of Heat: Instruments of Precision and Gestures of Accuracy in Early Victorian England." *Studies in the History and Philosophy of Science* 26 (1995): 73–106.
Siraisi, Nancy G. *The Clock and the Mirror: Girolamo Cardano and Renaissance Medicine*. Princeton: Princeton University Press, 1997.
———. *Medieval and Renaissance Medicine*. Chicago: The University of Chicago Press, 1990.
Slaughter, M. M. *Universal Languages and Scientific Taxonomy in the Seventeenth Century*. Cambridge: Cambridge University Press, 1982.
Slaughter, Sheila, and Gary Rhoades. "The Emergence of a Competitiveness Research and Development Policy Coalition and the Commercialization of Academic Science and Technology." *Science, Technology, and Human Values* 21 (1996): 303–39.
Slicher van Bath, B. H. *De agrarische geschiedenis van West-Europa (500–1850)*. Utrecht: Spectrum, 1960.
Sliggers, B. C., and M. H. Besselink. *Het verdwenen museum. Natuurhistorische verzamelingen 1750–1850*. Blaricum/Haarlem: V+K Publishing/Teylers Museum, 2002.
Sloan, Phillip R. "The Buffon-Linnaeus Controversy." *Isis* 67 (1976): 356–75.
Smith, Bernard. *European Vision and the South Pacific*. 2nd ed. New Haven: Yale University Press, 1985.
Smith, Crosbie, and M. Norton Wise. *Energy & Empire: A Biographical Study of Lord Kelvin*. Cambridge: Cambridge University Press, 1989.
Smith, Pamela. *The Business of Alchemy*. Princeton: Princeton University Press, 1994.
Snobelen, Stephen D. "Isaac Newton, Heretic: The Strategies of a Nicodemite." *British Journal for the History of Science* 32 (1999): 381–419.
Solomon, H. M. *Public Welfare, Science and Propaganda in Seventeenth Century France*. Princeton: Princeton University Press, 1972.
Spary, E. C. *Utopia's Garden: French Natural History from Old Regime to Revolution*. Chicago: The University of Chicago Press, 2000.
Stafford, Barbara M. *Body Criticism: Imaging the Unseen in Enlightenment Art and Medicine*. Cambridge, Mass.: The MIT Press, 1991.
Stafleu, Frans. *Linnaeus and the Linnaeans: The Spreading of Their Ideas in Systematic Botany, 1735–1789*. Utrecht: Oosthoek, 1971.
Stamhuis, Ida. *"Cijfers en Aequaties" en "Kennis der Staatskrachten". Statistiek in Nederland in de 19^e eeuw*. Amsterdam: Rodopi, 1989.
———. "Radeloos, redeloos, noch reddeloos: Jan de Witts lijfrenteberekeningen rond het Rampjaar." *Nieuw Archief voor Wiskunde*, fourth series, vol. 17 (1999): 439–52.
———. "An Unbridgeable Gap Between Two Approaches to Statistics." In *The Statistical Mind in a Pre-Statistical Era: The Netherlands 1750–1850*, edited by Paul Klep and Ida Stamhuis, 71–98. Amsterdam: Aksant, 2002.
Stamhuis, Ida, Onno Meijer, and Erik Zevenhuizen. "Hugo de Vries on Heredity: Statistics, Mendelian Laws, Pangenes, Mutations." *Isis* 90 (1999): 238–67.
"Statement of the Sizes of Men in Different Counties of Scotland, Taken from the Local Militia." *Edinburgh Medical and Surgical Journal* 13 (1917): 260–64.
Steenberghen, Fernand van. *Aristotle in the West*. Leuven: Nauwelaerts, 1954.
Stevens, Peter F. *The Development of Biological Systematics: Antoine-Laurent de Jussieu, and the Natural System*. New York: Columbia University Press, 1994.

Stolberg, Michael. "A Woman Down to Her Bones: The Anatomy of Sexual Difference in the Sixteenth and Early Seventeenth Century." *Isis* 94 (2003): 274–99.
Stroup, Alice. *A Company of Scientists*. Berkeley: University of California Press, 1990.
Struik, Dirk. *Geschiedenis van de wiskunde*. Amsterdam: Sua, 1977.
———. "Further Thoughts on Merton in Context." *Science in Context* 3 (1989): 227–38.
Takacs, David. *The Idea of Biodiversity: Philosophies of Paradise*. Baltimore: Johns Hopkins University Press, 1996.
Talbot, Charles. "Medicine." In Lindberg, *Science in the Middle Ages*, 391–428.
Tauber, Alfred I., and Leon Chernyak. *Metchnikoff and the Origins of Immunology: From Metaphor to Theory*. Oxford: Oxford University Press, 1991.
Taylor, Charles. *Sources of the Self*. Cambridge: Cambridge University Press, 1989.
Taylor, Peter. "Historical Versus Selectionist Explanations in Evolutionary Biology." *Cladistics* 3 (1987): 1–13.
Terrall, Mary. "Representing the Earth's Shape: The Polemics Surrounding Maupertuis' Expedition to Lapland." *Isis* 83 (1992): 218–37.
Thackray, Arnold. "Natural Knowledge in Cultural Context: The Manchester Model." *American Historical Review* 79 (1974): 672–709.
Theunissen, B. "Closing the Door on Hugo de Vries' Mendelism." *Annals of Science* 51 (1994): 225–48.
———. "Natuursport en levensgeluk. Hugo de Vries, Eli Heimans en Jac. P. Thijsse." *Gewina* 16 (1993): 287–307.
Theunissen, B., and R. P. W. Visser. *De wetten van het leven: Historische grondslagen van de biologie 1750–1950*. Baarn: Ambo, 1996.
Theunisz, J. *Carolus Clusius. Het merkwaardige leven van een pionier der wetenschap*. Amsterdam: Van Kampen, 1939.
Thijssen, J. M. "Bishop Tempier's Condemnation." In *Texts and Contexts in Ancient and Medieval Science*, edited by E. Sylla and M. McVaugh, 84–105. Leiden: Brill, 1997.
Thomson, William, and P. G. Tait. *Treatise on Natural Philosophy*. Oxford: 1867.
Toulmin, Stephen. *Cosmopolis: The Hidden Agenda of Modernity*. New York: The Free Press, 1990.
Truesdell, C. *Essays in the History of Mechanics*. New York: Springer, 1968.
Turner, John. "Random Genetic Drift, R. A. Fisher, and the Oxford School of Ecological Genetics." In Krüger, Gigerenzer, and Morgan, *Probabilistic Revolution*, vol. 2, 313–54.
Turner, R. Steven. "Justus Liebig versus Prussian Chemistry: Reflections on Early Institute-Building in Germany." *Historical Studies in the Physical Sciences* 13 (1982): 129–62.
Underwood, E. Ashworth. *Boerhaave's Men at Leyden and After*. Edinburgh: Edinburgh University Press, 1977.
Vermij, Rienk. *The Calvinist Copernicans*. Amsterdam: Koninklijke Nederlandse Akademie van Wetenschappen, 2002.
———. *Secularisering en natuurwetenschap in de zeventiende en achttiende eeuw: Bernard Nieuwentijt*. Amsterdam: Rodopi, 1991.
Vet, Paul van der. *The Aborted Takeover of Chemistry by Physics*. PhD dissertation, University of Amsterdam, 1987.

Vickers, Brian, ed. *English Science, Bacon to Newton.* Cambridge: Cambridge University Press, 1987.
———, ed. *Occult and Scientific Mentalities in the Renaissance.* Cambridge: Cambridge University Press, 1984.
Vincenti, Walter. *What Engineers Know and How They Know It: Analytical Studies from Aeronautical History.* Baltimore: Johns Hopkins University Press, 1990.
Visser, Rob. "Het Rijksmuseum van Natuurlijke Historie in de 19de eeuw." In Sliggers and Besselink, *Het verdwenen museum,* 175–85.
Vondel, Joost van den. "De Beemster." In *Volledige dichtwerken en oorspronkelijk proza.* Amsterdam: Becht, 1937.
Wachelder, Joseph. "Mirrors: Truth and Error." In *Theories, Technologies, Instrumentalities of Color,* edited by Barbara Saunders and Jaap van Brakel, 215–31. Lanham: University Press of America, 2002.
Wallace, William A. *Galileo and His Sources.* Princeton: Princeton University Press, 1984.
Wang, Jessica. *American Science in an Age of Anxiety: Scientists, Anti-Communism, & the Cold War.* Chapel Hill: The University of North Carolina Press, 1999.
Webster, Charles. *From Paracelsus to Newton: Magic and the Making of Modern Science,* Cambridge: Cambridge University Press, 1982.
———. *The Great Instauration: Science, Medicine and Reform 1626–1660.* London: Duckworth, 1975.
Wessely, Anna. "Transposing 'Style' from the History of Art to the History of Science." *Science in Context* 4 (1991): 265–78.
Westfall, Richard, *The Construction of Modern Science: Mechanisms and Mechanics.* 1971. Reprint, Cambridge: Cambridge University Press, 1977.
———. *Never at Rest: A Biography of Isaac Newton,* Cambridge: Cambridge University Press, 1980.
———. "Newton and Alchemy." In Vickers, *Occult and Scientific Mentalities,* 315–35.
Westman, Robert S. "The Copernicans and the Churches." In Lindberg and Numbers, *God and Nature,* 76–113.
———. "Magical Reform and Astronomical Reform: The Yates Thesis Revisited." In Westman and McGuire, *Hermeticism and the Scientific Revolution,* 5–91.
Westman, Robert S., and J. E. McGuire, ed. *Hermeticism and the Scientific Revolution.* Los Angeles: Williams Andrews Clark Memorial Library, 1977.
White, Hayden. *Metahistory: The Historical Imagination in Nineteenth-Century Europe.* Baltimore: Johns Hopkins University Press, 1973.
Wieland, Wolfgang, *Die aristotelische Physik.* Göttingen: Vandenhoeck & Ruprecht, 1962.
Wiesenfeldt, Gerhard. *Leerer Raum in Minervas Haus. Experimentelle Naturlehre an der Universität Leiden, 1675–1715.* Amsterdam: Koninklijke Nederlandse Akademie van Wetenschappen, 2002.
Wijnands, D. O., E. J. A. Zevenhuizen, and J. Heniger. *Een sieraad voor de stad. De Amsterdamse Hortus Botanicus 1638–1993.* Amsterdam: Amsterdam University Press, 1994.
Williams, Mary. "Deducing the Consequences of Evolution." *Journal of Theoretical Biology* 29 (1970): 343–85.
Wilson, Catherine. *The Invisible World: Early Modern Philosophy and the Invention of the Microscope.* Princeton: Princeton University Press, 1995.

Wise, M. Norton. "Work and Waste I." *History of Science* 27 (1989): 263–301.

———. "Work and Waste II." *History of Science* 27 (1989): 391–449.

Wittrock, Björn. "Dinosaurs or Dolphins? Rise and Resurgence of the Research-Oriented University." In *The University Research System: The Public Policy of the Home of the Scientists*, edited by B. Wittrock and A. Elzinga, 13–37. Stockholm: Almqvist & Wiksell International, 1985.

Worster, Donald. *Nature's Economy: A History of Ecological Ideas*. New York: Cambridge University Press, 1977.

Yates, Frances. *Giordano Bruno and the Hermetic Tradition*. Chicago: The University of Chicago Press, 1964.

———. "The Italian Academies." In *Renaissance and Reform: The Italian Contribution*, by Frances Yates, 6–29. 1949. Reprint, London: Routledge and Kegan Paul, 1983.

———. *The Occult Philosophy in the Elizabethan Age*. London: Routledge, 1979.

Yeo, Eileen Janes. "Social Surveys in Eighteenth and Nineteenth Centuries." In *The Modern Social Sciences*, edited by Theodore Porter and Dorothy Ross, 83–99. Vol. 7 of *History of Science*. Cambridge: Cambridge University Press, 2003.

Yoder, Joella. *Unrolling Time: Christiaan Huygens and the Mathematization of Nature*. Cambridge: Cambridge University Press, 1988.

Young, Robert M. *Darwin's Metaphor: Nature's Place in Victorian Culture*. Cambridge: Cambridge University Press, 1985.

———. "Malthus and the Evolutionists: The Common Context of Biological and Social Theory." *Past and Present* 43 (1969): 109–45.

Zevenhuizen, Erik. *Vast in het spoor van Darwin: Biografie van Hugo de Vries*. Amsterdam: Atlas, 2008.

Zilsel, Edgar. *Die sozialen Ursprünge der neuzeitlichen Wissenschaft*. Frankfurt: Suhrkamp, 1976.

———. "The Origins of William Gilbert's Scientific Method." *Journal of the History of Ideas* 2 (1941): 1–32.

———. "The Sociological Roots of Science." *American Journal of Sociology* 47 (1942): 245–79, reprinted in *Social Studies of Science* 30 (2000): 935–49.

INDEX

abacus/abacists, 40, 48, 59
Abélard, Pierre (1079–1142), 28, 30
Academia Secretorum Naturae, 101
Académie Royale des Sciences, 116, 118–19, 121, 148; on circulation in plants, 142–43, 184
academies, 47, 128; Bessarion's, 50–51; experimentation flourishing in, 49, 128; Sforza's, 58–59
Accademia dei Lincei, 70, 77, 101
Accademia del Cimento, 71, 109
Achenwall, Gottfried (1719–1772), 197, 208–9
action at a distance, 161, 230; Newton's use of, 151–52, 245; rejection of, 154–55
Adelard of Bath (1070–1146), 28
Aeschylus, 23–24
Agricola, Georgius (1494–1550), 99, 103
Agricola, Rodolphus (1433–1485), 51, 76, 181
agriculture, 219; Agricultural Revolution, 177; artificial selection in, 236, 242; botany and, 175–79, 210; statistical analysis in, 213–14
Agrippa von Nettesheim, Cornelius (1486–1535), 99
Alberti, Leon Battista (1404–1472), 55–58
alchemy, 29, 38, 51, 126; claims to have made gold from lead, 93–94, 96; drawbacks of, 94–96; growing irrelevance of, 93–94, 96; influence of, 96, 104, 117; interest in, 49, 94, 151–53; natural and technical curiosities and, 107–8; opposition to, 97, 103, 110; physicians' interest in, 97–98, 100; representations of, 94–95
Aldrovandi, Ulisse (1522–1605), 168–73
Alembert, Jean d' (1717–1783), 158–59
Alexander of Hales (1185–1245), 32
Alexander the Great (356–323 BC), 21
Alhazen (Ibn al-Haytham) (965–1040), 29, 36–39, 57, 137
Al-Khwarizmi (800–865), 29, 40–41
Al-Kindi (801–866), 29, 36
Ampère, André (1775–1836), 119
Amsterdam, botanic garden in, 183, 189, 224

analogies, 2, 135, 160; anatomical, 137–42; Aristotle and, 14–15, 24, 134; in biology, 162–63; Descartes's use of, 144–46, 148; figurative language vs., 136–37; models based on cybernetics, 162–63
anatomy, 101; analogies in, 137–42; botany and, 172–73; comparative, 233–34; of plants, 179, 181, 184, 192–93; in Plato's *Timaeus*, 225
Anaxagoras, 19
Andriesse, C. D., 148
Annales school, 5
anti-Cartesians, 123–24
apothecaries, 169–70, 172
Aquinas, Thomas. *See* Thomas Aquinas
Arab culture, 35, 107; influence of, 9–10, 29, 49, 103; mathematics and, 40–41, 50–51
Arabic, Greek writings translated from, 28–29
Archimedes (287–212 BC), 15, 46, 135; influence of, 48, 64; on mathematics, 39–40, 65–66; works of, 29, 52
Archytas, 18
Arianism, Newton's belief in, 153
Aristotelianism, 126, 139, 180, 192; classification schemes in, 181–88; experience in, 104–5; at University of Leiden, 121–22
Aristotle (384–322 BC), 33, 199; analogies and, 14–15, 134, 136; on arithmetic, 39–40; deductive style of, 13, 22, 25–26, 87, 256; Descartes and, 27, 30–31, 87; on ethics, 22–23; on experience, 38, 118; first principles of, 13, 15–17; influence of, 22, 27, 36–37, 50–52, 64, 82–83; influence on medical education, 30, 42; Lyceum of, 21; mathematics' influence on, 19–20; medieval period lacking direct link to, 27–28; metaphysics of, 13–14; methods of, 13, 22, 24–25, 110; on nature vs. technology, 134; observation in methods of, 15–17, 37, 46, 106; opposition to, 22, 31–32, 73; on physics, 55, 61, 63, 65; Plato and, 20–22, 35–36; reception

355

Aristotle (continued) history of, 25; taxonomy of, 181; translations of works of, 28–29, 52; works of, 13, 22–23, 25, 28, 36–37, 107
Arnold of Villanova (1240–1311), 42, 79
art, 56–58, 94, 188, 228; and crafts, 46, 53–54, 92, 96; history of, styles in, 6–8
artisan-engineers, 2, 47–49, 55–57, 126
artisans, 54, 127
Ashworth, William, 168
astrology, 29, 62, 94, 97–99, 105
astronomy, 17, 51, 69, 205; error analysis in, 206–7; as mathematics, 62, 80–81; Regiomontanus focusing on, 52–53. See also heliocentrism; Ptolemaic system
atomism, 117, 152, 153, 161; mechanism and, 144; opposition to, 19, 77
atoms, 144–45, 163, 255
Auerbach, Stanley, 267
Augustine (354–430), 29, 33, 35–36, 39, 97, 105, 107–8
Averroës (Ibn Rushd) (1126–1198), 22, 42, 78; influence of, 30, 32, 50
Avicenna (Ibn Sina) (980–1037), 22, 30, 42, 97; declining influence of, 49–50; translations of works of, 28–29

Babylonians, 17–18
Bacon, Francis (1561–1626), 102, 106–7, 109, 111, 116, 124, 136; on experiments, 46–47, 90, 93, 110; interest in occult, 2, 49, 93, 104; scientific reforms by, 112–14; on wonders of nature, 108, 117
Bacon, Nicholas, 111
Bacon, Roger (1219–1292), 32–33, 96, 105, 107; experimentation by, 36–39; influence of, 44, 56; optics and, 36, 44; on perspective, 37–39, 57
Baconianism/Baconians, 114–16, 127, 139; Dutch, 120–26; English, 122–23; French, 118–19
Bakhtin, Mikhail, 6
Baltimore, David, 265–66
Banks, Joseph (1743–1820), 191
Bateson, William (1861–1926), 213, 240
Bauhin, Gaspard (1560–1624), 169–70, 173, 177, 180
Bauhin, Jean (1541–1612), 173
Bauman, Zygmunt, 259, 275–76
Bayes, Thomas (1702–1761), 205, 213
Becher, Johann Joachim (1635–1682), 93–94
Becking, Lourens Baas, 249
Becquerel, Henri, 248
Beeckman, Isaac (1588–1637), 88, 116, 144
Beemster polder, 125–26
Belgium, 157, 198
Bellarmino, Robert (1542–1621), 70–71, 75, 82
Belousov-Zhabotinskii reaction, 249–50
Beneden, Edouard van (1846–1910), 241
Benedetti, Giovanni-Battista (1530–1590), 64
Bentley, Richard (1667–1742), 151, 154
Bergson, Henri (1859–1941), 249
Bernal, J. D., 254
Bernard, Claude (1813–1878), 128

Bernoulli, Daniel (1799–1782), 158–59
Bernoulli, Jakob (1654–1705), 203–4
Bernoulli, Johann (Jean) (1667–1748), 148, 158–59, 203
Bernoulli, Nicolaus, 203–4
Bessarion, Johannes (Basilius) (1403–1472), 50–51, 52
Biancani, Giuseppe (Josephus Blancanus) (1566–1624), 106
"Big Science," 90, 263, 266–68, 273
biology, 57, 143, 225, 232, 243, 247; Aristotle's works on, 22–24, 26; cell theory in, 162, 239; evolutionary theory in, 221, 228, 243, 250; as field, 3, 193, 260; metaphors and analogies in, 162–63; molecular, 250, 251; on production of new species, 239, 243
biotechnology, 250–51, 273
Bjerknes, Carl Adam, 161
Boerhaave, Herman (1668–1738), 94, 124–25, 187
Bohr, Niels (1885–1962), 163
Boltzmann, Ludwig (1844–1906), 204
Bombelli, Rafael (1525–1576), 61
Bonaventure (1221–1274), 32
Bonnet, Charles (1720–1793), 227, 229
Borgia, Cesare (1475–1507), 56
botanical gardens, 109, 172; in Amsterdam, 183, 189, 224; as collections, 168, 189; in Leiden, 122, 175, 224; in Padua, 171, 224; rearrangements of, 228–29
botany, 171, 174, 183, 193, 227, 241; agriculture and, 175–79, 210; collecting trips for, 189–90; drawing in, 57, 175, 178; medicine and, 172–73, 175. See also plants
Bouguer, Pierre (1698–1758), 157
Bourdieu, Pierre (1930–2002), 7
Bowley, Arthur (1869–1957), 216
Boyle, Robert (1627–1691), 89, 103, 115, 120, 135, 140–41, 152–53; contributions of, 90–91; interest in occult, 2, 93, 116–17
Bradwardine, Thomas (1290–1349), 41, 43–44
Brahe, Tycho (1546–1601), 77, 83
Brunelleschi, Filippo (1377–1446), 55, 57
Bruno, Giordano (1548–1600), 94
Buffon, Georges-Louis Leclerc, Comte de (1701–1788), 185, 191, 211, 224–25, 228–30
Burckhardt, Jacob, 51
Burgers, Jan (1895–1981), 249
Burgersdijk, Franco (1590–1635), 121
Buridan, Jean (1295–1358), 33, 41, 61, 134
Burman, Johannes (1707–1778), 189, 191
Burman, Nicolaas (1733–1793), 191
Burnet, Thomas (1635–1715), 224–25
Bush, Vannevar, 261–62, 266
business: college graduates entering, 132, 255; research and development in, 272–73; universities' links with, 259–62, 269–72; use of probability theories, 200–203
Byzantine culture, 49–50

Calvin, John (1509–1564), 73, 75, 123, 139
Calvinism, 78, 120, 123–26
Calzolari, Francesco (1521–1600), 169–72

Camerarius, Rudolph Jacob (1665–1721), 179, 183
Candolle, Alphonse (1806–1893), 210
Candolle, Augustin-Pyramus de (1778–1841), 191–93, 209–10
Cardano, Gerolamo (1501–1576), 59–60, 100, 108–9
Carnap, Rudolf, 255
Carnot, Sadi (1796–1832), 119, 247
Carson, Rachel, 265
Cartesianism, 121–22, 124, 126, 144, 152, 246; mechanism of, 161, 227; Newtonianism superseding, 154–55, 157; on vortex theory, 146–48
Casaubon, Isaac (1559–1614), 51, 96
Catholic Church, 28, 49, 51, 128; Copernican system and, 75–76, 79–82, 146; Council of Trent deciding dogma of, 75–78; Franciscans and Dominicans in, 29, 34; on magic and alchemy, 95–97; schisms in, 50, 74, 77–78; universities and, 29, 118. *See also* religion
Cauchy, Augustin-Louis, 254
Caus, Salomon de (1576–1626), 143
causality, 14, 35, 56, 83, 96, 145, 164; analysis of variation and, 219; correlation and, 213; hierarchical, 20, 134; probability and, 199–200, 204; in search for explanations, 153–54, 159; use of statistics in, 209–11
cell theory, 239, 250–51
CERN project, 264–65
certainty, 144, 161, 200; inductive arguments and, 206, 214; lack of, 88, 201
Cesalpino, Andrea (1519–1603), 139–40, 169, 173, 180–82
Cesi, Federico (1585–1630), 70
Chambers, Robert (1802–1871), 223
chaos theory, 198, 211, 247
character, formation of, 21, 54, 130–31, 257–58, 274
Charles I, King (1600–1649), 141
Charles II, King (1630–1685), 90–91, 116
Charles the Bold (Duke of Burgundy), 97
Charles V, Holy Roman Emperor, 103
Charles V, King (1338–1380), 135
Charles VIII, King (1470–1498), 54
Charleton, Walter (1620–1707), 102, 152–53
Chartier, Roger, 5
Chelsea Physic Garden, 172
chemistry, 100, 102, 122, 127; action in distance in, 151–52; mechanistic explanations in, 144–45; relation to physics, 254–55
China, 9–10, 40
Christianity, 35. *See also* Catholic Church; religion
Chrysippus (280–208 BC), 21, 34–35
Chrysoloras, Manuel (1353–1415), 50, 52
Cicero, Marcus Tullius (106–43 BC), 35, 51, 72, 76, 97, 199
circulatory system, 227; analogies of, 24, 138–42; of plants, 142–43, 184
civil engineering, 157
cladism, taxonomies of, 194
Clairaut, Alexis (1713–1765), 157
Clarke, Samuel (1675–1729), 154–56, 200
Clausius, Rudolf (1822–1888), 204, 247

Clavius, Christopher (1537–1612), 54, 58, 70, 80–82, 137–38
Clement VII, Pope (1478–1534), 80
Clifford, George (1685–1760), 189
climate, 229, 231, 235–36, 267–68
clocks, as allegorical, 134–35, 155
Clusius, Carolus (Charles de L'Ecluse) (1526–1609), 173 –176, 178, 180
Clutius, Theodorus, 175
Cocceius, Johannes (1603–1669), 124
Colbert, Jean-Baptiste (1619–1683), 118
Colding, Ludwig A., 246
Coleridge, Samuel, 254
collections, 170; arrangements of, 188; of data, 110, 166, 178–79, 198, 208–9, 212; of diseases, 187; of natural and technical curiosities, 107–9, 169–70; natural history, 188–92, 194; of objects with powers, 96–97; of plants, 169–71, 174; switch from bizarre to ordinary, 189; taxonomies stemming from, 167, 191–92
Collingwood, R. G. (1889–1943), 6
colonies, and spread of plants, 175–76
Commelin, Jan, 183
common-law system, English, 76, 111–13
comparative method, taxonomies in, 167
competition: in neoliberalism, 271–73; for resources, 234–36
Comte, Auguste, 254
Condillac, Etienne Bonot de (1714–1780), 245–46
conservation principle, 144, 151, 163, 246–47
Copernican system, 3, 146; efforts to prove truth of, 81–84; Galileo forced to abjure, 70–71; support for, 83–84, 103, 135
Copernicus, Nicolaus (1473–1543), 7–8, 51, 53; Church and, 79–80; influence of, 3, 77, 100
corpuscular philosophy, 182
Corvinus, Matthias (king of Hungary), 52
Cosimo de' Medici, the Elder (1389–1464), 51, 96
Cosimo II, de' Medici (1590–1621), 69–70, 106
cosmology, 34–35, 134–35
cosmos, 18, 20, 145–46
Council of Trent (1545–1563), 75–78, 82, 95, 107
Counter-Reformation, 78
Cousin, Victor (1792–1867), 207
Crick, Francis (1916–2004), 162
Croll, Oswald (1560–1609), 96, 100
Crombie, Alistair, 44–45, 132; on Aristotle, 13, 23; on styles of thinking, 1, 3–4; use of "style," 6, 8
Cromwell, Oliver (1599–1658), 110, 142, 154
culture, 6, 28; civilizations and, 228–29; relation of science to, 1–2
Curie, Marie (1867–1934), 258
Curie, Pierre (1859–1906), 248
Cusanus (Nicholas of Cusa) (1401–1464), 35, 134
Cusanus (1401–1464), 50, 58–59
Cuvier, Georges (1769–1832), 190, 192–93, 222, 233
cybernetics, models based on, 162–63

Dalton, John, 129
Damian, Peter (1006–1072), 31–32
Dante Alighieri (1265–1321), 32, 57–58

Danziger, Kurt, 218
Darwin, Charles (1809–1882), 119, 199, 211, 231, 248; forerunners in evolution theory, 221–23; as gentleman scientist, 129–30; on hybrids in evolution, 237–38; influences on, 210, 234–35, 238; on natural selection, 234–36, 243; publication of *On the Origin of Species* by, 221, 223
Darwin, Francis, 241
Darwin, George, 248
Daston, Lorraine, 201, 256
data, 253, 268; collections of, 166, 178–79, 198, 208–9, 212; reuse of, 268–69; statistics in analysis of, 197–98; trustworthiness of, 265–66; variability in, 210–11
deductive method, 39, 43, 76, 78, 87, 118, 162, 199, 204, 256; Aristotle and, 19, 22–26; criticisms of, 125, 253; developed by Greeks, 2, 9–10; experiments and, 66, 73; goals of, 27, 144; god's whims limited, 33; natural laws and, 44, 149, 208; opposition to, 22, 72–73, 110; status of, 2, 12, 252; usefulness of, 15, 186, 192
Dee, John (1527–1608), 51, 54, 94
della Mirandola, Pico, 34
della Porta, Giambattista (1535–1615), 101–2, 104, 169
Demiurge, in Plato's creation story, 16, 18, 31, 225–26
Democritus, 19, 22
demographics, 202, 207, 216
De Moivre, Abraham, 206
Desaguliers, Jean (1683–1744), 156–57
Descartes, René (1596–1650), 26, 81–82, 88, 110, 122, 140, 226; analogies of, 136, 145–46, 148; atomism of, 161; influence of, 125, 154; influences on, 30, 102; mechanical philosophy of, 96, 116, 143–47, 227; methods of, 84–85, 87, 136; on physics, 150–51, 246; on science, 27, 132; seeking certainty, 73, 78, 86
determinism, 204–5, 211, 219, 244
de Vries, Hugo (1848–1935), 193, 241–43, 248, 258
de Witt, Johan (1625–1672), 124, 202–3
dialectics, tradition of, 76
Diderot, Denis (1713–1784), 189, 229
Dijksterhuis, E. J. (1892–1965), 4, 37, 44, 66–67, 144–45
Diogenes Laërtius (ca. 220), 52, 72
Diophantus (ca. 250), 50–51, 52, 61
Dioscorides (ca. 50), 171, 176–77, 179
DNA, 162–63, 250–51
Dobzhansky, Theodosius, 243
Dodoens (Dodonaeus), Rembert (1517–1585), 122, 168–69, 173, 177, 179–80
Domela Nieuwenhuis, Ferdinand (1846–1919), 209
Dondi, Giovanni de' (1318–1389), 135
Drakenstin, Hendrik van Reede tot, 175
Driesch, Hans, 249
Duns Scotus, John (1265–1308), 32–34
Dürer, Albrecht (1471–1528), 57
Durkheim, Emile (1858–1917), 165–66

Dutch Republic, 157; Baconianism in, 114, 120–26; religion's influence on science in, 124–25
dynamics, 61; fluid, 158–59; statics vs., 65–66. *See also* motion

Earth, 14, 20; age of, 237, 248; history of, 222, 226. *See also* heliocentrism; solar system
ecology, 8, 235
economy, 271; law of conservation in, 246–47; science and technology powering, 115, 265
Eddington, Arthur, 256
Edgerton, David, 264
Edison, Thomas, 258
education, 59, 122; alternatives to universities for, 257; formation of character in, 54, 130–31, 274. *See also* universities
Einstein, Albert (1879–1955), 26, 256, 258
electricity: and magnetism, 160
electromagnetic field theory, 161
elementary particle physics, 166
elements: arrangement of, 20; in Stoic cosmology, 34–35
Elizabeth I (queen of England), 111
empiricism, 24–25, 78, 117–18, 192, 254
energy, 246–48, 255
engineering, 3, 46–48, 54, 56, 103, 156–57. *See also* artisan-engineers
England, 125, 183, 202; agriculture in, 177, 213–14; Baconians in, 114–16, 122–23; blend of intellectual schools in, 152, 154; collections in, 109–10; empiricism in, 78, 117–18; experimentation in, 46–47; industrialization in, 247–48; legal system of, 111–13; mathematics and alchemy in, 94–95, 103; Newtonianism in, 149, 154–56; politics in, 155–56; religion in, 110–11; research funding in, 264, 272
Enlightenment, 96, 128, 158, 190, 225; scientists holding values of, 258–59; utilitarianism in, 130
entropy, 247–49
environment, 231, 268; beings passively reacting to changes in, 229, 232, 236, 245; ecological flexibility in, 236–37, 245; organisms' relation to, 233–37, 243, 245; problems in, 265–67
Epicurus (341–270 BC), 19, 21, 152
epigenesis, 227, 230–31
epistemes, 188, 193
epistemology: and historic mode of thought, 232–3
Erasmus, Desiderius (1469–1536), 50–51, 76, 78, 99, 103, 107; Luther vs., 74–75; religious tolerance of, 73, 80
error analysis, Laplace's techniques for, 206–7
essences, and anti-essentialists, 182–83
ethics, 8–9, 21–24
Euclid (325–265 BC), 13, 15, 28–29, 39, 41
eugenics, Galton promoting, 211–12
Euler, Leonhard (1707–1783), 158–60
Europe: CERN project in, 264–65; research funding in, 264; science policies of, 274; technology transfers from university research in, 272–73. *See also* specific countries

evolutionary taxonomy, 194
evolutionary theory, 212, 224, 226; biology field and, 194, 243, 250–51; creation of new species in, 213, 228, 237–40, 242; Darwinian, 236, 248; Darwinian vs. non-Darwinian, 237, 239–40; Darwin's forerunners in, 221–23; Darwin writing up, 221, 223; gradual vs. by leaps, 248; hybrids and, 237–40; influence on taxonomies, 166, 194; mechanisms of change in, 236; mutations in, 242–43, 248; natural selection in, 234–35, 243–44; neo-Darwinian synthesis in, 243–45; niche construction in, 235–36, 245; opposition to, 222, 229, 241, 248; responses to, 231–33, 244; search for laws in, 243–45; as style of science vs. ordinary theory, 232; Wallace's, 223
experience, 41, 118; conceptions of, 38, 106; manifest vs. occult, 104–5, 116; role in science, 44, 56
experiments, 46, 109, 159, 161, 214; in biology, 193–94; Boyle's, 117; della Porta's, 102; design of, 213–15; De Vries's, 241; in dynamics, 66; Galileo's, 48, 66–68; goals of, 76, 88; Harvey's, 138–41; as intervention in nature, 92–93, 132; in medical science, 41–43; Mendel's, 238–40; methodology for, 47–48, 128; Newton's, 153–54; objections to, 13, 127–28; on projectile paths, 65; qualitative, 41; relation to theories, 12, 66–68, 124, 132–33, 253; requirements for, 89–90; skill in, 88, 102
eyes/sight, 36–39, 57–58, 89, 137–38

Fabricius, Hieronymus (1533–1619), 140
Faraday, Michael (1791–1867), 160–61, 254–55
Ferdinand I, Holy Roman Emperor, 103, 173
Ferdinand II de' Medici (1610–1670), 70–71
Fermat, Pierre de (1601–1665), 201
Fibonacci (Leonardo of Pisa) (1170–1240), 40–41, 58
Ficino, Marsilio (1433–1499), 51, 53–55, 136, 152; enhancing status of occult, 97, 99; translating Hermes Trismegistus, 96–98
first principles, 12, 33, 87, 126; Aristotle on, 15–17, 25; induction of, 37, 110; true knowledge from, 13, 15, 61
Fisher, Ronald A. (1890–1962), 212–15, 219–20
Flemming, Walther (1843–1905), 241
Flourens, Pierre-Jean-Marie (1794–1867), 119
Fludd, Robert (1574–1637), 95–96, 104
fluid mechanics, 159–60
force, 152, 160; natural laws and, 34, 246; Newton on, 150–51, 154–55; in Stoic cosmology, 34–35
Forster, Georg (1754–1794), 209
Forster, Johann (1729–1798), 209
fossils, 222, 226, 237
Foucault, Michel (1926–1984), 5, 187–88, 198, 232
Fourier, Jean-Baptiste (1768–1830), 119, 162
France, 87, 116, 123, 148, 198, 254; botany in, 175, 191, 210; deductive style in, 78, 118–19; Newtonianism in, 157–58, 159; religion in, 78, 118–19; universities in, 29, 128
Francis I, King (1494–1547), 56, 188

Fresnel, Augustin (1788–1827), 119
Frye, Northrop, 6
Fuchs, Leonhard (1501–1566), 169
Fugger, Anton (1493–1560), 174

Gaertner, Carl Friedrich von (1772–1850), 238–39
Gaertner, Joseph, 192
Gafori, Franchino (1451–1522), 51
Galen (129–199), 35, 52, 101, 120, 171, 179; on blood and circulation, 139, 141
Galilei, Galileo (1564–1642), 2, 7, 77, 96, 106, 110; background of, 49, 68; career of, 66–71, 101; condemnations of, 70–71, 74–75, 84, 86; efforts to classify style of, 82–83; experimental style of, 46–48, 66–68, 73, 88, 93; explanation for moon's dark spots, 57–58; Guidobaldo and, 65–68; on mathematics, 62, 65–68, 81; as philosopher, 62–63, 69, 71; on Skepticism, 73, 83; trying to prove Copernican system, 81–82
Galilei, Vincenzo (1520–1591), 68, 89
Galton, Francis (1822–1911), 197, 217, 237; promoting eugenics, 211–12; on statistics, 210–11, 213, 241–42
Gassendi, Pierre (1592–1655), 86, 152–53, 161, 230–31
Gauss, Carl Friedrich (1777–1855), 206
Gay-Lussac, Joseph Louis, 128, 131
Gell-Mann, Murray (b. 1929), 166
Gemma Frisius, Regnier (1508–1555), 80
genetics, 162, 213, 243; hybridization and, 238–40; population, 251; regression to mean in, 211–12, 242; statistics in, 212, 241–42; traits in, 237, 239–41
geology, 222–23, 235
geometry, 18, 20, 54; arithmetic and, 39–40; Euclid's, 15; plane, 39. *See also* mathematics
George I, King (1669–1727), 155
George III, King (1738–1820), 191
Gerard of Cremona (1114–1187), 28
Germany, 80, 164, 198; in CERN research group, 264–65; collections in, 109–10; metallurgy in, 93–94, 95; research institutes in, 257–58; universities in, 54, 128, 130–31, 257
Gesner, Conrad (1516–1565), 123
Ghiberti, Lorenzo (1378–1458), 55, 57
Ghini, Luca (1490–1556), 170, 176
Gibbon, Edward (1737–1794), 228
Gilbert, William (1544–1603), 103
Gini, Corrado, 216–17
God, 36, 75; activity in the world by, 154–55, 225; Aristotle's concept of, 14, 31; controlling nature, 113–14, 153, 155, 158; creation by, 16, 123, 125, 144, 230; existence of, 84–85, 87, 158; intentions of, 87, 106; knowability of nature and, 33–34, 134; Linnaeus's system approaching creation plan of, 167, 186–87; miracles of, 118; power of, 30–33, 44–45, 153; role in the world, 19, 134–35; as source of truth, 85–86, 97
Goethe, Johann Wolfgang von (1749–1832), 225
Gombrich, Ernst (1909–2001), 6–9

Gool, Jacob (Golius) (1596–1667), 50
Gottlob, Abraham, 224–25
Goulart de Senlis, Simon (1543–1628), 123
Grassi, Orazio (1590–1654), 70, 77
Graunt, John (1620–1674), 197, 216
Gravesande, Willem J. 's (1688–1742), 124–25, 156
gravity, 3, 151, 161; efforts to explain, 96, 159; Newton's theories of, 148–54, 157, 200; vortex theory on, 146–47
Greeks, 12, 17, 31, 33, 137, 171; cultural continuity from, 9, 28; deductive style developed by, 2, 9–10; efforts to recover works of, 52–54; on ethical decisions, 23–24; experimentation and, 13, 46; language of, 28, 50–51; mathematics of, 39–40; philosophical schools of, 19, 21, 34; translations of works from, 28–29, 49, 51, 152
Gregorian calendar, 80–81
Gregory XIII, Pope (1502–1585), 81
Grew, Nehemiah (1641–1712), 179, 183–84, 227
Gronovius, Jan Frederik (1686–1762), 191
Grosseteste, Robert (1168–1253), 36–37, 39
Grotius, Hugo (1583–1645), 78, 200
Guidobaldo, Marquis del Monte (1545–1607), 48–49, 62, 64–68, 71
Guidobaldo II (duke of Urbino) (1514–1574), 65

Haber, Fritz, 258
Hacking, Ian, 3–4, 132, 198, 201, 208
Hall, Rupert, 47, 54, 64, 68
Haller, Albrecht von (1708–1777), 161
Hamilton, William (1805–1865), 162, 246
Handler, Philip, 265
Hartlib, Samuel (1599–1670), 114–15, 177
Hartsoeker, Nicolaas, 127
Harvey, William (1578–1657), 27, 138–39, 227
Hasselqvist, Fredrik (1722–1752), 190
Hauksbee, Francis, 127
Hebrew traditions, 31, 33
Heereboord, Adriaan (1613–1661), 123
Hegel, Georg Friedrich Wilhelm (1770–1831), 254
Heimans, Jacobus (1889–1978), 240
Heisenberg, Werner (1901–1976), 264
heliocentrism, 53, 74, 80–81; Church's rejection of, 75–77; importance of, 3, 7–8
Helmholtz, Hermann (1821–1894), 246–47
Helvetius (Johann Friedrich Schweitzer) (1625–1709), 93
Henry IV, King (1553–1610), 60, 78
Heraclitus (5400475 BC), 14
herbals, 169, 178
Hermes Trismegistus, 51, 95–98
hermeticism, 95–96, 152
higher-order principles. *See* first principles
Hippocrates, Galen vs., 101
Hippocrates of Cos (450–370 BC), 18–19
historical thinking, 221; about Earth, 222, 226; applied to physical systems, 245–46; epistemology and, 232–33
historicism: vs. philosophy, 254
history, 4–6, 235; of Earth, 237; of nature, 231–32; of physical systems, 250
Hobbes, Thomas (1588–1679), 89–90, 117

Hooke, Robert (1635–1702), 91, 103–4, 109, 127, 222, 226
Hudde, Johan (1628–1704), 202
Huizinga, Johan (1872–1945), 2, 120
humanism/humanists, 62, 76, 88, 96, 103; Aldrovandi as, 170–71; classics and, 51–52, 97, 131; decline of, 87; of Gassendi, 152–53; Italian, 47, 50–51; on mathematics, 48–49, 53–55, 57; recovery of Greek works and, 52, 53–54; universities and, 47, 50–51, 53
Humboldt, Alexander von (1769–1859), 209–10, 233–35, 253
Humboldt, Wilhelm von, 130
Hume, David (1711–1776), 205
Hunt, Lynn, 5
Hutton, James (1726–1797), 224–25
Huxley, Julian, 258
Huygens, Christiaan (1629–1695), 88, 121, 127, 143–44, 149, 201; mechanical philosophy of, 96, 116; on vortex theory, 146–48
hypotheses, 144, 161; experiments to test, 12, 124, 132–33; Newton's use of, 150, 153–54; rejection of, 13, 110

Imanishi-Kari, Theresa, 265–66
impetus theory, 41, 63, 66–67, 139
inductive method, 15, 119, 149, 192, 204, 215; Bacon's, 110–11, 113–14; by Descartes, 84–85; first principles from, 37, 110; search for rational procedure for, 205–6; statistics and, 208–9, 214, 219–20; use of models as, 161–62
Industrial Revolution, 157, 247–48
industry, 131; research for, 256–58, 262–63, 265; universities' links with, 255–62, 269–72
Inquisition, 78, 82, 101–2
instruments, scientific, 126–27, 137–38, 206; objections to experiments requiring, 90, 92–93; refinements of, 37, 129
International Biological Program (IBP), 266–67
International Geosphere-Biosphere Programme (IGBP), 267–68
intuition, 15–17, 180, 256
Ionians, 17
Italy, 61, 96, 109; academies in, 47, 49; artisan-engineers of, 54–56; humanism in, 47, 50–51

Jacob, François (b. 1927), 163
Jacobins, 128
James I, King (1566–1625), 111, 173
James II, King (1633–1701), 155
Jasanoff, Sheila, 269
Jesuits, 88, 181; educational institutions of, 54, 62, 70; on heliocentrism, 74–75, 81; methods of, 76–77, 84
Johannsen, Wilhelm (1857–1927), 240, 243
John XXII, Pope, 97
Jonston, Jan (1603–1675), 168
Jordanus, Nemerarius (ca. 1220), 63–64
Joule, James Prescott (1818–1889), 128–29, 246
Jung (Jungius), Joachim (1587–1657), 178, 181–82
Jussieu, Antoine-Laurent de (1748–1846), 192–93, 225, 228–29

Jussieu, Bernard de (1699–1777), 193
Justinian, Emperor (482–565), 35

Kalm, Pehr (1715–1779), 190
Kant, Immanuel (1724–1804), 231–32
Kepler, Johannes (1571–1630), 44, 73, 82–84, 88, 95, 105, 135; on eyes and sight, 137–38; on planetary orbits, 18, 147, 149–50
Keverberg de Kessel, Charles-Louis de (1768–1841), 216
Keynes, John Maynard (1883–1946), 95
Kiaer, Anders Nicolai (1838–1946), 216
Kilgore, Harley, 261–62, 266
Kircher, Athansius (1602–1680), 71
knowability, 12; of the cosmos, 9, 16; of nature, 6, 33–34, 125, 134, 136; of technology, 134, 136
knowledge; acquisition of, 16–17, 35–36; assurances of truth of, 85–86; Baconian collections of facts, 110–11, 113–15, 119, 121; causal relationships in, 199–200; experiential, 104–5; hidden, 101, 106–7; sharing of, 103, 107, 127; sources of, 97, 105–6, 120
Kölreuter, Joseph (1733–1806), 238–39
Koyré, Alexandre (1892–1964), 44, 48
Kramers, Hendrik (1894–1952), 163
Kuhn, Thomas (1922–1996), 7–8, 92, 119

laboratories, 255; locations of, 127–30, 258; in universities, 130–31
LaCapra, Dominick, 5–6
Lactantius, 97
Lagrange, Joseph-Louis, 254
Lamarck, Jean-Baptiste (1744–1827), 193, 222–23, 229
languages, 14, 17, 28–29, 222
Laplace, Pierre-Simon (1749–1827), 119, 158, 205, 209–10, 216; error analysis by, 206, 213; on time, 245–46
Laqueur, Thomas, 184
Latin, 28, 51, 131; translations into, 52, 174–75; vernacular vs., 177, 201
Lavoisier, Antoine-Laurent, 254
Leclerc, Georges-Louis. *See* Buffon, Georges-Louis Leclerc, Comte de (1701–1788)
Leeghwater, Jan (1575–1650), 126
Leeuwenhoek, Antoni van (1632–1723), 227
legal system, 111–14
LeGoff, Jacques, 29
Leibniz, Gottfried Wilhelm (1646–1716), 159, 161, 222, 246; Newton vs., 124–25, 149, 154–56, 158; on statistics, 197, 202; transformism and, 229, 231
Leiden, 123; botanic garden in, 122, 175, 224; University of, 50, 54, 121–22, 172, 174–75, 187, 241
Leonardo da Vinci (1452–1519), 56–59
Leopold I, Emperor (1640–1705), 95
Leopoldo de' Medici (1617–1675), 69–71
Le Play, Frédéric (1802–1882), 216
Libavius, Andreas (1560–1616), 96
Liebig, Justus von (1803–1873), 131
Liender, J. van, 157

light, 36, 38, 89, 144, 151, 256
Linnaeus, Carl (1707–1778), 166–68, 172, 178, 225, 238; plant classification scheme of, 183–88, 190, 192; plant collection by, 189–91
literature: metaphors in, 135–37
L'Obel (Lobelius, or de Lobel), Matthias de (1538–1616), 168–70, 173, 180
Locke, John (1632–1704), 182, 232–33
Löfling, Peter (1729–1756), 190–91
logic, 14, 20, 23
Lorentz, Hendrik (1853–1928), 161, 258
Louis XIV, King (1638–1715), 116
lucerne, rediscovery of, 176–77
Lucretius (95–55 BC), 19, 52, 145
Ludovico Sforza (1452–1508), 56, 58–59
Luther, Martin (1483–1546), 74–75, 120
Lyell, Charles (1797–1875), 223, 234–35
Lyotard, Jean-François, 272–73

MacDougal, Daniel Trembley (1865–1958), 242–43
Machiavelli, Niccolò (1469–1527), 107
machines, 246–48
macrocosm-microcosm thesis, 34, 98–100
magic, 103; Bacon and, 2, 37, 104, 110; growing irrelevance of, 93–94; influence of, 96, 98; interest in Italian courts, 96–97; natural and technical curiosities and, 107–8; in objects, 96–99; relation to science, 38, 101; representations of, 94–95
magnetism, 116, 147, 160
Malpighi, Marcello (1628–1694), 179, 183–84, 227
Malthus, Thomas (1766–1834), 234–35
Mannheim, Karl (1893–1947), 6–9
manuscripts, medieval copying of, 176–77
Margaret of Austria (1480–1530), 99
Maricourt, Pierre de, 39
Marlowe, Christopher (1564–1593), 95
Martini, Francesco di Giorgio (1439–1501), 56
Marxism, 4–5
Massachusetts Institute of Technology (MIT), 259–60, 262
mathematics, 17–18, 39, 50, 110; algebra, 50–51, 59–60, 61; "applied," 103; arithmetic, 40, 210; authorities on, 65–66, 116; bookkeeping and, 48, 58, 60; calculus, 158–59; effects of fluid dynamics on, 158–59; in Egypt, 17–18, 54; experiments and, 66–68, 93; fields under, 48, 61, 62; Galileo as philosopher, not mathematician, 69, 71; humanism and, 48–49, 54; influence of, 19–20, 119; in logical positivism, 255–56; low status of, 55, 59, 61–62, 68–69, 79, 103; magic and, 94, 101, 103; mathematicians and, 59, 80, 103; meanings of, 37, 84, 162; numeral systems in, 40; philosophy and, 61–62, 81; of Phoenicians, 54; "pure," 254; of Pythagoreans, 17; relation to reality, 82–83; relation to science, 56, 119, 254; in statistics, 197–98, 212; in theory of transformation, 225–26; universities and, 52, 61–62; uses in understanding nature, 47, 148–54; uses of, 51, 66–68, 82–83, 106, 121, 127, 201; utility of, 53–57, 121

Index ⊙ 361

matter, 144, 249; activity of, 153, 155–56; properties of, 182, 230–31; types of, 145–48
Mattioli, Pier Andrea (1500–1577), 171, 173, 176–77
Maupertuis, Pierre-Louis Moreau de (1698–1759), 157, 230, 231
Maurice (prince of Netherlands) (1567–1625), 60
Maurolico, Francesco (1494–1575), 53, 137–38
Mauss, Marcel (1872–1950), 165–66
Maximilian I, Emperor (1459–1519), 99
Maximilian II, Emperor (1527–1576), 173–74
Maxwell, James Clerk (1831–1879), 160–61, 207, 256
Mayer, Julius, 246–47
Mayr, Ernst, 243–44
Mayr, Otto, 154
Mazarin, Cardinal (1602–1661), 75
measurement, 41, 47, 54, 83, 127; of Earth, 157; enthusiasm for, 208–9; Galileo's, 48, 67, 77; increasing precision of, 129, 255; of time, 77, 88
mechanics, 41, 52, 163; of animals, 96–97; fluid, 158–59; Guidobaldo's research in, 48–49; Newton's, 154, 161
mechanistic philosophy, 125, 148, 152, 230; in biology, 138–40; Cartesian, 123, 143–47, 150, 154–55; on gravity, 149, 151; on occult, 96, 116–17; popularity of, 86–87, 222, 226–27; rejection of, 123, 154–55, 255
medical-industrial complex, 272
medical schools, 30, 42, 49–50, 140; botany at, 172–73; curriculum of, 28, 30, 187
Medici family, 41, 69–71. See also Cosimo; Ferdinand, Leonardo
medicine, 42, 122, 187; biotechnology and, 250–51; Paracelsus's influence on, 99–100; plant uses in, 122, 168–71, 175, 179–80; recipes in, 107, 171; research in, 41–43, 264; statistics in, 208–9, 215
Melanchthon, Philipp (1497–1560), 54, 75, 76, 80, 103
Mendel, Gregor (1822–1884), 213, 238–40, 242
Mendeleev, Dmitri (1834–1907), 166
merchants, 10, 40–41. See also business
Mersenne, Marin (1588–1648), 48–49, 96, 102, 121, 136; experimentation and, 46–47, 88; Skepticism and, 73, 76, 78, 87–89; work on music theory, 88–89
Merton, Robert (1910–2003), 119, 259
metaphysics, 13–14, 144, 159, 254–55
Mettrie, Julien Offray de la (1709–1751), 232
microcosm-macrocosm thesis, 104, 144
microscopes, 178–79, 227
Middle Ages, 9, 35, 171, 226; continuity with Renaissance, 44–45; copying of manuscripts in, 176–77; scholasticism of, 51–52
military, 272; mathematics used by, 54, 60; scientific research for, 261–65
Milstein, César, 269
Milton, John (1608–1674), 154
mind, conception of, 85, 232–33
mining, 99, 103
Mirowski, Philip, 246

Monardes, Nicolás (1493–1588), 174, 175
Monod, Jacques (1910–1976), 162–63
Montaigne, Michel de (1533–1592), 73, 75–76
Montesquieu, Charles-Louis de Secondat (1689–1755), 229
moon, dark spots on, 36, 57–58
morality, 200, 207
Morange, Michael, 251
More, Henry (1614–1687), 152
Morgan, Thomas H. (1866–1945), 242–43
motion, 77, 144; of fluids, 159–60; Galileo's work on, 47–48, 66–71; impetus theory of, 41, 63, 66–67, 139; naturalness of circular, 14, 20, 146, 148; Newton's laws of, 150–51, 156; perpetual, 14, 151; science of, 62–65. See also dynamics
Mullis, Kary, 269–70
museums, 128, 169–70, 188, 190
music, 39; intervals in, 18, 68; in measurement techniques, 67, 77; theory in, 88–89
Musschenbroek, Petrus van (1692–1761), 124–25
mysticism, 10, 18, 55. See also magic

Nägeli, Carl (1817–1891), 239–41
NASA, 263, 267–68
National Academy of Science (U.S.), 265
National Defense Research Council (NDRC) (U.S.), 261–62
National Research Development (Britain), 269
National Science Foundation (NSF) (U.S.), 261–63, 266, 268–69
natural history, 4, 113, 136, 166, 232; collections in, 109, 169–70, 188–92, 194
natural laws, 38, 44–45; God and, 33–34, 114; science to discover, 113; as unchanging, 226, 230
natural philosophy, 17, 27, 37, 46, 108, 224; alchemy and, 94, 98; artisans and, 49, 127; definition of, 61
natural sciences, in philosophy, 61
natural selection: in evolutionary theory, 234–36, 243–44; mutation theory and, 243; niche construction vs., 245
natural theology (physico-theology), 120, 123, 125–26, 225, 234
nature, 22, 123, 204, 247; efforts to control, 11, 49, 99, 104, 110; efforts to explain, 27, 100–101; experimentation forcing actions of, 46, 92–93, 132; God's control over, 153, 155; history of, 231–32; knowability of, 6, 125, 134, 136; mathematics and, 18, 47; methods of understanding, 17, 47, 120, 135; observation of, 105–6, 120; regularity of, 199, 205; secrets of, 14, 19, 101; technology and, 56, 134, 136; variability in, 198–99, 210–11, 231–32, 236, 242–43; violence in, 234–35; wonders of, 105, 107–8, 117, 120
nature-nurture debate, 211
necessity, 19, 26; Aristotle on, 31, 33; rule of, 199–200
Needham, Joseph (1900–1995), 10
Ne'eman, Yuval (b. 1925), 166

neo-Darwinism, 234–36, 243–45
neo-humanism, 130
neo-Lamarckism, 237
neoliberalism, 271, 275
Neoplatonism, 55, 108; Aristotle and, 35–37; efforts to reconcile, 35–37, 152, 227; hermeticism and, 51, 98; influence of, 29, 117; popularity of, 26, 51
Netherlands, 78, 115, 125, 191; botany in, 183, 195; Cartesians in, 121–22; interest in science in, 46–47, 157, 206; natural history collections in, 188–89; uses of statistics in, 198, 202–3, 209, 216
Neumann, John von (1903–1957), 211
Newton, Isaac (1642–1727), 7, 101, 127, 224; on active principles, 151–52; gravitation theories of, 3, 147, 148–54, 200; influence of, 158–59; influences on, 152–53; interest in alchemy, 93, 95, 151–53; laws of motion by, 150–51; Leibniz vs., 124–25, 154–56, 158; mechanics of, 154, 158–59, 161; religion of, 153–54
Newtonianism, 154–56; Dutch, 124–25; in France, 157–58, 159; reception of, 157–58
New World, plants and animals of, 168, 190–91
Neyman, Jerzy (1894–1981), 215, 217
Nieuventijt, Bernard (1654–1718), 120, 125–26
Nordenskiöld, Eric, 140
Noyes, Arthur, 259
Nussbaum, Martha, 23–24, 24–25

objectivism, on chance processes, 198–99
objectivity, 255–56, 271
observation, 14, 84, 149; Aristotle's use of, 22, 37, 46; changing perception of, 105–6, 233; confidence in, 82–83, 255; deductions from, 43, 208; knowledge of nature from, 120, 125; limitations of, 15–17, 25; use of large numbers of, 206–7; uses of, 15–17, 85, 256
occult, 104, 116–17. *See also* alchemy; magic
Ockham, William of (1289–1349), 20, 32–34
Office of Scientific Integrity, 265–66
O'Hanlon, Redmond, 235
Olby, Robert, 240–41
optics, 39, 44, 61, 88, 137, 152–53
Oresme, Nicole (1320–1382), 41, 43–44, 61, 134–35
Orta, Garcia de (1500–1568), 174–75
orthogenesis, 237, 241
Osiander, Andreas (1498–1684), 80
Ostwald, Wilhelm, 255
Ovid (43 BC–18), 226

Pacioli, Luca (1445–1517), 51, 58–59
paleontology, 222–23, 237, 243
Paley, William, 234–35
Palladius, Rutilius (ca. 350), 176
Paludanus, Bernard (1550–1633), 167, 170, 174–75
pangenesis theory, 230–31, 237, 241
Paracelsus (Theophrastus Bombastus von Hohenheim) (1493–1541), 99–100, 120, 136, 152
Paradise, botanic gardens recreating, 168, 224
particles, taxonomy of, 166

Pascal, Blaise (1623–1662), 76, 118, 201
Pasquier, Etienne (1529–1615), 100
patents, 55, 69, 269–71
patronage, 49, 60, 71; experiments without, 128–29; Galileo's, 65, 69–70, 106; for humanists, 50, 52; for mathematicians, 52, 54; for physician-botanists, 173–175; by state, 110, 116; universities and, 175, 255
Paul III, Pope (1468–1549), 77–78, 80
Pearson, Egon (1895–1980), 212–13, 215
Pearson, Karl (1857–1936), 197, 212–13
Pels, Dick, 9
periodic table, as taxonomy, 165–66
Peripatetics, 120
Perrault, Claude (1613–1688), 143
perspectiva, 36, 37–39
perspective, linear, 57–58
Pestre, Dominique, 258, 264
Petty, William (1623–1687), 197
Peurbach, Georg von (1423–1461), 53, 79
pheneticism, 194
Philip II, King (1556–1598), 60, 173
Philip the Good (Duke of Burgundy) (1419–1467), 96–97
Philo Judaeus of Alexandria (20 BC-50), 33, 35
Philolaos, 18
philology, 51
philosophers, 78; mathematicians vs., 69, 71; scientists vs., 254
philosophy, 72, 124, 254; applied to evolutionary theory, 243–44; Desaguliers applying, 156–57; mathematics and, 61–62, 81; "modern," 85; Skepticism and, 73, 75
physicians, 102, 120, 122; apothecaries and, 170, 172; botanists as, 172–75; interest in alchemy and magic, 97–98, 100, 103; surgeons vs., 42, 99; techniques of, 30, 42, 98–100; use of astrology, 94, 98
physics, 3, 13, 23, 55, 81, 88, 199; chemistry and, 254–55; classical, 226; Cold War funding for research in, 262–64; Desaguliers applying, 156–57; experiments in, 47–48, 253, 255; mathematics and, 61, 254; modern, 255; Newtonian, 124–25; optics in, 36, 61; status of, 252, 273
"physics envy," 243, 252
Pickstone, John, and ways of knowing, 4
Pico della Mirandola, Giovanni (1463–1494), 51, 97
Piero della Francesca (1416–1492), 51, 57–58
Planck, Max (1858–1947), 164, 258–59
Plantijn, Christoffel (1520–1589), 174–75
plants, 129, 210, 231; anatomy of, 179, 181, 192–93; circulatory system of, 142–43, 184; classification schemes for, 168–69, 177–88, 192, 194, 228–29; collections of, 168–71, 174, 189–90; collections of dried, 187, 189; discovery of new species of, 169, 180, 192; hybridization of, 238–40; identification of, 178–79, 193–94; medicinal, 122, 168–72, 175, 179–80; names of, 176–78, 180; reproduction of, 183–88; sexuality of, 179, 185, 192; spread to new continents, 175–76. *See also* botany

Plato (429–347 BC), 16, 18, 73, 135; Aristotle and, 20–22, 35–36; creation story in *Timaeus* by, 13, 225–26; ethical theory of, 21–22; influence of, 13, 51; on mathematics, 19–20, 55, 65; works of, 28, 51
Platter (Plater), Felix (1536–1614), 137, 173
Plattes, Gabriel (1600–1644), 114–15
Pliny the Elder (23–79), 171
Plotinus (205–270), 35
Poincaré, Henri, 256, 258
Poisson, Denis (1781–1840), 119, 162
Polanyi, Michael, 265
politics, 107, 220; contract with science, 274–76; Dutch, 157; English, 154–56; Harvey's, 141–42; philosophy applied to, 156–57
polymerase chain reaction (PCR) technique, 250–51, 269
Pomian, Krzysztof, 189
Popkin, Richard, 85
Popper, Karl (1902–1994), 132, 243–44
positivism, 8–9, 254–56
preformation, theory of, 179, 184, 227–28, 230–31, 238
Prigogine, Ilya (1917–2003), 249–50
printer-publishers, 50, 174–75, 178
printing press, effects of, 52
probabilism (plausibility), 76, 81, 87
probability, 198–201, 206, 211, 244; Bernoulli's new model of, 203–4; inverse theorem of, 204–5; used in insurance business, 202–3
Protagoras (485–415 BC), 17, 19
Protestantism, 78, 119–20
Prussia, 130–31, 197
Pseudo-Aristotle, 63–65
psychology, effects of statistics in, 217–19
Ptolemaic system/geocentrism, 135; arguments against, 78–79; choosing between Copernican system and, 81, 84; usefulness of, 79–80
Ptolemy, Claudius (121–151), 57, 79; influence of, 52–53, 77; works of, 29, 52
public service, 50, 258–59, 275
Puritanism, 110–11, 120, 125
Pyrrhon, 72–73
Pythagoras (ca. 530 BC), 17–18, 37

quantification, 39, 47. *See also* measurement
quantum mechanics, 163, 198, 211
Quételet, Adolphe (1796–1874), 197–98, 205–6, 211; discovery of "average man," 207–8; influence of, 208–9, 235; use of statistics, 213, 216, 241–42
Quiccheberg, Samuel (1529–1567), 167–68
Quintillian, Marcus Fabius (35–100), 199

race, 228, 230, 231–32
radioactivity, 198, 243, 248
Ramus, Petrus (Pierre de la Ramée) (1515–1575), 54–55, 76, 100, 181
Ray, John (1627–1705), 168, 169, 182–83, 183–84
reality, 14, 25, 150, 161, 256; mathematics' relation to, 37, 81–83; representations of, 79, 82

recollection/recognition, in acquisition of knowledge, 16
Reformation, 73–74
Regiomontanus (Johannes Müller von Königsberg) (1436–1476), 50–54
Regius, Henri (1598–1679), 86, 122
Reinhold, Erasmus (1511–1553), 80
relativity theory, 26, 256
religion, 118, 126, 135, 222; Galileo's fall from grace and, 70–71; influence on development of science, 10, 70–71, 119–20, 124–25, 153–54; Newton and, 153–54; Skeptics and, 73–75, 78; splits in, 50, 77–78, 125; tolerance in, 73, 80, 155. *See also* Catholic Church
Renaissance, 35, 55, 96, 109, 136; botany in, 171, 176; continuity with Middle Ages, 44–45; experimental style developed in, 2, 104; hypothetical-analogical style developed in, 2–3; macrocosm-microcosm thesis in, 98–99; science of motion in, 62–63; taxonomic style developing in, 3, 167, 169; *virtù* in, 56–57
Renerius, Henri (1593–1639), 122
research institutes, 256–58
Rheticus, Joachim (1514–1574), 80, 135
Ricci, Ostilio (1540–1603), 68, 71
Riccioli, Giovanni Battista (1598–1671), 77
Richelieu, Cardinal (1585–1642), 75, 78
Riegl, Aloïs (1858–1905), 9
Ripa, Cesare, 136
Roberval, Gilles Personne de (1602–1675), 147–48
Rockefeller Foundation, supporting research, 260–61
Romans, 28, 46; cultural continuity from, 9, 55; numeral systems of, 40–41
Rondelet, Guillaume (1505–1566), 172–73
Rorty, Richard, 85
Rosenberg, Alexander, 244–45
Royal Society (England), 109, 115, 156, 179, 191, 255; mathematics in, 103, 116; self-limitation of, 90–91
Royen, Adriaan van (1704–1779), 191–92
Rudolf II, Emperor (1552–1612), 83, 95, 173
Rumphius, Georg (1626–1693), 175
Rutherford, Ernest, 248

Sachs, Julius (1832–1897), 129, 241
Sacrobosco (John of Holywood) (ca. 1250), 52–53
Saint Germain-en-Laye, gardens of, 143
scala naturae (ladder of nature), 179, 183, 187, 193, 229, 233
Scaliger, Joseph (1540–1609), 54
Scarperia, Giacomo d'Angelo da, 52
Scheiner, Christopher (1573–1650), 70, 77, 81, 138
Schindewolf, Otto (1896–1971), 237
Schleiden, Mathias (1804–1881), 193, 239
Scholasticism, 28, 30; decline of, 51–52, 110–11; development of, 33–34
schools, 21, 28. *See also* academies; medical schools; universities
Schooten, Frans van, Jr. (1615–1660), 121
Schrödinger, Erwin (1887–1961), 249
Schwann, Theodor (1810–1882), 193

science: basic, 260–62, 265; clusters of convictions about, 6; as completed, 32, 132; confidence in progress from, 265; cultural value of, 274–76; deductive vs. performative, 273–74; definitions of, 27; disciplines/specializations in, 13, 164, 254–55, 273–74; engineering and, 47–48; goals of, 21, 78; limitations on, 46, 255, 266; mission-oriented, 266–68; "pure," 254, 261–64; reforms of, 112–14; religion's relation to growth of, 119–20; research funding for, 261–64, 265; states' relations to, 274–76; styles of, 221, 273; technology and, 38–39, 43, 115, 253, 264, 274; utility of, 18, 115, 253–54, 256–59; values of, 259; Western vs. non-Western, 9–10
sciences, classical vs. Baconian, 92–93, 116, 119
sciences, higher vs. lower, 199–200
scientia, 37, 61, 105; loss of faith in, 205–6; *opinio* vs., 27, 161, 199–200; Skeptics challenging, 72–74
scientific community, 90–91, 259
scientific method, 96, 139, 161–62
Scientific Revolution, 3, 44–45, 104, 161
scientists, 223, 254, 260, 271; professional vs. amateur, 129–30; roles of, 258–59, 276; sharing knowledge, 103, 107; status of, 258, 265–66, 274–75
Scotland, 247–48, 264
Seba, Albert, 189–90
Second Law of Thermodynamics, 247–49
secrecy/secrets, 101, 127; in alchemy, 95–96; hidden knowledge and, 106–7; occult implying, 104–5
Seitz, Frederick, 265
senses, 88–89, 104–5, 122
Servetus, Michael (1511–1553), 139–40
Severinus, Petrus (1540–1602), 100, 152
Sextus Empiricus (ca. 180), 72–73, 87
Siemens, Ernst Werner von, 257
Siger de Brabant (1240–1284), 32
Skepticism, 79, 111, 205; challenging deductive style, 72–73; Descartes rejecting, 73, 84–85, 122; moderate, 76, 78, 87–88, 91; opposing syllogism, 73–74; opposition to, 83–84, 87–89; religion and, 74–75
Snellius (Snell), Willebrord (1580–1626), 54, 121
social sciences, statistics in, 217–19
Solander, Daniel (1736–1782), 191
solar system, 151, 157, 159, 248; history of, 222, 248; planetary orbits in, 18, 149–50, 158; stability of, 158, 205; vortex theory in, 144, 146–48, 156, 158
Sophism, 19
Sophocles, 23
Spallanzani, Lazzaro (1729–1799), 227
Spengler, Oswald (1880–1936), 164
Spinoza, Baruch (1632–1677), 125, 156
state, 124, 194; neoliberalism of, 271–73; patronage of science by, 110, 116; science policies of, 110, 116, 260–62, 266–68, 274–76; scientists as advisors to, 258–59; supporting research, 256–58, 270; supporting universities, 255,
270; uses of statistics by, 197–98, 209; war demanding strong, 272
statics, vs. dynamics, 61, 65–66
statistics, 167; in calculation of risk, 201, 215–16; correlations and, 212, 214–15; demographic uses of, 202, 207, 216; development of, 197, 212; error analysis in, 213; in eugenic movement, 211–12; in genetics, 241–42; as indices of greater wholes, 208–9, 216–17; individuals in, 196, 199, 208, 217–18, 220; inferential, 219–20; meaning of "average" in, 211; method of least squares in, 206; normal distribution in (bell curve), 206–7, 210–11; null hypotheses in, 215–16; to overcome chaos at micro level, 207–8; in public health, 208–9; random sampling in, 216–17; regression to mean in, 211–12, 237, 242; in social sciences, 217–19; states' uses of, 197–98, 208–9; test of significance in, 213–14, 218; types of, 198
status, social, 99, 265; of artisans, 48–49, 127; in hierarchy of sciences, 92, 164; of mathematicians, 54–55, 59, 68–69, 103; university education and, 30, 257
Stensen (Steno), Niels (1638–1686), 222, 226
Stevin, Simon (1548–1620), 60, 88, 246
Stoicism, 34–35, 52, 73, 98
Strasburger, Eduard (1844–1912), 241
Strasbourg, astronomical clock in, 135
Suringar, W. (1832–1898), 241
Swammerdam, Jan (1637–1680), 227
syllogism, Skeptics opposing, 73–74
Sylvius, Franciscus (1614–1672), 122
Synod of Dort (1618), 78
systematics, 194–95

Tartaglia, Niccolò (1499–1557), 59–60, 62, 64–65, 71
Tasso, Torquato (1544–1595), 48–49
taxonomies, 4, 178; based on Aristotelian logical structure, 181–88; biological, 183, 193–94; classification schemes in, 168–69; evolutionary, 194; for higher taxa, 168, 185–86; as keys, 166, 182; Linneaus's, 190–91; stemming from collections, 189, 191–92; theories' influence on, 166–67
Taylor, Charles, 119
technē/tuchē, 21, 23–24
technology, 1, 3, 88, 108, 137, 162, 265; biotechnology and, 250–51; nature vs., 56, 134, 136; relation to science, 38–39, 43, 274; research institutes advancing, 257–58; science and, 11, 115, 264; status of, 253–54, 264; transfers from university research, 269–73
Telesio, Bernardino (1509–1588), 35, 143–44
Tellier, Michel-François Le, 118–19
Tempier, Etienne (ca. 1277), 32–33
Theodoric of Freiberg (1250–1320), 39
theology, 124, 153
Theophrastus (372–288 BC), 21, 171, 179
thermodynamics, 207, 211, 247–50. *See also* conservation principle
Thirty Years' War, 78

Thomas Aquinas (1225–1274), 27, 34, 79, 97, 106, 134; on mathematics, 39–40, 61, 65; on religion, 32, 75–76
Thomson, D'Arcy Wentworth (1860–1948), 225–26
Thomson, William (Lord Kelvin) (1824–1907), 160–61, 237, 247–48
thought-experiments, Leonardo using, 56
Thunberg, Carl Peter (1743–1828), 191
time, 135, 228, 245, 247, 256
timelessness, Aristotle on, 13–14
Toland, John (1670–1722), 156
Toulmin, Stephen, 78
Tournefort, Joseph Pitton de (1656–1708), 143, 168, 172, 177, 180, 182, 184
transformation, theory of, 225–26, 229
transformism, 221–22, 231
truth, 77, 259; certainty of, 72–74, 78, 86; criterion for, 74–75; hypotheses vs., 85–86; literal vs. allegorical, 135–36; sources of, 80–81, 85, 88, 136

Uffenbach, Zacharias von (1683–1734), 109–10
Unger, Frans (1800–1870), 238–39
United States, 242, 258, 273; science policies of, 261–63, 266–68; universities in, 131–32, 259–61, 270
universities, 47, 49, 97, 194; botany at, 172–73, 175; competing for students, 131, 255, 260; curricula of, 28–30, 39, 41, 92; development of, 29; effects of neoliberalism on, 275; entrepreneurship at, 258, 269–71; expansion of, 30, 257; faculty of, 62, 68–69, 255, 269, 273–74; humanists' separation from, 50–51, 53; jobs for graduates of, 132, 255; laboratories in, 130–31, 255; links with private sector, 259–63, 269–72; mathematicians' independence from, 52, 61; mathematics at, 54, 61; mathematics' low status in, 55, 61–62, 103; reforms of, 54–55, 115, 127–32; research at, 130–32, 256–57; research funding for, 264, 272–73; teaching methods in, 131; in United States, 131–32
Unmoved Mover, 14, 35, 134
Urbanus VIII, Pope (Maffeo Barberini), 70–71, 81
utilitarianism, in universities, 130–31

Valla, Lorenzo (1407–1457), 51, 56–57, 76
van Helmont, Jan Baptista (1579–1644), 102, 117, 122, 141, 152–54
van Helmont, Franciscus Mercurius (1614–1699), 102
Vasari, Giorgio (1511–1574), 68
Vergil, Polydore (1470–1555), 54
Vesalius, Andreas (1514–1564), 101, 137, 139–40, 184

Viète, François (1540–1603), 60
virtù, 9, 56–57
virtues, Aristotle on, 23–24
virtuosi. See artisan-engineers
Vissering, Simon (1818–1888), 208–9
vita activa, 9, 50, 57
vitalism, 152, 153; Harvey's, 139–42; mechanism and, 145, 230; Second Law of Thermodynamics and, 248–49
Vitruvius, 55
Voetius (Voet), Gijsbert (1589–1676), 123–24, 125–26
Volder, Burchard de (1643–1709), 121–22, 149
Voltaire (1694–1778), 157
voluntarism, 32, 34, 44–45, 154
vortex theory, 144, 146–48, 156, 158
Vosmaer, Arnout (1720–1799), 190
Vossius, Gerard (1577–1649), 121

Walaeus (Jan de Wale) (1573–1639), 122, 140
Walker, William, 259
Wallace, Alfred Russell (1823–1913), 221, 223, 235
Watson, James (b. 1928), 162
Watt, James (1736–1819), 157, 246–47
Weaver, Warren (1894–1978), 250, 260
Weber, Max, 259
Westfall, Richard, 151, 153
Weston, Richard (1591–1642), 177
Weyl, Hermann, 255
Whewell, William, 254
Whiston, William (1667–1752), 224
White, Hayden, 6
Wickelmann, Johann, 228
Wieland, Wolfgang, 17, 25
Wilkins, John (1614–1672), 115–16, 200
William III of Orange, 124, 154, 155, 173
William of Moerbeke (ca. 1270), 29
William the Silent (1533–1583), 121
William V, 190
Wilson, Edward O. (b.1929), 195
Witelo (1230–1275), 39, 44, 58
Witsen, Nicolaes (1640–1717), 189
Wölfflin, Heinrich, 7
World Soul, 98
worldviews: influences on, 7–8, 19–20; styles vs. schools of thought in, 6–7
Worsley, Benjamin (1618–1677), 114–15
writing style, 118; of natural history books, 168–69; for plant descriptions, 187; for reporting experiments, 90

Yates, Frances (1899–1981), 6, 98, 104

Zarlino, Gioseffo (1517–1590), 89
Zilsel, Edgar (1891–1944), 47, 49
Zola, Emile, 130